T0138251

Darwin's Orchids

DARWIN'S ORCHIDS

Then and Now

Edited by Retha Edens-Meier
and Peter Bernhardt

The University of Chicago Press
Chicago and London

Retha Edens-Meier is associate professor in the College of Education and Public Service at Saint Louis University and a research associate with the Missouri Botanical Garden in St. Louis and the Kings Park and Botanic Garden in Perth, Western Australia. **Peter Bernhardt** is professor of biology at Saint Louis University and a research associate with the Missouri Botanical Garden and the Royal Botanic Gardens and Domain Trust in Sydney, Australia.

Funding for the publication of all color plates was graciously provided by The Missouri Botanical Garden, St. Louis, Missouri, USA, The Royal Botanic Gardens and Domain Trust, NSW, Australia, and Peter and Armalinda Bernhardt, St. Louis, Missouri.

MISSOURI
BOTANICAL
GARDEN

The Royal
BOTANIC GARDENS
& Domain Trust

The University of Chicago Press, Chicago 60637
The University of Chicago Press, Ltd., London
© 2014 by The University of Chicago
All rights reserved. Published 2014.
Printed in the United States of America

23 22 21 20 19 18 17 16 15 14   1 2 3 4 5

ISBN-13: 978-0-226-04491-0 (cloth)
ISBN-13: 978-0-226-17364-1 (e-book)
DOI: 10.7208/chicago/9780226173641.001.0001

Frontispiece *Ophrys insectifera* (*muscifera*) was a favorite flower of Charles Darwin, who knew it as the "Fly Orchis." He recorded the natural removal of its pollinaria over several seasons near his home in Kent, England (see Chapters 1 and 3). Photographer: John Palmer. By courtesy of Irene Palmer.

Library of Congress Cataloging-in-Publication Data

Darwin's orchids : then and now / edited by Retha Edens-Meier and Peter Bernhardt.
     pages   cm
     Includes bibliographical references and index.
     ISBN 978-0-226-04491-0 (hardcover : alk. paper) — ISBN 978-0-226-17364-1
(e-book) 1. Orchids.   2. Botany—History.   3. Darwin, Charles, 1809–1882.
I. Edens-Meier, Retha.   II. Bernhardt, Peter (Botanist)
QK495.064D292   2014
584'.4—dc23
                                                            2014003532

♾ This paper meets the requirements of ANSI/NISO Z39.48–1992 (Permanence of Paper).

We dedicate this book to Carlyle Luer, whose taxonomic studies on the orchids of the Western Hemisphere have inspired pollination studies for decades. His classifications, careful observations, and excellent photographs of the orchids of North America brought attention to the roles of bumblebees in the pollination of *Spiranthes* and butterflies in the pollination of *Platanthera*. Thanks to interest in Dr. Luer's original segregation of tropical orchids in the genus *Dracula*, a generation of researchers learned that the labellum petals of these flowers mimic mushrooms to attract fungus flies in the genus *Zygothrica*. Meticulous research in taxonomy leads to great research in pollination biology.

CONTENTS

Color plates follow page 228.

Great scientific research is a paradox, because initial publications in its wake invariably stimulate more questions than they answer. Often these questions develop into new branches of research. Surely the works of Charles Darwin (1809–1882) are prime examples of this paradox, as new disciplines were produced almost every time he wrote a book. In particular, *On the Origin of Species* (Darwin 1859) is one of the greatest examples of how one piece of research and observation profoundly influenced and impacted scientific thought even to the present day. We remind the reader that after 1859, Darwin wrote another 10 books, 6 of which emphasize evolution and adaptation in plants. How have these books left their impact on the development of scientific disciplines and the history, nature, and philosophy of science?

Historians of science often refer to Darwin's first publication after 1859 as "the little book with the long title": *On the Various Contrivances by Which British and Foreign Orchids Are Fertilised by Insects, and on the Good Effects of Intercrossing*. Originally published in 1862, the book reached the 150th anniversary of that first edition in 2012. Darwin's work on orchids was meticulous as well as stimulating—*On the Various Contrivances* compelled people from all over the world to send him correspondence and specimens. As a true gentleman, Darwin acknowledged these individuals in his books. Accordingly, this orchid book enjoyed a second and much expanded edition in 1877 (see Chapter 1).

On 23–30 July 2011, the City of Melbourne, Australia, hosted the

18th International Botanical Congress. As always, the congress was divided into a series of symposia and guest lectures. Two symposia honoring the work of Darwin (1862) were held on 28 July. We decided to honor Darwin's memory by publishing the proceedings of these symposia in the pages that follow.

We wanted a published account of the proceedings that would be most accessible to a historian, scientist, and/or orchid lover interested in reading or rereading Darwin. That explains why, following Chapter 1 (a brief historical account of Darwin and his orchid work), the arrangement of chapters in this book parallels the order of topics presented in the second edition of *On the Various Contrivances*. However, we must emphasize that we are not trying to recreate either edition of Darwin's book. Instead, we simply want to make it easier for readers to compare his original contributions with the studies published afterward.

Therefore, we would like the reader to think of each orchid genus studied by Darwin as a model system. As in Darwin's (1877) second edition, Chapters 2 and 3 begin in the English and Eurasian countrysides with *Orchis* and *Ophrys*, respectively. Next we visit the Southern Hemisphere to study the orchids of southern Africa (Chapter 4), followed by the terrestrial orchids of Australasia (Chapters 5–7). We then plunge into the New and Old World tropics, examining advances in the study of orchids with the largest and most fragrant flowers (Chapters 8 and 9). We end this tour of orchid lineages as Darwin did, with a treatment of pollination in the extravagant architecture of the slipper orchids (Chapter 10).

Of course, Darwin's books end with a long, provocative discussion synthesizing and expanding on the previous chapters. We felt that our last 2 chapters should synthesize what we now know about how orchid pollinators perceive color, and how a variable climate influences orchid flowering periods. Please note that our original symposium was unable to fit the schedules and/or finances of researchers studying either the Australian sun orchids (*Thelymitra*; see Chapter 7) or the vast number of Neotropical species dependent on insects we now call orchid bees (Apidae: tribe Euglosini). Those 2 chapters were inserted after the symposia, and for Chapter 9 we are particularly grateful to Dr. David Roubik of the Smithsonian Institution.

We must warn the reader that just as the study of orchid evolution has expanded and changed over the past 150 years, so have the scientific names and classification of many familiar and beloved species. In fact, orchid classification remains a dynamic study to the present day; orchid taxonomists often disagree about the names and number of species, because they use different tools to investigate their collections. That explains why we can no longer rely exclusively on all the scientific names employed by Darwin in either edition of his orchid books. Additionally, the family Orchidaceae is now subdivided into subfamilies, but the concept and number of subfamilies have changed radically since 1981. We've encouraged our coauthors to stick to preferred but modern classification systems that best allow them to help their readers understand major advances in the study of orchid pollination. Yet we must also warn that the literature on orchid classification remains in great flux in certain parts of the world. Orchid collectors and amateur naturalists may be quite surprised by recent changes in, say, the names of orchids native to Madagascar and particularly Australia.

What, then, is the purpose of this book? A need exists for more than one review emphasizing how Darwin's 2 editions (1862, 1877) continue to influence botanical research. Of course, Darwin addresses interesting topics in both editions, but we have focused on the main question that he addresses in the title (slightly modified) of his second edition—Just what *are* the various contrivances by which orchids are "pollinated" by insects? Darwin (1877; p. 288) poses a related question: "Why do the Orchideae exhibit so many perfect contrivances for their fertilisation?"

These same questions have been addressed and answered, in part, by the contributing authors. We comment on how their review and personal research help answer these questions while suggesting new inquiries. Please join us again in the summary after you have read the following chapters.

Peter Bernhardt and Retha M. Edens-Meier

Saint Louis University

St. Louis, Missouri 63103

1 September 2013

PART I

Darwin Shares His Orchids

ONE

Darwin's Orchids (1862, 1877): Origins, Development, and Impact

Peter Bernhardt and Retha Edens-Meier

INTRODUCTION: WHY DID DARWIN STUDY ORCHIDS?

In 2009, scientists, naturalists, and journalists worldwide celebrated the bicentennial of the birth of Charles Darwin (1809–1882) and the 150th anniversary of the first edition of a book now known simply as *The Origin of Species* (Darwin 1859). If botanists, horticulturists, and plant conservationists wish to continue this happy tradition, we should recognize that first editions of Darwin's "plant books" appeared between 1862 and 1880, which means that the final 150th anniversary will be in 2030 (see Darwin 1880). The first book Darwin published following *On the Origin of Species* interpreted the morphology and biomechanics of flowers in the orchid family (Orchidaceae; Darwin 1862). Why did he select these flowers as model systems to expand on such concepts as "descent through modification" and adaptation, first addressed in *On the Origin of Species*?

We will probably never know for certain. Did Darwin's early love of flowers (Siegel 2011) in general predetermine a long-term, and unusually personal, inquiry on orchids decades later? Allen (1977; p. 45) noted that Darwin left a notebook from his university days that shows dissection of the anther and pollinia of *Anacamptis morio* (syn. *Orchis morio*). Yet it's most unlikely that Darwin's botany professor at Cambridge University, the Reverend John Stevens Henslow (1796–1861), turned his student into an enthusiastic orchidophile; the surviving Henslow-Darwin correspondence fails to mention orchids. It is far more likely that Henslow influenced Darwin's interests in the leaves of carnivorous plants and flower forms in *Primula* spp. (see Barlow 1967). Moreover, the orchid species Darwin encountered and collected during his 5 years on the HMS *Beagle* never evoked comments in his recounting of the voyage (Darwin 1845). In fact, *orchid* is not even an entry in the book's index.

Darwin's collections of orchids during that voyage were few, although Hooker (1860) credited him with the collection of the terrestrial and temperate *Chloraea magellanica*, a species that produces shiny white flowers that appear to be etched with bold green veins. Darwin was clearly unable to recognize the unique floral features of the Orchidaceae during this voyage, as 1 of the 2 "orchids" he collected in Tierra del Fuego and preserved in wine turned out to be a *Calceolaria* sp. (letter to Hooker; 19 May 1846). The story that the great botanist Robert Brown (1773–1858) stimulated Darwin's interest in orchid pollination by convincing him to read Sprengel (1793) did not convince Ghiselin (see foreword in Darwin 1984; pp. xvii–xviii), who insisted that Darwin was experimenting on a range of flowers as early as 1839.

In fact, Darwin's notes and correspondence with J. D. Hooker (1817–1911), the influential botanist and Director of the Royal Botanic Gardens at Kew, show they began discussing irritability in orchid flowers as early as 8 December 1844. In a letter to W. D. Fox (1805–1880) dated 8 February 1857, Darwin asked his cousin if he would observe any *Mormodes* spp. in bloom in a private collection at Oulton House, to see "which eject their pollen masses when irritated"(see Chapter 9). There is no surviving record of a reply from Fox.

Of course, the mere presence of existing correspondence can't prove Darwin's intense interest in any topic. For example, while his published and/or online letters from 1858 to 1859 fail to mention orchid flowers, it's obvious he recorded natural rates of insect-mediated removal of pollinia from flowers in populations of *Ophrys muscifera* (syn. *O. insectifera*) in 1858 (Darwin 1862). Therefore, the most plausible explanation for Darwin's orchidophilia remains with his seventh son, Francis (1848–1929): "He [C. Darwin] was probably attracted to the study of Orchids by the fact that several kinds are common near Down" (Darwin F 1896; p. 303). More recently, Boulter (2008; p. 157) reported that he read a notebook that C. Darwin had written in during his early days at Down House in the 1840s. Boulter insisted that Charles and his wife Emma transplanted orchids from the wild into their hothouse to better observe them.

THE VICTORIANS AND THEIR BELOVED ORCHIDS
(DARWIN AS ORCHID COLLECTOR)

Whatever the case, Darwin's scientific interest in orchids parallels Victorian sentimentality toward flowers in general, rising literacy with respect to education in botany (see Scourse 1983), and the 19th-century fad for privately owned, living collections of exotic plants (Tyler-Whittle 1970). In England in particular, greenhouse technology improved significantly during the 19th century as iron became cheaper and the tax on window-panes was repealed (Tyler-Whittle 1970; Woods and Warren 1988). A mania for tropical, epiphytic-lithophytic orchid species was just one of many plant-collecting passions of the Victorian era that also included tropical/temperate species of ferns, aquatic plants, *Rhododendron* spp., and palms, among others (Scourse 1983). However, the craze for wild-harvested, epiphytic orchids proved to be a longer-lasting and resilient mania (Bernhardt 1989b; Darwin 1868; Siegel 2011) for two overlapping reasons. First, most of England's expanding middle class did not own enough land to support the culture of choice groves of giant conifers and *Magnolia grandiflora* from North America or the arborescent *Rhododendron* spp. of the Himalayas. They could not afford to build either the huge tropical pool houses required to sustain the South American *Victoria amazonica* (Coats 1970), or the palm conservatories associated with Kew Gardens and the great private estates. Instead, small conservatories and glasshouses attached to urban and suburban homes (Woods and Warren 1988) housed dozens of orchid specimens from Old and New World tropics. Some popular writings of the day offered stories about how orchids were collected, bringing mild excitement to armchair travelers. A number of these authors insisted that orchid collections conferred status on their owners, as their cultivation showed good taste and horticultural expertise (see Boyle 1893). Second, Victorian collectors found orchids unique, as mass propagation of these plants from either seeds or meristem tissues was impossible until the 20th century (Bernhardt 1989b).

DARWIN'S ORCHID STUDIES BEFORE 1862

By 1861, Darwin had amassed a sizable collection of orchids from collections at Kew and the commercial nurseries of James Veitch and sons as well as from gifts given by private collectors (Siegel 2011). Addressing his father's research on orchid flowers, Francis Darwin (1896; p. 303) wrote that "in 1861 he [C. Darwin] gave part of summer and all of autumn to the subject." Some people have misread this sentence and come to the conclusion that *all* Darwin's orchid research for the first edition of *On the Various Contrivances* began and ended in 1861. While it is reasonable to assume that it took Darwin 10 months to write the book, and that most of his dissections of exotic species were performed and recorded in this space of time, his correspondence indicates that his work on British species was older. In particular, there is Darwin's famous letter to the *Gardener's Chronicle* (4–5 June 1860) discussing his observations, before and during 1858, recording (insect-mediated) pollinia removal in flowers of *Ophrys muscifera*. Within the same letter, he contrasts these results with the absence of pollinia removal and self-pollination in *Ophrys apifera* (see Chapter 3). It was not until the second edition of the orchid book that Darwin (1877) finally released his data on poor pollinia removal rates in flowers of *Orchis morio* during the cold and wet season of 1860 (see Chapter 2).

Based on collected correspondence, it was in 1860 that Darwin first attempted to enlist other naturalists to make observations on orchid flowers native to Britain. He wrote Alexander Goodman More (1830–1895) on 24 June 1860, but More would not respond until the following year. So the belief that Darwin began and completed his entire study on orchid floral biology pollination in 1861 is a nice story, but it tends to fall apart after reading the first half of one sentence in Darwin (1862; pp. 34–35): "I have been in the habit for twenty years of watching Orchids."

BUT WAS DARWIN A POLLINATION BIOLOGIST BEFORE 1862?

But this sentence is a double-edged sword. If we complete it, we have to wonder if Darwin was much of a pollination biologist by modern standards: "[I] have never seen an insect visit a flower, excepting butterflies twice sucking *O. pyramidalis* and *Gymnadenia conopsea*" (Darwin 1862; pp. 34–35). Therefore, in the first edition of his book, Darwin's firsthand descriptions of orchid flower-insect interactions are few and usually credited to other people. In the second edition, Darwin (1877) generously credits the observation of the pollination of *Herminium monorchis* by a parasitic wasp and moth pollination of *Gymnadenia conopsea* to his son George (1845–1912). Müller (1871) also complained that visitors to the flowers of *Orchis* were infrequent, but it was Müller, not Darwin, who observed and collected bees visiting several *Orchis* spp. (Müller 1883; see Chapter 2). It is ironic to think that Charles Darwin, a great collector of beetles in his youth and a hunter of so many different, and far larger, animals in South America (Darwin 1845), was inept with a butterfly net. Twenty years of bad luck, season after season, seems unlikely even if we emphasize a combination of personal ill health and bad weather in Kent from 1844 to 1861. Grant Hazlehurst, assistant warden of Downe Bank for the Kent Wildlife Trust, noted that members of the trust continue to revisit Darwin's favorite orchid sites each spring. They spend some time catching and photographing male wasps pollinating *Ophrys insectifera* on sunny, cloudless days (Grant Hazlehurst, personal communication).

We suspect that much of Darwin's lack of field observation was based on two personal limitations. First, his correspondence shows him to be an extremely busy man who was always juggling several lines of research at the same time. It is unlikely he could spend hours in the field every day covering the full flowering seasons of each orchid species as fieldworkers do, or should do, today. We presume that then as now, each population of orchids remained in bloom for 2 or 3 weeks. Second, Darwin never accepted K. C. Sprengel's (1730–1816) interpretation of the false nectar flower (the *scheinsaftblumen*; sensu Sprengel 1793). That is, Sprengel noted that even though the flowers of some species always fail to secrete nectar, they do produce appropriate visual and/or scent advertisements and re-peatedly attract insect pollinators. These insects probe for a nonexistent reward and serve as passive pollen taxis. While Darwin read Sprengel's treatise on pollination, he confided in a letter to Harvard's professor of botany, Asa Gray (1810–1888), that it was "a curious old book full of truth with some little nonsense" (19 January 1863).

DARWIN ON NECTAR SECRETION IN ORCHIDS

Ironically and consequently, the nectar glands Darwin describes in some of his orchid flowers in both the first and the second edition of his book have since been reinterpreted as floral sculptures that fail to secrete nectar but have other functions during the act of insect-mediated pollination. Darwin's predilection for interpreting small, novel floral structures as functional nectar glands was accepted with enthusiasm by Müller (1871) and others. However, few botanists today regard the hammer glands on or in the flowers of most lady's slipper orchids (*Cypripedium*) as nectar glands (see Chapter 10), breaking with Darwin (1862, 1877) and Müller (1871). We think it's notable that following the publication of his first edition, Darwin netted insects visiting the flowers of *Spiranthes autumnalis* and *Epipactis latifolia* (Darwin 1877), both nectar-secreting species. Had he understood that entire lineages in the orchid family fail to produce any reward (Tremblay et al. 2005; Bernhardt and Edens-Meier 2010; see Chapters 2, 3, 4, 5, 7, 10) and remain pollinator limited (sensu Committee on the Status of Pollinators in North America 2007), he might have spent more time in the field or even reinvested his time in another diverse angiosperm lineage with bilaterally symmetrical flowers.

Would the book have had the same general appeal and long-term impact if, for example, he compared flower and pollinator interactions in the pea family (Fabaceae)? We think not, considering the specific plant collection fads during the Victorian age. Ignorance was bliss in Darwin's case. At least he never experienced the disappointment of spending 3 seasons at the same population of *Cypripedium reginae*, only to capture 6 insects carrying the pollen of the large, colorful but nectarless flowers (Edens-Meier et al. 2011).

DARWIN'S ORGANIZATION OF THE FIRST EDITION (1862): *ORCHIS* AS A CELEBRITY

Therefore, the first edition of Darwin's book would seem to be an unlikely place for a breakthrough in evolutionary botany. It is a rather short study in the comparative floral morphology and biomechanics of, at the time, 28 species (in 15 genera) distributed throughout Britain. The flowers of an additional 43 genera represent 42 species with Neotropical and Paleotropical distributions. The outline of the book follows the classification of the orchid family by John Lindley (1799–1865), then editor of the *Gardner's Chronicle*. Most historians of orchidology treat Lindley's classification as the first real attempt to subdivide all species into tribes (see Dressler 1981) based primarily on variation in the architecture and fusion of the floral column and the alignment of the anther(s) to the stigma (Lindley 1826).

To make the book's topic more familiar and appealing to a British audience, Darwin employed an introductory technique he used in *On the Origin of Species*. He would repeat this technique in all his plant books. He began the book with a simple and familiar example. For instance, Darwin (1859) introduces the new and complicated concept of natural selection by beginning the book with a survey of popular breeds of domesticated pigeons produced by artificial selection. It was, after all, an era in which people ate, bred, raced, exhibited, and shot pigeons according to breed.

Therefore, although Darwin (1862) follows Lindley's much earlier subdivision of orchids into tribes (Lindley 1826), he begins his book with

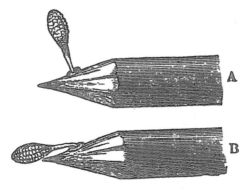

Fig. 1.1 The famous pencil experiment. The position of the pollinarium changes from a subvertical to a horizontal position as the caudicles dry (Sowerby from Darwin 1877; p. 12).

those British species that produce the largest and prettiest flowering stems (*Orchis* s.l.; see Chapter 2). These are the same wildflowers avid naturalists would recognize easily while taking a spring botany walk through a moist countryside. In fact, the first flowers described in the book are those of *Orchis mascula*, and that is probably not a coincidence either. An important insight is offered by a long-standing member of the Kent Wildlife Trust. Specifically, Irene Palmer informed us that *O. mascula* remains the earliest flowering orchid species in Kent. It sometimes blooms as early as 3 April, but a few flowers linger until 13 May (Irene Palmer, personal communication). We also wonder whether Darwin knew that his educated readers would also recognize *Orchis mascula* as the long purples in *Hamlet* (Shakespeare, 4.7.168–71). While the identification of those flowers in Ophelia's suicidal garlands will always remain open to debate (Wentersdorf 1978; Otten 1979), they are arguably the only orchids mentioned in the entire works of William Shakespeare.

Of greater importance, Darwin (1862) uses *O. mascula* in the first chapter to illustrate his famous "pencil experiment" (Figure 1.1). In it, he removed the pollinarium of *O. mascula* with the tip of a sharpened pencil and watched each pollinia pair change position as their connective stalks (caudicles) dried. Darwin argues that this is an adaptation encouraging cross-pollination (see below, entry 2). While the pollinaria of *Orchis* spp. bear long, easily observed caudicles, these same structures are often much reduced or absent in many other flowers in the orchid family (Dressler 1981). *Orchis* spp., then, gave Darwin a familiar yet flamboyant model system to begin a book that emphasizes repeatedly how unfamiliar flower organs encourage cross-pollination while reducing the frequency of self-pollination.

ABOUT THE ILLUSTRATOR

Darwin (1862) contains 34 illustrations, reproduced as woodcuts, by George Brettingham Sowerby *fils* (1812–1884), the third generation of an artistic family of naturalists specializing in works of botany and conchology. Was the remainder of Sowerby's illustrations in this first edition as novel and convincing to the first-time reader as the graphic depiction of the pencil experiment? Perhaps not. After all, by 1862 there wasn't anything unique about illustrations that magnified and identified the fine details of an orchid's floral architecture. The finest were executed as color plates, and were essential tools for plant taxonomists and morphologists. Consider, for example, how Robert Brown (1773–1858) benefited from the opus of Ferdinand Bauer (1760–1826; see Mabberly 1999).

NOVEL CONTRIBUTIONS IN THE FIRST EDITION (1862)

Therefore, we must now emphasize which components of Darwin (1862) represent novel contributions to orchidology, plant evolutionary biology, and the history of science beyond the mere study of comparative floral anatomy. We see 8 outstanding novelties.

1. By publishing his table showing the natural removal of pollinia in flowers of *O. muscifera*, and by covering flowering stalks in bloom of *Orchis morio* with a bell jar to prevent insect visitation, Darwin (1862) emphasized the potential for vector-mediated cross-pollination in a flower with irregular (bilateral) symmetry. This was essential in a century when some botanists still believed that individual orchid flowers inseminated themselves by some internal process, because male and female organs always fused together to form an interconnected column (see Bateman 1837–1843). Darwin developed and much expanded this study later in his career, performing and/or supervising actual crossing experiments on orchids and many other flowering plants (Darwin 1876). Based on a far earlier letter to J. D. Hooker (19 July 1856), it's clear he was thinking of the importance of cross-pollination years in advance (see Chapter 2).

2. Using needles and pencils, Darwin (1862) showed how the orchid column releases and receives a pollinarium. He made readers understand that the whole pollinarium is deposited on the insect as it leaves the flower, and paired pollinia are deposited on receptive stigma lobes as the same insect enters a second flower. The novelty is that Darwin's "pencil experiments" showed that the pollinia of most species change their angle or position. Those pollinia hadn't moved by themselves. The connective

structures (caudicles or stipes) attached to the pollinia dried and bent once they were removed from the column. Darwin interpreted this predictable period of caudicle-stipe desiccation as selectively advantageous to cross-pollination. An insect foraging on more than one flower on the same stem removes whole pollinaria from each flower, and each pollinarium releases 2 pairs of pollinia. Each pair of pollinia stands upright until its connective stalks shrivel and bend over several minutes later (Figure 1.1). Freshly removed, upright pollinia are far less likely to contact receptive stigma lobes on flowers borne on the same stem, provided pollinators move quickly from plant to plant. This prevents self-pollination between flowers on the same stem or on the same flower if the insect probes them repeatedly before leaving. In contrast, the longer a pollinarium remains on an insect, the more time the connective stalks have to dry out completely so that their pollinia lay down flat. At this new angle, the pollinia could be inserted onto the stigma lobes of a flower found on the second, third, or fourth plant visited by the same pollinator, thus encouraging cross-pollination.

3. What use is this process of pollinia movement if deposition of the sticky disc (viscidium) on the pollinator is such a haphazard affair that the whole pollinarium is deposited at a site on the insect's body that never contacts the stigmatic lobes in the flower? An important novelty in Darwin (1862) is his close examination of pinned specimens of insects that had been caught by colleagues and family while these insects were foraging on orchid flowers. Darwin noted that the disc of a pollinarium is, in most cases, deposited with monotonous and predictable regularity on the same part of the pollinator's body. We learn in Darwin (1862) that butterflies and moths regularly carry pollinaria of *Orchis* (syn. *Anacamptis*) toward the bases of their proboscides. One of Sowerby's illustrations depicts the head of a moth wearing the greatest number of pollinaria toward the base of its proboscis (Figure 1.2).

It is clear that Darwin recognized the importance of combining his own observations and experiments on floral organs with collections of insects observed as they foraged on the same flowers. For example, after 1863, Darwin (1877) reported he saw common wasps (*Vespa sylvestris*) carrying pollinaria on their heads after consuming nectar in flowers of *Epipactis latifolia*. During a trip to Torquay, he caught two unidentified humble bees (*Bombus* sp.) carrying whole pollinaria and viscidia on their proboscides after witnessing them foraging on flowers of *Spiranthes autumnalis*.

4. Examination of flowers of Neotropical orchids, now placed in the genus *Catasetum*, produced 2 novelties reported by Darwin (1862). First,

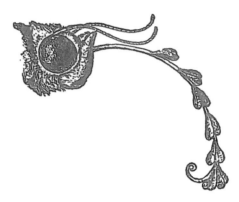

Fig. 1.2 Specimen from the box sent by Mr Bond. Proboscis of *Acontia luctuosa* wearing pollinaria of *Orchis* (*Anacamptis*) *pyramidalis* (Sowerby from Darwin 1877; p. 31).

Darwin showed that plants placed originally in separate genera, due to radically different floral architecture, were actually males and females of the same species. This was the first record of dioecy (two genders, male and female) in the family Orchidaceae. *Catasetum* spp. and related genera were also the first orchids Darwin described as having irritable columns that "shoot" their pollinia after certain structures are touched (see Chapter 9).

5. The original prediction, that *Angraecum sesquipedale* could be pollinated by moths with "proboscides" 10 or 11 inches in length, appeared in the first edition (Darwin 1862). The significance of this prediction will be treated fully in a later chapter. However, it is clear here that Darwin was beginning to think that some orchid species had canalized pollination systems. That is, only a limited and closely related lineage of insect species was the pollinator of flowers having such exaggerated floral organs with such elongated nectar spurs (see Chapter 8).

6. Darwin examined the living flowers of diandrous orchids (Cypripedioideae) of Asian origin. He listed them all as *Cypripedium* spp., although all were reclassified as *Paphiopedilum* spp. by 1876 (Bernhardt 2008). Darwin believed that the pollinator contacted a dehiscent anther accessible through 1 of 2 basal and lateral openings formed by the column interconnecting with the much-inflated labellum. He argued that when the insect flew to a second flower, it inserted its head in the same basal and lateral opening, transferring the pollen smear onto the undersurface of the large stigma. As early as 1863, Darwin admitted this was a blunder (letter to Roland Trimen; 27 August 1863), but his published correction did not appear until much later (Darwin 1869 and see Chapter 10).

7. In his concluding remarks in the first edition, Darwin asks, why do orchids release their pollen grains as conjoined pollinia? His answer

is simple enough: to prevent waste and exhaustion (i.e. of energy and resources) in the act of transportation (pp. 355–356). As all the orchid ovaries he dissected contained hundreds, if not thousands, of unfertilized seeds (ovules), Darwin believed that if modern orchids produced an "extravagant amount" of dust-like pollen grains, like their ancestors, the result would have been "exhaustion." Instead, he discusses the number of pollen grains in pollinia required to match ovules in an ovary required in turn to maximize seed set following cross-pollination. Orchid pollen grains, he explains, are produced in packets or masses (pollinia) to prevent waste in the act of transportation. We argue that this passage shows the germ of both a modern approach to plant reproductive ecology (e.g. mate selection; sensu Willson 1983) and the importance of pollen/ovule ratios (Cruden 1977) in determining adaptations for breeding systems (Richards 1986).

8. We argue that the most important novelty in the first edition remains the floral homology at the end of the book. It is the unique and overpowering "punch line" uniting all the preceding pages' descriptions and comparisons of organ morphology in all orchid flowers that produce only one fertile stamen. Like most botanists of his day, Darwin used the standard, schematic *bau plan* to plot the relative number and location of each organ (Figure 1.3). The model showed the bilateral symmetry expected for both the corolla and the androecium, but there was something quite different in Darwin's diagram, its caption, and its accompanying text. The complete suite of organs regarded as unique to the flowers of monandrous orchids, including the labellum, the 2 wings (or lobes) forming the clinandrium (sensu Dressler 1981), and the rostellum, were reinterpreted. The labellum was reinterpreted as a unit of fused petals. The 2 wings or lobes were reinterpreted as 2 infertile stamens (staminodia). The rostellum was reinterpreted as a stigma lobe now adapted for the release of a sticky disc instead of the reception of pollen (Darwin 1862).

To Darwin, these organs were evidence of descent by modification. He argued that living orchids descended, ultimately, from a common ancestor with a flower that was complete (had 4 different kinds of organs), expressed radial symmetry, and had 6 fertile stamens arranged in 2 rings (whorls). He bolstered his argument by addressing the location of vessels (veins) in the flower, connected to what he considered rudimentary (vestigial) organs. This included the reduced and membranous lateral lobes (sterile stamens) in the clinandrium of *Malaxis paludosa* (Figure 1.3) and other species. For Darwin, bilateral symmetry in the monandrous orchid was also based, in part, on organ reduction. The now modified flower

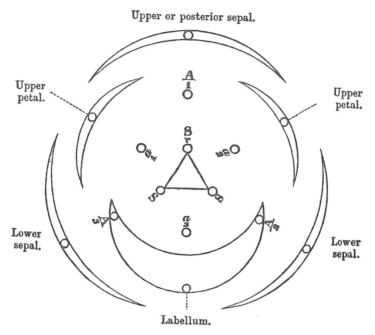

Upper or posterior sepal.

Upper petal.

Upper petal.

Lower sepal.

Lower sepal.

Labellum.

Fig. 1.3 Darwin's schematic diagram of the flower of a monandrous orchid. Small circles represent conducting vessels; S = stigma; A = fertile anther; $a1$ and $a2$ = rudimentary anthers (staminodia) (Sowerby from Darwin 1877; p. 236).

also lacked 3 free, fertile stamens (formerly located directly opposite the column) and at least 3 free petals.

Studies in developmental botany have never confirmed Darwin's interpretation of the 3-lobed labellum petal as a compound organ produced by the lateral fusion of 3 petals (coalescence). Sometimes one lobed, hypermorphic petal is just one lobed, hypermorphic petal. Vestigial veins linking to sites where the 3 missing fertile stamens developed in the extinct ancestor have not been found either. There is no reason to presume that vessels (veins) in plants must show the same predictable development as vestigial bones in vertebrate skeletons (Stebbins 1974). However, no authority has successfully refuted Darwin's interpretation of column structures (see above) in monandrous and diandrous orchids, or his derivation of the rostellum from a pollen-receptive stigma lobe. The stigmas of most monandrous orchids remain both a pollen-release and a pollen-reception organ based on the fusion (adnation) of the androecium (fertile and infertile stamen) to the dorsum of a pistil neck (style) in a flower in which the bases of all organs ultimately fuse to the ovary (epigyny).

The floral homology in the final chapter makes it easier to explain why Darwin rushed to study the organs and biomechanics of dozens of exotic orchid flowers in a single season, only to produce a rather small book written in what he admitted was a "semi-popular form." The orchids provided Darwin with model systems to expand on theories published in Darwin (1859). Orchid work produced additional evidence refuting the critics who attacked *On the Origin of Species*. How can we be so sure about this? Darwin gives the game away in one of his typically chatty letters to Asa Gray: "Of all the carpenters for knocking the right nail on the head, you are the very best: no one else has perceived that my chief interest in my orchid book has been that it was a 'flank movement' on the enemy" (23–24 July 1862).

While the last chapter in Darwin (1862) brought interpretation and argument together, Darwin persists in tantalizing and/or teasing Gray further, in that same letter of 23–24 July 1862. Gray was a proponent of the new natural teleology, and so Darwin wrote, "I should like to hear what you think about what I say in last Ch. [chapter] of Orchid book on the meaning and cause of the endless diversity of means for same general purpose. It bears on design—that endless question." Does this mean that both editions of this book (1862, 1877) represent works of teleology (see below)?

THE SECOND EDITION (1877): PREPARATIONS AND EXPANSIONS

Darwin went on to other projects in 1862, and the morphology and mechanisms of orchid flowers stopped dominating his correspondence in favor of simple, schematic illustrations of flowers from other families. However, Darwin (1862) had an international impact on botanists, entomologists, and natural historians. Gray obeyed Darwin's request for input from other naturalists by performing the same observations and experiments on orchid flowers belonging to North American congeners of British species (Gray 1862; Darwin 1877). Darwin now had several long-term correspondents who studied the morphology and pollination mechanisms of orchid flowers. These new colleagues published their results, addressing orchid species on 5 continents (see Chapters 4, 6, 7, 9). By the time Darwin published the second edition of his book, he was ready to incorporate his later research and reinterpretations (Darwin 1869) in conjunction with an additional 38 publications, written by other authorities. How, then, does Darwin (1862) differ from Darwin (1877)?

1. Biodiversity is much increased in the second volume, although G. M. Sowerby added only 4 new illustrations. Retaining the system of

Lindley's classification (see above), Darwin (1877) added information on the flowers of approximately 23 species from North America and Europe, 4 species from southern Africa, 10 species from Australia and New Zealand, and 3 species from tropical Asia.

2. Darwin finally published his data on the insect-mediated loss of pollinia in 5 *Orchis* spp. and in *Aceras anthropophorum* (now *Orchis anthropophora*) that he had compiled over many years. Once again, he notes that rates of pollinia removal declined in some years due to climatic conditions (see Chapter 12). Furthermore, in the last chapter of the book (pp. 280–281) Darwin compares low fruit set in European *Ophrys* spp. with observations his colleagues made in South America, Australasia, and southern Africa, noting that the conversion of flowers into capsules was extremely low in *Vanilla* sp., *Epidendrum* sp., *Dendrobium speciosum*, *Coryanthes triloba* (misprint; actually *Corysanthes* = *Corybas*), and *Disa grandiflora* (compare with Chapter 10). The rate of capsule set was as low as 5/200–1/1000. Darwin notes (pp. 281–282) that mechanisms for cross-pollination were "so elaborate" that "they cannot yield seeds without the aid of insects." Darwin (1877; p. 281), like Müller (1871), then offers some of the earliest published information that cross-pollinated orchid species are often pollinator limited (sensu Committee on the Status of Pollinators in North America 2007).

3. As mentioned above, the second edition added additional data on flower-pollinator interactions. In the final chapter of this edition, Darwin (p. 282) wrote, "We know that certain Orchids require certain insects for their fertilizations," and mentions members of the genera *Vanilla*, *Sarcochilus*, and *Angraecum* as primary examples (see Chapter 8). Thanks to his own observations, and those of German colleagues, he wrote, "in Europe *Cypripedium calceolus* appears to be fertilized only by small bees of the genus *Andrena* [see Chapter 10] and *Epipactis latifolia* only by wasps" (p. 282). This suggests the development of a modern trend in evolutionary ecology, indicating that some plant species (especially orchids) have canalized and specialized modes of floral presentation (Gurevitch et al. 2006), thereby limiting the spectrum (sensu van der Pijl and Dodson 1966) of prospective pollinarium vectors.

4. Just as Darwin expanded and added data on insect-mediated cross-pollination, so he added new information on species with obligate-facultative self-pollination systems. While "Nature tells us, in the most emphatic manner, that she abhors perpetual self-fertilisation" (Darwin 1877; p. 293), he noted that the phenomenon is not restricted to *Ophrys apifera* and a few allied European species. In fact, Darwin (1877) compared self-pollination in the British *O. apifera* with self-pollination in the

bud (e.g. cleistogamy) in two Australian *Thelymitra* spp. (see Chapters 5 and 7).

5. Darwin's second edition (1877) contains the corrected and revised interpretation of pollen transfer and pollen reception in the flowers of *Cypripedium* spp. This reinterpretation was based on a combination of factors, including correspondence (see Darwin's letter to Asa Gray, 20 April 1862), dissections by Gray (1862), field observations by Müller (1871), and Darwin's own experimentation (Darwin 1869 and see Chapter 10 this book).

IMPACT OF THE SECOND EDITION (1877) ON NATURALISTS AND BOTANISTS

Consequently, the second edition offered what we now regard as the modern approach to the pollination ecology of orchids. While naturalists, entomologists, and botanists continued to read Darwin long after his death, few depended on the first edition after 1877 (Müller 1883). Preferences for Darwin (1877) are overwhelmingly exemplified as citations in books and papers published in the 20th century, especially in Australia. There, orchid pollination studies were published primarily in journals devoted to the studies of amateur naturalists through the early 1980s (e.g. Bernhardt and Burns-Balogh 1983). Most of these Australian field naturalists read Darwin (1877) in association with Fitzgerald (1875–1895). They emulated the work of these men by collecting flowers of native species to repeat those earlier experiments with needles, probes, and pencils, noting which flowers regularly dropped their pollinia onto stigmas in the absence of insect visitors (Rogers 1913; Erickson 1965b). Some observed, collected, and documented pollinia-carrying insects and/or recorded conversion rates of flowers into viable capsules (Rayment 1935; Erickson 1965a; and see review in Armstrong 1979; Cady and Rotherham 1970). In particular, the contributions made by Edith Coleman (1874–1951) were especially detailed and informative (Coleman 1927, 1928a,b, 1929a,b, 1930a,b, 1931, 1932, 1933a,b, 1938). She made the camera an important recording tool in her study of pseudocopulation in *Cryptostylis* spp. (Coleman 1929a,b; 1930a,b). Such publications would have amazed Darwin, as they were among the earliest studies of pseudocopulation paralleling the same discoveries in Eurasian-African *Ophrys* spp. (see review by Proctor et al. 1996 and Chapters 3 and 11).

Most important of all, during the second half of the 20th century, the works of these naturalists were read by a new generation of Australian academics and botanical curators. Just as Darwin encouraged Fitzgerald, so the orchid-pollination papers of the early 20th century formed, and continue to form, a foundation for new field and lab studies on the flowers of Australian species through the present day (see review by Armstrong 1979 and Chapters 5–7).

AVOIDING TWO MISINTERPRETATIONS OF DARWIN (1862, 1877)

Before any final conclusions, we should address 2 interpretations of Darwin (1862, 1877) that now persist on the Internet and, therefore, will enjoy a dubious immortality regardless of what is written here. First, and most important, we must ask: are both editions ill-disguised works of natural teleology in which natural selection replaces the wisdom of what the Victorians called Providence? Once we excuse Darwin for writing in a semipopular style (letter to Gray; 10–20 June 1862), teasing Gray (letter to Gray; 23–24 July 1862), and using words like *contrivances*, it's very hard to read the text in both books and accuse him of insisting there is "directed or inevitable design" in nature. Darwin did not argue that the structures he studied evolved *toward* a predictable and/or final cause. Surely there is a distinct difference between a philosophy that organic structures are shaped for a purpose versus reports, based on firsthand observations, that certain structures perform predictably upon maturity. If this is not the case, then we must conclude that virtually every textbook on vertebrate anatomy/physiology is a work of teleology. To some extent, it is unfortunate we now live in an era in which some historians of science have made biologists shy about using the word *adaptation* within its evolutionary context. No one wants to be identified as either a "vulgar adaptationist" or a "Panglossian adaptationist," and those pejoratives, alas, are also all over the Internet.

Before forming an opinion, the reader is directed to a foreword written by Ghiselin for a facsimile of the second edition of Darwin (1877). We must concur with Ghiselin when he traces support for this "natural teleology" back to Asa Gray and Darwin's son Francis. Ghiselin notes that while Darwin described adaptations in orchid flowers, he "did not assert that *all* adaptation was perfect" (p. xv). It is also reasonable to agree that both editions of Darwin (1862, 1877) show early applications of the principles of preadaptation and function shift. One must also concur that Darwin didn't attribute functions to floral parts without introducing the evidence of experimentation or field observations (see Ghiselin in Darwin 1984).

Finally, it is hoped that this chapter allows us to refute a comment by Gould (1978) that the 1862 edition was Darwin's "most obscure work." This statement damned with faint praise. Indeed, we note that a recent biographer (Quammen 2006) also failed to grasp the impact and value of either volume within the context of the voluminous Darwin canon. Correspondence between Darwin and Gray (10–20 June 1862 and 23–24 July 1862) indicates that the first edition triggered much the same range of exchanges between academics, naturalists, and aristocrats as did the publication of Darwin (1859), albeit on a smaller, far less heated scale. If a first edition of a scientific work goes on to a second and expanded edition without obvious objection and rejection by the scientific community, it can't be called obscure. If the revised edition continues to be read and cited by laypersons and professional scientists for seminal information, including some protocols still in use today, it can't be called obscure. If the book provides repeated inspiration to authors of imaginative fiction, it can't be considered obscure. Yes, at least one of the editions of the book was read by the young Herbert George Wells (1866–1946), who used Darwin's comments on the relative absence of known pollinators in *Cypripedium* (*Paphiopedilum*?) to write the short story "The Flowering of the Strange Orchid," published in 1895. Since then, at least 5 more authors have written variations of the story of orchids, using their flowers to prey on humans at least through 1970 (see Bernhardt 1989b).

CONCLUSION

Charles Darwin, through his detailed research on orchids and resultant publications (1862, 1869, and 1877), served and continues to serve as a global role model in the field of evolutionary botany for over 150 years. Consequently, scientists today are still asking some of the same questions Darwin asked: What pollinates this flower? How does the pollination mechanism work? Will a flower self-pollinate or is an insect vector required? How has this lineage changed over time?

PART II

Darwin's Orchids of the English and Eurasian Countrysides

Darwin on the Pollination of *Orchis*: What He Taught Us and What We Can Tell Him Today

Giovanni Scopece, Salvatore Cozzolino, and Amots Dafni

INTRODUCTION

In Charles Darwin's pioneering book, *The Various Contrivances by Which Orchids Are Fertilised by Insects* (1862, 1877), at least 10% of the text is devoted to the genus *Orchis*. Here, Darwin's observations focus mainly on the floral functional morphology of this large and heterogeneous genus and on its role in ensuring cross-pollination, while considering the floral adaptations to be a result of natural selection (see Chapter 1). Darwin found K. C. Sprengel's interpretation of the deceptive nature of orchid flowers unacceptable (Chapter 1). Like other contemporary biologists, he also supposed that the visiting insects pierced the cell wall while looking for "honey" and extracted cell sap. It is noteworthy that during 20 years of observation, Darwin had never seen even a single pollinating insect under field conditions, except for 2 butterflies visiting *Orchis* (=*Anacamptis*) *pyramidalis* and *Gymnadenia conopsea* (Darwin 1862; pp. 34–35). In the second edition of his book (Darwin 1877; p. 15), he notes, "When the first edition of this book was published, I had not seen any insects visiting the flowers of the present species (*Orchis mascula*), but a friend watched some plants and saw them visited by several bumble-bees, apparently *Bombus muscorum*." Thus, it is admirable how, with so little real evidence on active pollination, he ingeniously interpreted the functional adaptations of orchid floral morphology as "contrivances" of promoting cross-pollination.

This chapter discusses the post-Darwinian developments in the following aspects: the taxonomic delineation of the genus *Orchis* s.l. and its relatives; the pollination systems; the *Orchis*'s floral functional morphology; the pollinator's behavioral and cognitive aspects; the effects of pollinator behavior on the population ecology of deceptive species; the effects of pollinator behavior on the floral traits of deceptive species; and the role of the genus *Orchis* in testing the hypotheses on the evolution of deception.

In this chapter, we focus on the genus *Orchis* as it was circumscribed at Darwin's time. Concordantly, together with the 3 recently identified monophyletic groups (*Anacamptis*, *Neotinea*, and *Orchis* s.s.), *Orchis* s.l. also encompassed, at that time, the more genetically distant *Himantoglossum* and *Dactylorhiza*. From the viewpoints of pollination and floral adaptations, the species grouped in these genera closely resemble the nectarless *Orchis* and *Anacamptis* and are therefore included in this review, as previously done by Dafni (1987). In particular, previous studies related to *Dactylorhiza* may contribute in elucidating some ecological and evolutionary aspects also common to the deceptive species of *Orchis* s.l. In the present review, we conform to the nomenclatural treatment proposed by Bateman et al. (1997, 2003).

THE TAXONOMIC DELINEATION OF THE GENUS *ORCHIS* S.L. AND ITS RELATIVES

The origin of the taxonomical category *Orchis* dates back to Linnaeus (1737, 1753), who grouped a number of heterogeneous species (actually placed in 10 different genera). Afterward, numerous re-circumscriptions were implemented (Brown 1810) until Vermeulen (1947), who delimited a still heterogeneous but well-circumscribed *Orchis* genus. Ever since, the taxonomical controversy has focused mainly on sectional or subsectional levels (cf. Landwehr 1977; Delforge 1994).

With the advent of modern molecular phylogenetics, the taxonomy of the genus *Orchis* s.l. has undergone critical changes in the tribe Orchideae (Bateman et al. 1997). Molecular analyses showed polyphylies in previously supposed monophyletic groups, which suggested 2 possible nomenclatural changes: either to include the genera *Aceras*, *Anacamptis*, *Barlia*, *Chamorchis*, *Coeloglossum*, *Comperia*, *Dactylorhiza*, *Gymnadenia*, *Himantoglossum*, *Neotinea*, *Platanthera*, *Pseudorchis*, *Steveniella*, and *Traunsteinera* under the wide genus *Orchis* or to split the existing genus *Orchis* into 3 clades. Bateman et al. (1997) chose the second approach, which required fewer taxonomic changes. Following Bateman et al. (1997), 3 genera are now recognized: *Orchis*, *Anacamptis*, and *Neotinea* (Kretzschmar et al. 2007).

Far from being widely accepted, this molecular-based taxonomy has prompted great debate, and many treatments have been proposed since the first well-sampled DNA-based phylogenies of the subtribe were published (Bateman et al. 1997, 2003; Pridgeon et al. 1997; Aceto et al. 1999; Cozzolino et al. 2001). This controversy is mainly due to the sharply contrasting outcomes of morphological and molecular-based analyses. Morphological traits are exposed to the action of natural selection, and in plants pollinated by the same set of pollinators, they often represent parallelisms and convergences that blur the interpretation of ancestry. This could have been particularly true in the case of genus *Orchis* s.l., where several genetically distinct deceptive species (now separated in different genera) converged into very similar flower morphologies as a consequence of a deceptive pollination strategy that exploits the perceptual biases of a common set of pollinators.

POLLINATION SYSTEMS

Nectariferous Flowers

In the genus *Orchis* s.l., only a few species, namely *Anacamptis coriophora*, *A. fragrans*, *A. sancta* (Eberle 1974), and the autogamous *Neotinea maculata* (Duffy et al. 2009), produce a nectar reward for their pollinators (see Table 2.1). The presence of nectar in the spurs of these species attracts a

Table 2.1 Overview of genera once included in the genus *Orchis* (during Darwin's time) and their floral characteristics with respect to nectar production, spur, and papillae

Genus	No. Species producing nectar vs. total species	No. Species bearing spur vs. total species	No. Species bearing papillae vs. total analyzed species	Reference
Anacamptis	3/16	16/16	3/4	Bell et al. 2009
Dactylorhiza	1/28	1/28	2/3	Bell et al. 2009
Himantoglossum	0/6	6/6		Delforge 1994
Neotinea	1/4	4/4		Duffy et al. 2009
Orchis s.s.	0/26	26/26	2/5	Bell et al. 2009

wide variety of pollinators (Vöth 1975; Peisl and Forster 1975; Dafni and Ivri 1979) and, as a consequence, leads to the production of higher levels of fruit set (Smithson and Gigord 2001; Claessens and Kleynen 2011). When approaching the inflorescences of these nectariferous species, pollinators visit several flowers on the same inflorescence, spending considerable time on each of them. This pollinator behavior greatly increases the chances of pollinator-mediated self-pollination (geitonogamy; Dafni 1987).

The evolutionary significance of the nectariferous species in a genus that mainly encompasses deceptive flowers is a subject of debate (see Chapter 6). Based on phytogeographic arguments, Dafni (1987) considered the loss of nectar as a recent characteristic within the genus *Orchis*, suggesting that food-deceptive species evolved from nectariferous ancestors. However, the mapping of pollination strategies on phylogenetic trees suggests that deception might be the ancestral character, while nectar presence would be a derived development (Cozzolino et al. 2001; Cozzolino and Widmer 2005; Smithson 2009).

Deceptive Flowers

Almost all species belonging to the genus *Orchis* do not offer any nectar reward to their pollinators and thus attract insects through deceptive strategies (see Table 2.1). The concept of deception traces back to Sprengel (1793), who was the first to notice the empty flowers of *Orchis* species as "scheinsaftblumen" (flowers with sham nectar). Concerning *Orchis*, Darwin (1862; pp. 44–53, 279–285) claimed that the reward in *Orchis* flowers was hidden as a sap within the intercellular spaces. He asserted that while visiting the flowers, the insects would suck the sap by breaking up the spur's walls, with the pollinaria glued to their bodies. Later, in his second edition, he (1877; p. 33) commented, "He who believes in Sprengel's doctrine must rank the sense or instinctive knowledge of many kinds of insects, even bee, very low in scale."

Müller (1871; p. 281) followed Darwin's idea of "hidden sap" and supplied detailed observations on the visitors' behavior. He also criticized Sprengel:

> This absence of honey is a phenomenon without parallel in the vegetable kingdom. Sprengel on that account called them plants with "false nectaries," imagining (sic!) that the insects which visit them are deceived by the colours and form of spur into inserting their heads into the faces of the flower with the expectation of finding honey.

He continued:

> It is clear that Sprengel himself was conscious of not having completely deciphered the enigma. Darwin, too, as we read in his work on the Orchids, never succeeded in surprising insects in the field *Orchis*, although he had observed them diligently not less than twenty years.

While Darwin ignored the possibility of deception, Müller (though he did add detailed observations on pollinators of *Orchis* spp.) was wondering:

> But on the whole it seems improbable to me that the spurs, although lymphatic, yet not honey-bearing of *Orchis morio, latifolia, maculata*, etc., should attract bees.

Finally, Daumann (1971) provided the first experimental evidence that *Orchis*'s flowers are in fact rewardless.

Some studies have shown that even in the absence of nectar production, the presence of a spur significantly increases the chances of a flower being pollinated, whether in orchidaceous or non-orchidaceous plants (Bell 1986; Nilsson 1988; Hodges and Arnold 1995; see Chapter 4). Following these generalized observations, Bell et al. (2009) studied the micromorphology of the inner surface of the spurs of species of *Gymnadenia, Dactylorhiza, Anacamptis,* and *Orchis* s.s. in relation to nectar production and readsorption. They suggested that the orchid species that attract pollinating insects by food deception require these insects to probe the spur for as long and in the same manner as they would if being provided with a nectar reward. Thus, these authors hypothesized that when a potential pollinating insect encounters a spur with the papillate texture—a texture that the insect associates with a genuine reward in a nectariferous orchid species—this encourages the insect to anticipate a similar reward in species possessing morphologically similar, but deceitful, flowers. This hypothesis is based on the underlying assumption that the pollinators of the food-deceptive orchids have previous experience with other coflowering rewarding orchids.

The resemblance of the deceptive flowers to (genuinely) rewarding ones has been termed as "generalized food deception" (Nilsson 1992; Steiner 1998). Generalized food-deceptive species mimic typical floral traits of nectariferous species. The flowers of deceptive species generally appear equally large, with wide labella for pollinator landing, showy colors (see Chapters 4 and 11), and spurs, as well as nectar guides. Most species in the genus *Orchis* s.l. (and of *Dactylorhiza*) are regarded as being "generalized food-deceptive," though this assertion has been demonstrated for only a few species, such as *A. papilionacea* ssp. *palaestina* (=*Orchis caspia*; Dafni 1983), *A. morio* (Nilsson 1983a, 1984; Johnson et al. 2003b; Cozzolino et al. 2005), *O. boryi* (Gumbert and Kunze 2001), *O. mascula* (Nilsson 1983b), and *O. spitzelii* (Fritz 1990), as well as for *Dactylorhiza sambucina* (Juillet et al. 2007; Jersáková et al. 2008; Pellegrino et al. 2008), *D. romana* (Salzmann et al. 2007b), and *D. incarnata* (Vallius et al. 2007).

TRENDS TOWARD A MORE SPECIALIZED DECEPTION

Batesian Mimicry

When a food-deceptive species mimics a specific rewarding model, it is recognized as a case of "Batesian mimicry" (Brown and Kodric-Brown 1979; Bierzychudek 1981; Dafni and Ivri 1981; Little 1983; Dafni 1984, 1986). One may consider Batesian mimicry as species-specific generalized food deception, which implies that all the selective forces exerted by the pollinators are almost the same and can thus be discussed in the same terms (Jersáková et al. 2009). To establish the existence of Batesian mimicry, it must be demonstrated that the mimic resembles the model from the shared pollinators' perceptive point of view to such an extent that the conditioned signal receiver is unable to discriminate between mimic and model and moves freely between them (Wiens 1978; Dafni and Ivri 1981; Jersáková et al. 2009; see Chapter 5). The resemblance should be evolutionarily adaptive so that the mimic receives more visits and has higher fitness in the presence of the model than in its absence (Dafni 1984).

Several authors have mentioned the importance of a close match between the reflectance spectra of the flowers of Batesian mimics and their models (Nilsson 1983b; Johnson 1994a, 2000; Johnson et al. 2003a,b; Anderson et al. 2005; Peter and Johnson 2008; see Chapter 11), suggesting that visual signals may be essential for floral mimicry. All these stages were so far shown in the genus *Orchis* s.l. only for *Anacamptis israelitica* (Dafni and Ivri 1981). In this species, Galizia et al. (2005) showed that the rewarding *Bellevalia flexuosa* and the rewardless *Anacamptis israelitica* are optically similar from the honeybee's perspective. Partial evidence of higher reproductive success associated with the presence of a supposed model has been found also for *Orchis pallens*, which apparently imitates *Lathyrus vernus* (Vöth 1982), and for the color-dimorphic *Dactylorhiza sambucina* with *Mimulus guttatus* (Gigord et al. 2002). In the same color-dimorphic species, it has been shown that the fitness of each of the 2 morphs is strongly increased by the presence of a *Viola* morph of similar color, suggesting a double mimetic effect (Pellegrino et al. 2008). Jersáková et al. (2009) suggest that the rarity of Batesian mimicry in European Mediterranean orchids in comparison with South African species is related to the proliferation of bees in the former region (compare with Chapter 4).

Sexual Attraction

Sexual attraction is a highly specific pollination system that has evolved multiple times in different orchids' lineages (Schiestl 2005; see Chapters 3–6, 11). In the orchid subtribe Orchidinae, it characterizes the wide and diverse genus *Ophrys*. In the *Orchis* genus, increasing evidence suggests that some species may exploit the sexual behavior of insects in order to entice them to visit the flowers. Bino et al. (1982) found that the East Mediterranean species *Orchis galilaea* is pollinated only by males of the bee *Lasioglossum marginatum*. The authors further showed that as in the well-documented sexually deceptive systems, a key role in male attraction is likely played by volatile compounds that induce sexual behavior.

Valterová et al. (2007) found that the pollination of the generalized food-deceptive *Orchis pauciflora* is mainly performed by *Bombus terrestris* queens. The floral bouquet of this species contains constituents of male marking pheromones of many *Bombus* species. Analysis of *O. pauciflora* scent and *B. terrestris* male marking pheromone revealed the presence of the (S)-isomer of (E)-2,3-dihydrofarnesol in both samples. Electrophysiological experiments showed that mainly the (S)-isomer activates the antennal receptors. When flowers were enriched in field experiments with the main compound (E)-β-farnesene, the result was significantly increased pollinaria export. The occurrence of these chemical compounds is supposed to increase plant fitness by attracting *B. terrestris* females as pollinators. However, the role of (E)-β-farnesene in the pollination of *O. pauciflora* remains unclear. Nevertheless, this case might be an interesting example whereby different behavioral aspects of the pollinators' foraging and communication are exploited at the same time.

Another clue for exploitation of male sexual behavior comes from the Mediterranean orchid *Anacamptis papilionacea*. In an insect capture survey, Scopece et al. (2009) found that in a southern Italian population, this species is pollinated by 5 different bee species, but with an unusually high frequency of males in comparison with other food-deceptive orchid species that are mainly, if not exclusively, pollinated by females. These data were interpreted as an indication that *A. papilionacea* is also pollinated (in addition to generalized food deception) by male bees due to nonspecific male sexual attractants. This species is an early flowering orchid that demands pollination when male insects, which emerge first, are more abundant than females (Heinrich 1975). Therefore, the prevalence of male-mediated pollination might simply be explained as the sexual imbalance of the pollinator set available when this orchid is in flower. In this context, the production of potentially active scent compounds, such as those found by Schiestl and Cozzolino (2008), may represent a response to the selective pressure produced by the mainly male pollinator set to increase floral attraction for male pollinators through a still unclear sexual mechanism (Scopece et al. 2009).

Although to date still in a speculative form, it seems possible that complex fragrance patterns use multiple communication channels in order to exploit pollinators' different behavioral aspects to attract and manipulate those (Jersáková et al. 2009). Such a complexity in pollination strategies suggests that the traditional distinction between food-deceptive and sexually deceptive species may be an oversimplification, and that increased efforts should be made to describe these intricate strategies and understand the selective forces behind their evolution.

ORCHIS'S FLOWER FUNCTIONAL MORPHOLOGY

Pollinaria Bending

Darwin (1862; p. 31) was the first to explain the adaptive value of several floral mechanisms of the genus *Orchis* s.l., with special attention paid to the pollinarium-removal mechanism as a "contrivance" (see Chapter 1) to promote cross-pollination. Darwin noted:

> The firmness of the attachment of the cement is very necessary, for if the pollinaria were to fall sideways or backwards they could never fertilize the flower. From the position in which the two pollinaria lie in their cells, they diverge a little when attached to any object. Now suppose that the insect flies to another flower, or let us insert the pencil, with the attached pollinium, into the same or into nectar. . . . How then can the flower be fertilised? This is effected by a beautiful contrivance: though the viscid surface remains immovably affixed, the apparently insignificant and minute disc of membrane to which the caudicle adheres is endowed with a remarkable power of contraction (as will hereafter be more minutely described), which causes the pollinium to sweep through an angle of about ninety degrees, always in one direction, viz., towards the apex of the proboscis or pencil, in the course of thirty seconds on an average. . . . After this movement (of the pollinium), completed in an

interval of time which would allow an insect to fly to another plant . . . the thick end of the pollinium now exactly strikes the stigmatic surface.

Müller (1871; p. 282) realized that the removal of the pollinarium

required the precise time necessary for the viscous stalks of the pollen-masses to attach themselves firmly upon the heads of the insects; and that the time occupied by the pollen-masses securely attached to the insects in becoming depressed upon their stalk so as to be able to rub against the stigma, corresponds nearly to the time employed by the insects in visiting one plant and passing to another. In this way, intercourse between two individuals would necessarily take place.

He was the first to realize that if the pollinarium-bending duration has to be longer than the duration of the visit at the same inflorescence in order to ensure interplant pollinarium transfer, then

it will also be found that in about forty seconds after drawing it out, these masses will have completed that movement of declination by virtue of which they can come in contact with the stigma. Now, as a bee, from what we observed, does not remain on a given spike longer than twenty or twenty-two seconds, it is clear that it cannot fecundate it with its own pollen, but only with that of spikes previously visited (for *Orchis latifolia* and *O. mascula*).

In more recent times, studies by Johnson and his coworkers (Johnson and Nilsson 1999; Johnson and Edwards 2000) estimated the time of pollinarium bending and correlated it with the visiting time of pollinators. These studies elegantly confirmed the Darwinian ideas by highlighting the decisive role of the pollinarium bending as a mechanism to promote the high rate of outcrossing in the genus *Orchis* s.l., as well as in other genera.

Large and Showy Flowers and Inflorescences

One strategy for overcoming avoidance of the rewardless flowers is to provide superior floral signals that elicit higher spontaneous responses in pollinators (see Chapter 11). Some deceptive orchids actually provide very showy floral displays that do attract bees as they search for new food plants (Nilsson 1980, 1983a; see Chapter 7). Jersáková et al. (2009) argued that large and showy flowers of deceptive species display in order to out-

compete other rewarding plants and benefit from the presence of rewarding species of similar color (see below). A comparative study performed by
Huda and Wilcock (2008) confirmed that tropical nectarless orchids have
displays with larger flowers than their rewarding counterparts, irrespec-
tive of their habitat (see Chapter 9).

Inflorescence size is, however, constrained by the availability of resources, and plant species experience a trade-off between the production
of a showy floral display and the resources allocated to other life-history
traits. Both male and female reproductive success can be a decelerating
or an accelerating function of inflorescence size (Dafni 1987; Dafni and
Kevan 1996; De Jong and Klinkhamer 1994). In the former case, the evolution of pollination strategies that increase the efficiency of one or both
reproductive functions can allow a reduction of investment in inflorescence size. This resource saving may represent a significant improvement
in allocation strategy by releasing energy for other life-history traits, particularly in resource-limited habitats.

A recent survey of orchid reproductive performance in groups characterized by alternative pollination strategies (which also includes rewarding and deceptive *Orchis* species) showed dramatic differences in terms of
pollen fate. Indeed, food-deceptive species are found to export less pollen
than rewarding ones, and most of this pollen is lost. Rewarding species,
on the other hand, show high rates of pollen export and high rates of
pollination, indicating that less pollen is lost (Scopece et al. 2010). These
findings suggest that in a less efficient pollination strategy, a higher number of flowers should be produced in order to ensure pollination, and that
there may be a selection toward the production of large inflorescences in
the *Orchis* species (Scopece et al. in preparation). This latter outcome was
also directly demonstrated by Johnson and Nilsson (1999), who showed
that individuals of *A. morio* with more flowers attain higher levels of fruit
set, thus suggesting the existence of a selection for greater floral display in
this generalized food-deceptive species.

FLOWER LONGEVITY

Dafni and Bernhardt (1989) proposed that food-deceptive orchids, which
usually experience lower pollinator visitation rates than rewarding orchids,
may evolve into long-lived flowers in order to maximize the opportunities
for mating (see Chapter 12). Based on this hypothesis, Ruxton and Schaefer (2009) argued that the evolution of long-lasting flowers may at the
same time cause deceptive orchids to flower earlier than rewarding orchids.

The duration of flowering also depends on the time required for the development of the mature gametophyte. In tropical orchids, female gametophyte development is triggered by the arrival of pollen on the stigma (Zhang and O'Neill 1993), and the process that follows pollination can last more than 3 months. However, due to seasonality, temperate species, such as those included in the genus *Orchis*, have less time for completing this process. As a result, the female gametophyte is already partially developed before the arrival of the pollen (Barone-Lumaga et al. 2010), and the flowering time is comparatively shorter when compared with tropical groups.

POLLEN LONGEVITY

Dafni and Firmage (2000) proposed the hypothesis that the longer the pollen travel distances, the higher the pollen longevity. This concept is in line with the suggestion that when the average transit time between anther and stigma is longer, long-lived pollen may be favored (Proctor 1998). Such a phenomenon would be expected in rare and highly dispersed species, as it is for many Euro-Mediterranean orchids (Neiland and Wilcock 1995). Neiland and Wilcock (1995) found that the pollen of orchid species (*Gymnadenia conopsea*, *Dactylorhiza maculata*, and *D. purpurella*) was still able to germinate after a storage period of 37–51 days.

Bellusci et al. (2010) demonstrated the existence of a relationship between pollen viability duration and the type of deceptive pollination syndrome. The food-deceptive species (*A. papilionacea* and *O. italica*) showed significantly higher values of pollen viability than both the sexually deceptive (*Ophrys*) and shelter-mimicking (*Serapias*) species, indicating that their pollen can be better adapted to overcome longer traveling distance and low likelihood of pollinaria removal (Dafni and Firmage 2000). Shortened pollen viability duration, like that exhibited in their study by sexually deceptive and shelter-mimicking species, could limit the number of interspecific mates. In contrast, species with active postzygotic barriers, such as food-deceptive species (Scopece et al. 2007, 2008), may have longer pollen viability as a consequence of eliminating genetic contamination. A remarkable variation in pollen exine sculpturing, with a different level of variation within species groups, has been detected in microscopic analyses (Barone-Lumaga et al. 2006). In particular, a larger variety of exine conditions was found in the genera *Dactylorhiza* (psilate, psilate-scabrate, and reticulate) and *Orchis* s.s. (psilate, reticulate, perforate-rugulate, and baculate). This variation has no unequivocal correspondence to phylogenetic patterns and more likely reflects some ecological adaptations. This indirect evidence further supports the idea that the high viability and longevity of orchid pollen may be an adaptive response to species-specific pollination strategies.

POLLINATOR BEHAVIORAL AND COGNITIVE ASPECTS: NAÏVETÉ, LEARNING, AND AVOIDANCE

Delpino (1868–1875) was the first to mention that the pollination of deceptive *Orchis* is carried out by inexperienced bumblebee queens, and that after visiting empty flowers, they learn to avoid the nonrewarding ones. He argued that deceptive orchids are visited only during the first 2 or 3 days of flowering, when visiting bumblebees have newly emerged. At that stage, the bees have not become acquainted with appropriate nectariferous flowers, but soon recognize the deception. Nowadays, it is widely accepted that deceptive flowers exploit inexperienced pollinators that have not yet learned to avoid empty flowers (Vogel 1972, 1983; Nilsson 1980, 1983a; Ackerman 1981, 1986; Little 1983; Dafni 1984, 1987; Fritz 1990; Herrera 1991, 1993; Dressler 1981, 1993b; Smithson and Macnair 1997; Ferdy et al. 1998; Gumbert and Kunze 2001; Smithson 2002; Gigord et al. 2002; Smithson and Gigord 2003; but see Johnson et al. 2003a; Juillet et al. 2011; and see Chapters 7, 10, 11).

Dafni (1983) initially suggested that generalized food-deceptive pollination might rely on pollinators (mainly solitary bees, in *Orchis*) with poor memory or discriminatory abilities. However, this simplistic assumption was found to be more complicated, with further visits to a deceptive flower by bees dependent on several factors related to neurophysiological capacities, previous experience of the individual bee, and time elapsed from the last visit to a rewarding flower (Gumbert and Kunze 2001).

According to Chittka and Raine (2006), the short-lived memories usually last only a few seconds and rapidly decay, even without interference, being erased with relative ease by competing information. They found that bees do not easily forget the image of the previously encountered flower within the first few seconds (0–2 s), and that if newly incoming stimuli match the signal, then the bees visit another flower of the same species. After longer intervals (3–6 s), the bees tend to switch to different flower species and are more likely to choose those with similar coloration (Chittka et al. 1997). These factors may explain the frequency-dependent selective forces that are supposed to mold the efficiency of deception. A particular plant's success thus depends not only on the quality of its own signal but also on the efficiency of the signals of other species in the vicinity, as well as on their relative abundance, distribution, and degree of spatial intermixing (Chittka and Raine 2006).

Renner (2006) noted that while considering the role of naïve pollinators in the success of deceptive species, it is important to consider other factors as well, such as the individual bee's experience, densities of the deceptive and rewarding plants, yearly variation in emergence time, the existence of color morphs (see below), and the locality. These factors may vary greatly and may thus influence the process of learning and avoidance, which in turn may affect the visitation rates for the deceptive flowers. This variability may explain some conflicting results. For example, in *Orchis dinsmorei*, the pollination rate was found to be 70% in its first week of flowering, but decreased to 10% in the last week (Dafni 1986). Contrasting results were found for the spring-blooming orchid *Dactylorhiza sambucina*, in which pollinaria export proportions of flowers on the upper part of the inflorescence, which open late in the season, were significantly higher than those of flowers farther down, which flower early in the spring (Kropf and Renner 2008).

Ruxton and Schaefer (2009) mentioned several factors that may have an impact on learning and avoidance, including temporal variation in total pollinator availability, pollinator longevity, and unlearned response; the stability of plant communities over time; and the learned responses of individual pollinators. Learning alone would not necessarily select for early flowering by nonrewarders if temporal variation in pollinator numbers was strong or if naïve pollinators consistently appeared throughout the flowering season. Internicola et al. (2009) suggested that the reproductive success of deceptive species may be affected by temporal variation in the plant community composition with respect to the similarity in floral cues to rewarding coflowering species and the pollinators' experience (see Chapters 4, 7).

When bumblebees must divide their attention among an increasing number of rewarding and nonrewarding flower types, their ability to discriminate rapidly decreases (Dukas and Real 1993a,b). Bees usually do not learn the position of individual flowers, but instead memorize the location of rewarding and nonrewarding plants (Menzel 1985). A study on *Anacamptis* (=*Orchis*) *morio* found that *Bombus lapidarius* queens, both the experienced as well as the naïve ones, continue to probe *A. morio* as long as these empty flowers are at low density (Johnson et al. 2003b).

Ruxton and Schaefer (2009) emphasized the fact that later emerging bees may have a better chance to learn and avoid deceptive flowers than early emerging ones, because deceptive and rewarding plant species are more likely to coflower later rather than early in the season. This process speeds up the rate of bee learning (Dyer and Chittka 2004a,b). Another factor that may enhance the learning of naïve bees is that they learn more quickly when foraging in the company of experienced conspecifics (Leadbeater and Chittka 2007), which are less frequently encountered. These 2 factors hold only for social bees, which may transfer information between individuals, while many solitary bees are involved in food-deceptive pollination.

Gumbert and Kunze (2001) studied the individual bumblebee experience in relation to the frequency of visits to the deceptive *Orchis boryi*. They found that bumblebees approach and visit mainly orchids that more closely resembled the nectar-supplying species which they just visited. If being naïve were an important factor, then the bees would be expected to visit deceptive flowers in proportion to their abundance in the experimental populations. Thus, visitation rates seem to depend more on the individual bee's previous experience. As such, the mechanism of pollinators' learning must be understood within the context of the plant communities in which deceptive and rewarding species co-occur in space and time (Fritz 1990; Internicola et al. 2008, 2009).

Decreasing visitation rates, as a consequence of pollinator learning, may exert sufficient selective pressure on flowering phenology to explain why, in many temperate deceptive orchids, individuals that flower in early spring are pollinated at a higher rate than late-flowering ones (Tremblay et al. 2005). Higher reproductive success for early flowering individuals has been observed in *Anacamptis morio* (Nilsson 1984), *Orchis mascula* (Nilsson 1983b), *Orchis spitzelii* (Fritz 1990), *Orchis boryi* (Gumbert and Kunze 2001), *Orchis dinsmorei* (Dafni 1986), and *Dactylorhiza sambucina* (Nilsson 1980). Due to a general trend of low reproductive success in food-deceptive orchids in comparison with rewarding ones (regardless of their flowering season; Dafni and Ivri 1979; Gill 1989; Neiland and Wilcock 1998; Tremblay et al. 2005), selective pressure exerted on deceptive species is expected to enhance their floral advertisement as an adaptation in order to exhibit stronger signals and thus attract more pollinators (Jersáková et al. 2009).

EFFECTS OF POLLINATOR BEHAVIOR ON DECEPTIVE SPECIES' POPULATION ECOLOGY

Early Flowering

Heinrich (1975) was the first to hypothesize that the early flowering periods of rewardless species may represent an adaptive strategy to benefit from the higher abundance of naïve pollinators and to avoid competition with later-flowering rewarding species. Avoidance learning slows down when pollinators are unlikely to encounter deceptive and rewarding plants in short sequences (Dukas and Real 1993a,b; Internicola et al. 2006). Consequently, deceptive species may benefit from flowering at a time when few or no rewarding sympatric species do so. Jersáková et al. (2009) considered early flowering an adaptation to receive exploratory visits of inexperienced naïve pollinators that have just emerged from hibernation (Nilsson 1980, 1983a, 1984; Fritz 1990). In addition, pollinator avoidance learning usually increases with foraging experience (Internicola et al. 2007, 2009).

Kindlmann and Jersáková (2006) showed that the peak of flowering among deceptive orchids may occur earlier than that of rewarding orchids. Based on the onset of flowering among 233 European orchid species, Pellissier et al. (2010) found that food-deceptive orchids start flowering significantly earlier than rewarding orchids. Thus, they suggest that early flowering may be a common strategy among generalized food-deceptive orchids and hence a key factor in the success of generalized food deception.

While working on artificial models, Internicola et al. (2008) found that deceptive inflorescences received the most pollinator visits in the early phenology treatment, strongly suggesting that under natural conditions, the reproductive success of deceptive plants will be affected by the phenology of rewarding sympatric species. However, Ruxton and Schaefer (2009) mentioned that early flowering could simply be a natural corollary of the longevity of flowers needed to combat negative frequency-dependent selection and low overall visitation rates by pollinators, rather than a trait that has been specifically selected to reduce temporal overlap with competing rewarding species. In *Orchis dinsmorei*, for instance, the pollination rate of 70% in its first week of flowering was found to drop to 10% in the last week (Dafni 1986). Contrasting results were gained by Kropf and Renner (2008) for *Dactylorhiza sambucina*. Two interacting factors may be at work at different times, most important among them the density of other rewarding flowers, annual/seasonal variation in the emergence of bees, and for species with color morphs, such as *D. sambucina*, also the frequencies of yellow and purple morphs, with the more common morph being avoided more rapidly (Gigord et al. 2001).

Competition between Deceptive and Rewarding Coflowering Species

Several authors assumed that the early flowering of deceptive orchids may allow them to escape the potential competition from rewarding coflowering species in early spring (Heinrich 1975; Internicola et al. 2008; Jersáková et al. 2009; Pellissier et al. 2010), as well as from other conspecific deceptive plants which flower in dense populations (Internicola et al. 2009). The evidence for such a possibility is, so far, only circumstantial. Many species of *Orchis* and *Dactylorhiza* flower gregariously in early spring and are thus subjected to increased intraspecific competition for pollinators (Jersáková et al. 2009). Although the number of pollinaria removed and the level of fruit set have a tendency to increase with the number of plants, the proportion of pollinaria removed per plant decreases (Fritz and Nilsson 1996). Therefore, individuals flowering early or late in relation to population peak usually have higher pollination success than those flowering in the middle of the blooming period (Fritz 1990; Parra-Tabla and Vargas 2007).

THE RELATIVE DENSITY OF REWARDING AND DECEPTIVE POPULATIONS

Internicola et al. (2008) found that when deceptive and rewarding inflorescences are simultaneously present, bumblebees learn to avoid the deceptive ones more efficiently when these are offered after, rather than before, the rewarding inflorescences. The authors proposed that early flowering might be additionally advantageous to deceptive species, because the bumblebee's learning depends on the frequency with which rewarding plants are encountered when it acquires experience. This hypothesis is congruent with that of Smithson and Gigord (2003), who found that bumblebees avoid the deceptive inflorescences more efficiently when their relative frequency is high, which, practically, is related not to the flowering period but rather to the density of other rewarding species. By using potted plants in the field, Johnson et al. (2003a) found that the visitation rates for rewardless orchids depend not only on the relative abundance but also on the absolute abundance of empty and rewarding flowers, which in turn affect the pollinator density in a patch and movement patterns between patches.

EFFECTS OF POLLINATOR BEHAVIOR
ON DECEPTIVE SPECIES' FLORAL TRAITS

Color Variation and Polymorphism

In food deception, and especially in Batesian mimicry, there is a critical constraint of learning and avoidance imposed by the pollinators that recognize the deceptive plants and ignore them (Dafni 1984; Nilsson 1980,

1992). Some authors (Johnson et al. 2003a,b; Internicola et al. 2006, 2007) have mentioned that the pollinators exert selective pressure on the plants toward adaptations which may impede the learning and avoidance process that can reduce the mimics' pollination success. To overcome the pollinators' learning ability, many orchids display a variable set of floral cues, which, together with shape, size, scent, and color, are usually associated by the pollinators with a reward. These floral mimic signals utilize the innate and/or acquired preferences of pollinators and antagonize their inhibitory learning (Schiestl 2005; Internicola 2007, 2008, 2009).

Accordingly, in the genus *Orchis*, as in other generalized food-deceptive orchids, high levels of variability in floral traits, such as color (Nilsson 1980; Ackerman 1981; Cropper and Calder 1990; Nilsson 1992; Moya and Ackerman 1993; Pettersson and Nilsson 1993; Aragón and Ackerman 2004) or scent (Moya and Ackerman 1993; Andersson et al. 2002; Salzmann et al. 2007a,b), have been regarded as a means to disrupt the associative learning of pollinators through inhibition of their ability to recognize and avoid nonrewarding flowers (Heinrich 1975; Nilsson 1980; Dafni 1987; Pettersson and Nilsson 1993). In the recent review by Ackerman et al. (2011) on temperate and tropical orchid species, the finding that floral traits are significantly more variable in deceptive species than in rewarding ones supports the observed pattern of higher polymorphism of deceptive floral traits.

Variation in flower color can be expressed as a continuous variation (in *A. morio*, Nilsson 1984; in *A. papilionacea ssp. palaestina*, Dafni, personal observation) or as a distinct polymorphism (see Chapters 7 and 11). The recent monograph on the genus *Orchis* s.l. (Kretzschmar et al. 2007) recognizes only one *Orchis* species with "polychromy," namely *O. laeta*, which shows intrapopulation variation from whitish to purple, but not a real discrete polymorphism. A few species of *Dactylorhiza* exhibit discontinuous, genetically determined variation in flower color (Nilsson 1980; Gigord et al. 2001; Koivisto et al. 2002; Pellegrino et al. 2005; Jersáková et al. 2006b) and were used to test the hypothesis of reducing learning and avoidance. The experimental evidence failed to corroborate the hypothesis on the benefit of color polymorphism (Aragón and Ackerman 2004; Smithson et al. 2007; Juillet et al. 2011). Juillet and Scopece (2010) suggested that polymorphism might be maintained by a relaxed selection of floral traits due to weak flower constancy among deceptive species, rather than by an advantage in delaying pollinators' avoidance learning ability.

Some studies relate the maintenance of color polymorphism in deceptive orchids to a negative frequency-dependent selection exerted by pollinator behavior (Smithson and Macnair 1997; Ferdy et al. 1998; Gigord et al. 2001; Internicola et al. 2006; but see Pellegrino et al. 2005). However, in sharp contrast to the negative frequency-dependent hypothesis, the results of Dormont et al.'s (2010) study on natural and artificial color-dimorphic populations of *Orchis mascula* showed that the presence of a few white mutant individuals led to an increase in the pollination success of the more abundant purple-flowered morph. Several other studies relate the maintenance of polymorphism to different selective pressures on the color morphs (Jersáková et al. 2006b for *Dactylorhiza sambucina*; see also Tremblay and Ackerman 2007). An alternative view has been proposed by Internicola et al. (2009), who suggest that deceptive species may benefit from being polymorphic for floral cues because different morphs may be favored by different pollinators at different times.

Color Similarity

Pollinator behavior can also affect the color similarity between deceptive and rewarding species. Indeed, if deceptive and rewarding coflowering plants have similar flower colors, then potential pollinators may be less efficient in their discrimination due to a delay in their avoidance learning (Dafni 1984; Ackerman 1986; Nilsson 1992; Dyer and Chittka 2004a,b), which would result in higher visitation rates to the deceptive plants (Gumbert and Kunze 2001; Gigord et al. 2002; Johnson et al. 2003a,b; Dyer and Chittka 2004a,b; Internicola et al. 2007). Furthermore, even highly experienced pollinators (that are supposed to play a limited role in deceptive pollination systems) may tend to visit more deceptive inflorescences when their color resembles that of nectariferous plants previously visited (Gigord et al. 2002). To be sure, both experimental and field observations indicate that bees' foraging decisions are strongly influenced by their previous experience, and that bees choose deceptive species more frequently if they forage on more similarly colored species (Gumbert and Kunze 2001; Gigord et al. 2002).

Color similarity between the flowers of rewarding and deceptive species may affect pollinator visitation later in the season due to long-term memory effects. Gumbert and Kunze (2001) found that the pollination success of the food-deceptive orchid species *Orchis boryi* and *O. provincialis* is strongly correlated with the degree of similarity in their color to particular rewarding species known to individual pollinators. Some other studies (Johnson 2000; Anderson and Johnson 2006; Juillet et al. 2007; Peter and Johnson 2008) showed that the coflowering of deceptive species with more or less similarly colored nectar-producing species may increase the reproductive success of deceptive orchids. Likewise, an increase in fitness of each of the 2 *Dactylorhiza* morphs is correlated with the presence of a *Viola* morph of similar color (Pellegrino et al. 2008).

In a series of studies on color polymorphism and learning in bumblebees, Internicola et al. (2007) showed that over time, bumblebees increasingly avoid the deceptive plants, but at a significantly faster rate when deceptive and rewarding plants have dissimilar flower colors. This study also suggested that the spatial distribution of the plants may have a strong influence on the rate of learning and avoidance. Indeed, deceptive plants receive more visits when mingled in heterospecific patches with rewarding plants of similar flower color than when mingled with dissimilar ones.

Pollinator avoidance learning is affected by both the similarities of corolla color and the temporal flowering order of rewarding and deceptive species. Thus, it should be beneficial for deceptive plants to flower before rewarding plants. Indeed, early flowering may ensure that visitation rates are independent from the similarity in corolla color to any coflowering rewarding species (Internicola et al. 2008), which suggests that both pollinators and sympatric plant species are likely to exert strong selective pressure on the flowering phenology of deceptive orchids. Internicola et al. (2009) showed that deceptive inflorescences experience an increased number of exploratory visits by bumblebees when they are similar rather than dissimilar to the rewarding ones. The authors thus concluded that deceptive species may benefit from changes in cue-reward association in the long term as the similarity in floral cues between the rewarding and the deceptive species increases. All these studies show the tight interconnections among learning processes, color similarity, presence of other rewarding plants, and seasonal changes.

Odor Similarity and Variation

Given that scent is learned by bees quicker than color (Menzel 1985; Chittka and Raine 2006), it is expected that deceptive orchids would be scentless (Dafni 1987) or have high variability (Nilsson 1980; Moya and Ackerman 1993; Andersson et al. 2002; Salzmann et al. 2007a) in order to reduce pollinators' chances for learning and avoidance (Dafni 1987; Jersáková et al. 2009). The absence of scent is thus considered a selective advantage for increasing orchid fitness (Valterová et al. 2007). Jersáková et al. (2009) argued that because scent acts as a diffuse and long-distance signal, mimics growing near scented models might not need to produce any scent at all.

A study by Galizia et al. (2005) found no evident similarity in the scents between the rewardless *Anacamptis israelitica* and the rewarding *Bellevalia flexuosa*, suggesting that similarity in floral scent may not play a primary role in this mimicry system. Nevertheless, the odor of *O. israelitica* was found to be weaker than that of its model. In behavioral experiments with bumblebees foraging on artificially scented flower dummies, Gumbert and Kunze (2001) found that *Orchis boryi* produces a faint scent that is almost imperceptible to humans. Head-space analysis showed a low concentration of volatiles in deceptive flowers as compared with rewarding ones. Moreover, discrimination was poorest if the floral scents of the model and the mimic were similar.

Jersáková et al. (2009) suggested negative frequency-dependent selection as a possible mechanism that might maintain scent polymorphism. Accordingly, rare floral morphs would be expected to experience a selective advantage in comparison with common floral morphs, because pollinators are not able to learn to avoid them (Smithson and Macnair 1997; Ferdy et al. 1998). However, the positive impact of scent polymorphism on plant pollination success was not supported by a manipulative experiment in *Anacamptis morio* (Salzmann et al. 2007b; Juillet et al. 2011).

THE ROLE OF THE GENUS *ORCHIS* S.L. IN TESTING THE HYPOTHESES ON THE EVOLUTION OF DECEPTION

Most of the first studies on deception in orchids were completed in Europe and around the Mediterranean, where *Orchis* s.l. is a very common and widespread genus. Therefore, it is not surprising that several hypotheses on the possible evolution of floral deception have been examined while using *Orchis* s.l. and its closely related genera (especially *Dactylorhiza*) as a model system.

The Outcrossing Hypothesis

One hypothesis on the evolution of deception posits that deceptive species have an evolutionary advantage, because they experience higher rates of outcrossing than rewarding species. Dafni and Ivri (1979) compared visitor frequency on *Orchis* (=*Anacamptis*) *collina* (a nectarless species) with *Orchis* (=*Anacamptis*) *coriophora* (a nectariferous species) in the same season and habitat. Their results showed that pollinators visit fewer flowers on rewardless plants than on the nectariferous congeneric species and then switch to another conspecific plant. It has been proposed that deception serves as an adaptation to reduce geitonogamy (Dafni and Ivri 1979; Dressler 1981, 1993b; Dafni 1984, 1986; Ackerman 1986; De Jong et al. 1993; Klinkhamer and De Jong 1993; Johnson and Nilsson 1999; Smithson 2002, 2006; Johnson et al. 2003a,b; Johnson et al. 2004; Jersáková and Johnson 2006; Jersáková et al. 2006a, 2009) due to the visit to only a few flowers on the same nectarless plant. This hypothesis was confirmed by genetic analyses showing lower levels of genetic differentiation between populations and higher levels of genetic diversity within populations among deceptive species as compared with rewarding ones (Tremblay et al. 2005; Cozzolino and Widmer 2005).

The advantages of reduced pollinator-mediated self-pollination may compensate for the low visitation rate and level of fruit set (Gill 1989; Neiland and Wilcock 1998; see Chapter 10), which may confer an advantage to deceptive species. In this context, nectar-producing species should be less prone to inbreeding depression than deceptive ones, because repeated geitonogamous selfing could purge genetic load (Juillet et al. 2006). This hypothesis was tested by comparing the levels of geitonogamy in nectar-supplemented and nectarless flowers (Johnson and Nilsson 1999; Smithson and Gigord 2001; Smithson 2002; Johnson et al. 2004; Jersáková and Johnson 2006), which yielded equivocal results. Johnson and Nilsson (1999) showed that the experimental addition of nectar to the rewardless flowers of *Orchis mascula* and *Anacamptis morio* resulted in significant increases in the number of flowers probed by each pollinator, as well as the total time spent on plants. However, they did not obtain empirical evidence to support their prediction that nectar production would result in higher levels of geitonogamy. The results of Johnson et al. (2004) support the hypothesis that deception in orchids reduces geitonogamous self-pollination. Smithson (2002) found that nectar-supplemented *Anacamptis morio* flowers had more visits per plant, but not higher fruit set.

The Magnet Species Hypothesis

The magnet species hypothesis (Laverty 1992) posits that the presence of different coflowering rewarding species increases the pollination success of neighboring plants with inferior rewards. At the basis of this hypothesis is the idea that the magnet species increases the local abundance of pollinators, and that neighboring plants (both rewarding and deceptive) may gain a net benefit from the greater abundance of pollinators around the more attractive magnet species (Thompson 1978; Johnson et al. 2003b; Pellegrino et al. 2008). Several studies have shown that generalized food-deceptive species have higher reproductive success when growing close to rewarding plants, even though they do not resemble them (Dafni 1983; Laverty 1992; Pellmyr 1986; Johnson et al. 2003b). In this case, the rewarding plants act as a magnet species from which the deceptive species benefit, even though the pollinators may show relative constancy to the magnet species and cause reproductive interference (Laverty 1992; Alexandersson and Ågren 1996).

Remote Habitat Hypothesis (=Competition Hypothesis)

The remote habitat hypothesis predicts that generalized food mimics have better performance in habitats with few rewarding plants, because rewarding and deceptive plants compete for pollinators (Lammi and Kuitunen 1995). The common occurrence of deceptive orchids in habitats lacking many rewarding plants, such as marshes, has been considered as consistent with the idea of relaxed competition (Heinrich 1975; Boyden 1980; Nilsson 1980; Dafni 1984; Firmage and Cole 1988; Lammi and Kuitunen 1995; but see Johnson et al. 2003b). Supporting evidence is supplied by results showing decreased pollination success in patches of the deceptive marsh orchid *Dactylorhiza incarnata* following the experimental addition of nectar-producing *Viola* flowers (Lammi and Kuitunen 1995; but see Pellegrino et al. 2008).

The Pollinaria Removal Hypothesis

Harder (2000) argued that deception is successful in orchids because the male function is not compromised as much by low rates of visitation as it is in plants with granular pollen. Following this notion, Smithson and Gigord (2001) proposed that deception might actually increase the likelihood of pollinaria removal from flowers. Support for this rather counterintuitive idea came from a reduction in pollinaria removal from flowers of *Himantoglossum robertianum* following experimental supplementation with nectar (Smithson and Gigord 2001).

However, in a similar experiment conducted on *Anacamptis morio*, Johnson et al. (2004) revealed contrasting results that do not support the pollinaria removal hypothesis for the evolution of deception. The addition of nectar to spurs of *A. morio* resulted in significant increases in both the total number of pollinaria removed from inflorescences and the rate of removal of pollinaria from individual probed flowers. The pollinaria removal hypothesis was also not supported by previous studies of *A. morio*, which showed either a significant increase in pollinaria removal (Johnson and Nilsson 1999) or nonsignificant effects on overall pollinaria removal from inflorescences (Smithson 2002). Other contrasting results emerged from a comparative study based on direct observation of pollinaria removal in food-deceptive and rewarding species, demonstrating a higher proportion of pollen export in rewarding species (Scopece et al. 2010).

CONCLUSIONS

In this chapter we described the post-Darwinian advances in several aspects of research on orchid ecology and evolution. Since the publication of Darwin's pioneering book (1862), the orchid family has become a model for studying some of the most fascinating processes of flowering-plant evolution. The extraordinary species richness of this plant family is mirrored in the numerous astonishing adaptations of flowers to their pollinators, many of which still remain poorly understood and call for additional research.

The floral functional morphology of orchids represented the core of Darwin's interest. He meticulously described orchids' floral traits and used his natural selection theory as a key to understanding their evolutionary significance. With this vision, he was able to predict precisely the role of several "contrivances," notwithstanding the remarkable dearth of background information at the time. Thereafter, the discovery of floral deception offered a new interpretative approach to the interaction between flowers and pollinators, and generations of post-Darwinian researchers have contributed to the understanding of this peculiar pollination strategy.

Despite the fact that orchids in general, and the members of the genus *Orchis* in particular, have been at the center of pollination biologists' attention for over 150 years, we are still far from discovering all the underlying reasons for the surprising abundance of deceptive species in orchids; nor have we gained a satisfactory understanding of the evolution and function of floral deception. For instance, while sexual deception is the rule in the genus *Ophrys* (see Chapter 3), there is growing evidence of sexually based pollination in *Orchis*, which is still poorly understood.

It is expected that a refined chemical analysis, coupled with neurophysiological examinations and behavioral bioassays, will reveal the role of specific chemicals in the selective attraction of male bees. Such an approach may elucidate the question of the existence of a continuum between generalized food deception and specific sexual deception (see Chapters 4 and 5) and the possible transition from generalized to specialized pollination strategies, which may impose different pollinators' mediated selective forces on the plants. Moreover, while the role of visual cues has been deeply investigated, the role of scent in food-deceptive pollination strategies and whether its contribution may differ between generalized and more specialized (Batesian) deceptive systems remain unresolved.

Indeed, these 2 food-deceptive strategies are often interpreted as a continuum, and similarities rather than differences are normally emphasized. However, they are likely to differ in their evolutionary trajectories because of the potentially different actions of pollinator-mediated natural selection. In generalized food mimicry, pollinator selection can even facilitate the evolution of extreme traits (as supertraits) that enhance the innate pollinator response. In contrast, in Batesian mimicry, a strong evolutionary constraint on the trait selection of deceptive orchids is imposed by the similarity to the rewarding model and by its innate floral attraction limits.

Even after more than a century since Darwin's work, we should still refer to the master's guidelines and perform targeted studies on the strength and direction of pollinator-mediated natural selection in the 2 deceptive systems (and their models) in order to better understand their differences. The genus *Orchis*, one of Darwin's preferred pet systems, may still represent the best model for tackling these new research avenues.

THREE

Ophrys Pollination:
From Darwin to the Present Day

Nicolas J. Vereecken and Ana Francisco

INTRODUCTION: DARWIN'S CONTRIBUTIONS TO *OPHRYS* POLLINATION

> If the Orchid book . . . had appeared before the "Origin," [Darwin] would
> have been canonised rather than anathematized by the natural theolo-
> gians.
> —Asa Gray, in *The Life and Letters of Charles Darwin* (F. Darwin 1887;
> p. 274)

This quote from the eminent Harvard botanist Asa Gray (1810–1888),
chief American Darwinian advocate, clearly illustrates the intellectual
impact of Charles Darwin's *On the Various Contrivances by Which British
and Foreign Orchids Are Fertilised by Insects, and on the Good Effects of Intercrossing*, published in 1862. At the time of its publication, skepticism and
rationalism had already irreversibly changed the worldview of Darwin,
who regarded his "orchid book" as something that would "do good" to
his *Origin of Species*, published only 3 years earlier, in 1859. The main goal
of Darwin's treatise on orchids was to expand his argument that most,
if not all, hermaphrodite organisms require at least occasional crosses
with conspecifics. Several important details of the morphology of orchid
flowers had already been described by Sprengel (1793; English translation
in Sprengel 1996) and Brown (1833), who also provided an account of
the key role of insects in removing the orchids' pollen masses. With its

wealth of details and evidence on reciprocal morphological adaptations between insect pollinators and orchid flowers, Darwin's volume on orchids remains a classic for modern pollination biologists (see Chapter 1).

At the time of Darwin's observations, the field of orchid pollination biology was still in its infancy. Several authors before him had already gathered pollinator records and performed comparative morphological measurements on various organs of orchid flowers suspected to influence the insects' ability to transfer pollen from one flower to the next, but the details of the orchids' pollination strategies were still poorly understood. At the turn of the 18th century, Sprengel (1793) described in his book *Discovery of the Secret of Nature in the Structure and Fertilization of Flowers* that the intact, fresh, and unpollinated flowers of several orchid species were invariably devoid of nectar. Several of these species, such as *Dactylorhiza majalis* (syn. *Orchis latifolia*) and *Anacamptis morio* (syn. *Orchis morio*), which had also been examined by Darwin, were coined "sham nectar producers" by Sprengel (1793). This phenomenon perplexed Darwin to a great extent, as witnessed in a section of his orchid book where he claims he could not believe in "so gigantic an imposture" (Darwin 1862; p. 46) that plants attract their pollinators through an "organized system of deception" (Darwin 1862; p. 45). For Darwin, the reciprocal adaptations between plants and insects had to be based on mutual benefits at least to some extent, and he could hardly imagine that insect-pollinated deceptive plants could persist in natural habitats, regularly contributing to the next generations over vast periods of time by consistently attracting insects incapable of breaking this floral masquerade (see Chapter 2). In today's terms, Darwin regarded these orchids as having engaged in an evolutionary dead end.

Among the various deceptive orchid species investigated by Darwin in England are the *Ophrys* orchids (Figure 3.1). To the human eye, the flowers of these orchids resemble insects by color, shape, and hairiness, a phenomenon that has inspired a fanciful taxonomy tracing back to Linnaeus, who described the type species of this genus as the fly orchid, *O. insectifera* (syn. *O. muscifera*), since its overall habitus is reminiscent of insects perching on a grass stem (Figure 3.1). Darwin knew very well that this orchid does not produce nectar. In addition, he did not notice any obvious floral scent. However, after having found "abundant pollen on the stigmas of flowers, in which both their own pollinia were still in their cells" (Darwin 1862; p. 57), he became quite convinced that the activity of insects was essential if the fly orchid was to set seeds at all. Nevertheless, the raison d'être of the *Ophrys* insectiform flowers remained an open

Fig. 3.1 Three *Ophrys* species investigated by Darwin over the years in Kent and neighboring areas of England: (*top left*) *Ophrys apifera*, the bee orchid; (*top right*) *Ophrys insectifera*, the fly orchid, with a pseudo-copulating male wasp of *Argogorytes mystaceus* (Hymenoptera, Crabronidae); (*bottom*) *Ophrys fuciflora*, the late spider orchid, with a pseudocopulating male of *Eucera longicornis* (Hymenoptera, Apidae). Photos by N. J. Vereecken.

question for Darwin, because he never observed any pollinator in action. Over the years, he tried to gather more clues on the adaptive significance of the bizarre insectiform flowers of the fly orchid, and his field observations revealed that a great proportion of flowers in this species systematically fail to produce seeds. Darwin found out that less than 50% of the flowers examined had been visited (Table 3.1), and out of 11 plants with

Table 3.1 Observations on the rate of pollinia removal by insects in *Ophrys insectifera* over successive years in England

Year of observation	Total (*N*) of plants	Total (*N*) of flowers	Both pollinia or one removed by insects	Both pollinia remain in anther (not removed)
			(*N*) of flowers	
1858	17	57	30	27
1858	25	65	15	50
1860	17	61	28	33
1861	4	24	15	9
Total	63	207	88	119
		Total (%)	42.5	57.5

Source: After Darwin 1862; p. 59.

49 flowers investigated in 1861, an "extraordinarily favourable [year] to this species in this part of Kent" (Darwin 1862; p. 60), only 7 flowers on 5 plants produced capsules.

A low number of insect visits was also inferred from the observations of some flowers of the early spider orchid, *O. sphegodes*. Darwin was obviously puzzled by these results, and realized how difficult it was to come up with any reasonable explanation for this particular phenomenon. It all boiled down to 3 major issues: (1) How could such an apparently deceptive pollination strategy be maintained?; (2) Why have these flowers not been "rendered more attractive to insects" (Darwin 1862; p. 59)?; and (3) Could the pollinators not remove the orchids' pollen masses efficiently? All these questions are briefly encapsulated in a very concise statement that says it all: "Something seems to be out of joint in the machinery of [the Fly orchid's] life" (Darwin 1862; p. 60).

Darwin was even more perplexed when he noticed the deep contrast occurring between the insect-pollinated flowers of *O. insectifera*, which present a low fruit set of about 14%, and those of the bee orchid, *O. apifera*, the only species in the genus that has evolved morphological adaptations to self-fertilization, in which almost every flower produces a capsule, resulting in an extremely high fruit set. The rare case of "apparently perpetual self-fertilisation" of *O. apifera* (Darwin 1862; p. 71) seemed to challenge the idea conveyed in the subtitle of his orchid book, *On the Good Effects of Intercrossing*, encouraging Darwin to further investigate this phenomenon. Although the autogamy of *O. apifera* had already been reported by Smith (1829) and Brown (1833), Darwin was the first to recognize the morphological adaptations of its flowers to self-fertilization. He showed that like all orchids of the same tribe, the *Ophrys* flowers exhibit a remarkable contrivance for cross-fertilization: once out of its protective pouch (the bursicula) and exposed to the air, the sticky viscidium contracts and hardens, and the thin caudicle folds forward in a 90-degree movement, placing the pollinia in the exact position adapted to touch the stigmatic surface of another orchid flower upon the next visit by the insect (see Chapter 1). In *O. insectifera*, which presents distinctive double-bent-shaped caudicles, this bending movement of the pollinaria was found to occur at an unusually slow rate, being completed only after 6 minutes from their removal from the flower by the insect. When observing the flower structure of *O. apifera*, Darwin found some slight modifications in the configuration of the anthers, particularly in pollinaria, which he saw to be enough to assist self-pollination of the flowers. Compared with the other *Ophrys* species, *O. apifera* presents pollinaria with longer and thinner caudicles as well as heavier pollinia (pollen masses), which, together with the peculiarity of the widening anther, curve naturally downward, hanging in front of the stigmatic surface of the flower. Just a breeze is required for the pollinia to attach themselves to the receptive stigma, and self-pollination thus occurs (Figure 3.1).

On the other hand, Darwin found out that besides these important morphological adaptations for self-fertilization, the pollinaria of *O. apifera* retain the ability to bend forward after removal from the anther, and the viscidia maintain their viscosity and adhesive properties to the stigmatic surface. These perplexing observations showed that contrivances adapted to remarkably contrasting pollination strategies come together in a single flower, leading Darwin to predict that the autogamous *O. apifera* should receive sporadic insect visits which culminate in occasional events of cross-pollination. Therefore, Darwin's main finding that "some great good is derived from the union of two distinct flowers, often borne on distinct plants" (Darwin 1862; p. 71)—inferred from the enormously diverse and sophisticated floral adaptations to cross-fertilization that he found throughout the orchid family, and later in other plant families (Darwin 1876)—remains unquestionable (see Chapters 4, 5, and 7).

Most questions raised by Darwin reflect his lack of knowledge about the insects' mode of attraction to the *Ophrys* flowers. He knew that in obligate insect-pollinated plant species, "unless the flowers were by some means rendered attractive, most of the species would be cursed with perpetual sterility" (Darwin 1877; p. 275), but because he never had the opportunity to witness any pollination event in *Ophrys* flowers, he could only make predictions on the potential roles of floral traits in pollinator attraction. His observations on *Ophrys* species other than the self-pollinated *O. apifera* led him yet to dismiss Brown's (1833) hypothesis that the insectiform flowers of *Ophrys* are intended to deter rather than attract insects. Darwin noticed 2 small, shining protuberances with "an almost metallic luster, appearing like two drops of fluid" (Darwin 1862; p. 57), near the base of the *Ophrys* labellum and regarded them as a likely example of sham nectaries (Sprengel 1793). Their location close to the stigmatic surface of the flower led him to suppose that these shining protuberances, today known as pseudo-eyes and interpreted as mimics of the eyes of insects or the tegulae of their wings (Devillers and Devillers-Terschuren 1994; Paulus 2006), were the attractive feature that would draw the insects' attention.

But what he did not know was that other sophisticated contrivances of *Ophrys* flowers are responsible for the highly specific interactions of these orchids with their pollinators. In his orchid book (1862; p. 68), Darwin included a short footnote mentioning the description by Smith (1829) of some interesting observations by Mr. Price—perhaps referring to the Reverend Ralph Price, Rector of Lyminge and Paddlesworth, Kent, to whom Smith dedicated his book—who "has frequently witnessed attacks made upon the Bee Orchis by a bee, similar to those of the troublesome *Apis muscorum* [syn. *Bombus muscorum*]" (Smith 1829; p. 25). Curiously, the observations of Mr. Price were carried out in the autogamous *O. apifera*, which is visited by insects (in particular Eucerini solitary bees) even less frequently than all the other *Ophrys* species. This description of the "attacks" of a bee on the *Ophrys* flower, whose meaning Darwin could not understand, is probably the very first allusion to the pseudocopulation attempts that hymenopteran insects perform on the labellum so that pollination can occur, a phenomenon which remained unknown until the beginning of the 20th century.

THE DISCOVERY OF *OPHRYS* POLLINATION

By the early 20th century, Darwin's orchid book had already been edited several times in different languages. His observations caused a heightened interest in naturalists, who set out to investigate their native orchid flora in detail (see e.g. Camus 1885; Camus et al. 1908; see Chapter 1). The eminent Swiss botanist Henry Correvon (1854–1939) (Figure 3.2), creator of the first alpine botanical garden of Switzerland, was particularly interested in the fertilization of orchids. He described facts of their natural history, including the pollination of *O. apifera* and the floral structure of *O. insectifera*, in his *Album des orchidées d'Europe centrale et septentrionale* (Correvon 1899). Correvon lived in southern Switzerland, in an area where some of the orchids he was interested in were not native. Consequently, while preparing a new edition of his *Album*, he urged one of his many correspondents, Maurice-Alexandre Pouyanne, to gather more facts on the biology of the *Ophrys* orchids. Pouyanne was an amateur botanist who also happened to be the President of the Court of Appeal at Sidi Bel-Abbès in Algeria. He had meticulously investigated the native orchids in coastal Algeria for the best part of the early 20th century, and he decided to share his findings on the interactions between the *Ophrys* orchids, particularly *O. fusca*, *O. lutea*, and *O. speculum*, with Correvon:

> Sit down, in the sun, a small bouquet of *Ophrys speculum* in your hands, on a slope upon which males of *Colpa aurea* [syn. *Dasyscolia ciliata*] execute their evolutions. It won't take long before you will notice that they have sniffed, so to say, that they have located the flowers that you are handling. They start turning around them and soon you see 1, 2, 3 and more of them swooping down on your bouquet. They rush towards the *Ophrys*; sometimes two or three bump into each other on the same flower until one of them . . . takes possession of the place. It then lands on the labellum, its head towards the stigma, just below the pollinia, and its abdomen dives at the bottom in the long red hairs that look like the labellum's bearded crown. The abdomen tip becomes agitated against these hairs with messy, almost convulsive movements, and the insect wiggles around; its movements, its attitude are all similar to those of insects attempting copulation. (Correvon and Pouyanne 1916; p. 42; translated from the French)

This remarkably detailed excerpt from Pouyanne's findings was published in 1916 along with more observations with the help of Correvon, who introduced Pouyanne's findings to the National Botanical Society of France in a threefold, coauthored paper (Correvon and Pouyanne 1916). Both Pouyanne and Correvon were very much aware of Darwin's findings and, most of all, of what Darwin had *not* discovered on *Ophrys* pollination. Pouyanne had also read Jean-Henri Fabre's accounts on the instinct of insects (Fabre 1879–1907), which helped him interpret the behavioral sequences he observed upon contact between the hymenopterans and the *Ophrys* orchids.

Pouyanne's first report is extremely rich in details, and in it he also confirms Darwin's observations that on average, less than 25% of the available flowers of *O. fusca*, *O. lutea*, and *O. speculum* are pollinated (Correvon and Pouyanne 1916). He also provides the first adaptive explanation for the *Ophrys* insectiform flowers, which he suggests were the result of a complex sexual mimicry involving visual, tactile, and chemical signals. Pouyanne provided 2 additional reports on *Ophrys* pollination in 1917 (Pouyanne 1917) and in 1923 (Correvon and Pouyanne 1923). His observations were later confirmed by the British naturalist Masters John Godfery (1856–1945) in a series of papers on the species investigated earlier by Pouyanne (Godfery 1925a,b, 1927, 1928, 1929, 1930) as well as on other representatives of the genus *Ophrys* from southern France, including *O. insectifera*, which was found to be pollinated by males of the digger wasp *Argogorytes mystaceus* (syn. *Gorytes mystaceus*) (Figure 3.1). Godfery's classification of the genus *Ophrys* introduced the section *Pseudophrys* (Godfery 1928), still valid today and proved to be monophyletic (Bateman et al. 2003; Devey et al. 2008), which encompasses species that deliver the pollinaria on their pollinators' abdomen tip during pseudocopulation. It did not take long before the extraordinary early findings of Pouyanne and Godfery were extended overseas and confirmed by the Australian (but British by birth) naturalist Edith Coleman (1874–1951), who reported in a series of papers (Coleman 1927, 1928a,b, 1929a,b, 1930a,b, 1931, 1938) on her discovery of the pollination of 3 *Cryptostylis* orchid species through pseudocopulation by the Ichneumon wasp, *Lissopimpla semipunctata* (see Gaskett 2011; Chapters 1 and 11).

Although the theory of insect-flower mimicry became very much en vogue after the publication of works describing this newly identified pollination mechanism, a significant proportion of the academics and naturalists remained highly skeptical about the nature of this interaction and claimed that the phenomenon was nothing but improbable, based on interpretations that smacked of anthropomorphism. The stumbling block to a wider acknowledgment of this pollination strategy was of course not the fact that orchids require the assistance of insect vectors to reproduce. Instead, it was the behavior the male bees and wasps display upon contact with the flowers that was subject to debate, along with these orchids' exclusive reliance on sexual mimicry for their reproduction.

BERTIL KULLENBERG AND HIS INTELLECTUAL LEGACY

Besides a review by Ames (1937), the field of orchid pollination by sexual deception was left unexplored until 1945, when the Swedish entomologist Bertil Kullenberg (1913–2007) (Jonsson 2008; Vereecken et al. 2009) initiated a long series of experimental studies on the interactions between the *Ophrys* orchids and their pollinators (Figure 3.2). His first experiments

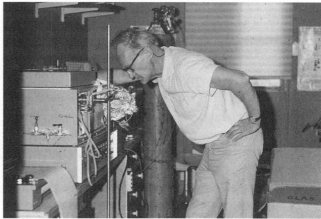

Fig. 3.2 Henry Correvon (1854–1939) and (*right*) Bertil Kullenberg (1913–2007), 2 major actors in the field of *Ophrys* pollination. Photograph of Kullenberg by G. Bergström; used by permission.

in Morocco and Sweden (Kullenberg 1950a,b,c), performed in situ as well as with plants transferred in pots to places with high insect activity, largely confirmed the hypothesis formulated earlier by Pouyanne and Godfery on the pollination of *Ophrys* orchids, including the absence of nectar, the very weak floral scent perceptible by the human nose, and the fact that these flowers are completely ignored by foraging insects. Kullenberg also published the first photograph of a male scoliid wasp *Dasyscolia ciliata* (syn. *Campsoscolia ciliata*; syn. *Colpa aurea*) pseudocopulating on a flower of *O. speculum* (Kullenberg 1949).

Kullenberg published a series of introductory papers on the subject (Kullenberg 1951, 1952a,b, 1953, 1956a,b) as a sort of prelude to his magnum opus: *Studies in* Ophrys *Pollination* (Kullenberg 1961), a 340-page monograph lavishly illustrated with line drawings, photographs of pollinator behavior, and numerous quantitative analyses on the floral colors and morphology. His works attracted considerable interest in the scientific community (see Ferlan 1957 for a review of what was known at the time) owing to his unique multidisciplinary approach to the study of orchid pollination by using advanced analytical tools to characterize the colors (relative reflectance curves measured with a spectrophotometer) and the scents (gas chromatography of volatiles emitted by the flowers), all investigated with an insect perspective on the interactions and incorporating the latest advances in the experimental study of animal instinct and ecology.

Kullenberg was particularly interested in the chemical ecology of the

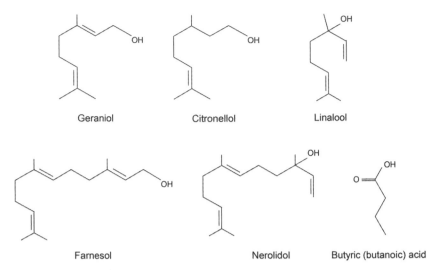

Geraniol   Citronellol   Linalool

Farnesol   Nerolidol   Butyric (butanoic) acid

Fig. 3.3 Some of the key volatile compounds used by Kullenberg (1956b) in field bioassays. Except for butyric acid, all the other compounds are also commonly found in floral scents of many other angiosperms. N. Vereeken. Used by permission.

insects, and he understood that the key to reaching a better understanding of *Ophrys* pollination was to gain insights on the volatile compounds mediating the targeted insects' sexual behavior. One of his first important contributions was his *Field Experiments with Chemical Sexual Attractants* (Kullenberg 1956b), where he investigated the chemical ecology of several species of wild bees and wasps, including the 2 *Argogorytes* species involved in the pollination of *O. insectifera*. His results provided evidence that certain mono- and sesquiterpene alcohols like citronellol (and the aldehyde citronellal), farnesol, linalool, geraniol, and nerolidol (Figure 3.3), which are also common constituents of floral scents and are widely used in the fragrance industry, are utilized by these insects as male sexual attractants (Kullenberg 1953, 1956a,b; Jacobson 1965; see also Knudsen et al. 2006 and Schiestl 2010 for reviews on floral scents and on the cross talk between plant volatiles and insect pheromones, respectively). It also became obvious to Kullenberg that a tactile stimulation was often required to release a genuine copulation attempt (albeit without ejaculation, *contra* Coleman 1928a and Gaskett et al. 2008) when testing scented dummies on the patrolling flight routes of male bees and wasps.

Although Pouyanne suspected that several sensory modalities were acting in concert to attract the male insects and release their copulatory behavior on the orchid flowers (Correvon and Pouyanne 1916, 1923;

Pouyanne 1917), Kullenberg was the first to experimentally quantify the hierarchical importance of the different floral stimuli (olfactory, visual, and tactile), and to show that the floral scent is the key to the attraction of a narrow range of male insects as pollinators (Kullenberg 1951, 1956a,b, 1961, 1973b). While the complex visual stimuli provided by the shape, colors, and contrast of the flowers against the background were found to have only secondary importance as key stimuli for close-range approach to the flowers, the micromorphological pattern of the labellum tactilely stimulates the male insects (thigmotaxis) and plays a primary role in their correct positioning for the uptake of the orchid's pollinaria (Kullenberg 1961). Once combined with a continuous emission of odor, the specific arrangement of the stiff trichomes of the labellum surface appears to provide tactile sensations to the males similar to those evoked by the hairiness of the females, encouraging them to perform copulatory attempts with the labellum in a certain position (abdominal vs. cephalic pseudocopulation), essential for pollination to take place (Kullenberg 1961; Kullenberg and Bergström 1976b; Ågren et al. 1984).

Haas (1946, 1949, 1952) had already reported that volatile odorant compounds are released by several *Bombus* and wild bee species from their mandibular glands and used to deposit chemical landmarks on their patrolling flight routes. As early as 1956, Kullenberg hypothesized that at least for the *Argogorytes-O. insectifera* interaction, it is "the [female] abdominal scent that [has] the same irresistible attraction and exciting effect for males on copulation flight as the scents of *Ophrys* flowers" (Kullenberg 1956a; p. 29). Yet most of the subsequent investigations carried out by Bergström, Tengö, and colleagues on *Andrena-, Colletes-,* and *Eucera*-pollinated *Ophrys* have focused on the chemistry of highly volatile compounds produced by the flowers and by the mandibular and the abdominal Dufour glands of the female insects (e.g. Bergström 1974; Bergström and Tengö 1974, 1978; Tengö and Bergström 1975, 1976, 1977; Tengö 1979; Cane and Tengö 1981; Francke et al. 1984; Borg-Karlson and Tengö 1986; Borg-Karlson 1990). In behavioral tests carried out with patrolling male pollinators in the field, these highly volatile compounds were shown to trigger approaching mating flights to the scented dummies, but failed to elicit copulation attempts by most male insects, a finding that led researchers to believe that *Ophrys* flowers emit only "second-class attractivity compounds" (Borg-Karlson 1990; p. 1382).

Kullenberg went on with his field studies on *Ophrys* pollination (Kullenberg 1973a,b; Kullenberg and Bergström 1973, 1976a,b), and he reported on the fact that solvent (*n*-hexane) extracts of both the orchid labella and the female body odor, which contain low-volatile compounds such as cuticular hydrocarbons and their derivatives, had "a full sexual exciting effect" on the targeted *Eucera* males (Kullenberg 1973b; 36; Kullenberg et al. 1984). He also noted that the thorax + abdomen extracts of *Eucera* females were highly attractive and capable of triggering sexual behavior in patrolling males, but he suggested that these extracts had been contaminated by the highly volatile compounds found in cephalic secretions, which he thought were the key attractants (Kullenberg 1973b). Meanwhile, significant advances had been made in the field of analytical studies of insect pheromones (Ställberg-Stenhagen et al. 1973), and the development of electro-antennographic detection using *Ophrys* scents with antennae of wild bees (Priesner 1973; Ågren and Borg-Karlson 1984) paved the way for a new generation of studies.

The results obtained by Kullenberg since the late 1940s were remarkable, not only because the experiments were well designed for the time, but also because they were conducted several years before the German Nobel Prize laureate Adolf Butenandt (1903–1995) published his landmark study on the identification of the first insect sex attractant, bombykol (named for *Bombyx mori*, the China silkworm, used as a study organism; Butenandt 1955; Butenandt et al. 1959), and also even before the term *pheromone* was proposed by Karlson and Lüscher (1959) to refer to this specific class of biologically active compounds.

MODERN STUDIES IN *OPHRYS* POLLINATION

Several remarkable reviews of the "post-Kullenberg" advances in chemical ecology of the *Ophrys* genus have emerged in the literature in the past decade (Schiestl 2005; Paulus 2006, 2007; Cozzolino and Scopece 2008; Vereecken 2009; Ayasse et al. 2010, 2011; Dötterl and Vereecken 2010; Vereecken and McNeil 2010).

The hypothesis proposed by Kullenberg (1961) that the scent of *Ophrys* flowers is highly attractive to male insects due to its similarity to the sex pheromone of their conspecific females has only been confirmed some 4 decades later, when Schiestl and collaborators (1999) conclusively identified the components of the female sex pheromone of the solitary bee *Andrena nigroaenea*, the pollinator species of *O. sphegodes*. Using the innovative analytical technique of coupling electro-antennographic detection with gas chromatography (GC-EAD) for identification of the full range of physiologically active scent compounds for males, and then testing their behavioral activity in field bioassays using dummies impregnated with synthetic mixtures of these compounds, Schiestl and collaborators (1999, 2000) demonstrated that the female sex pheromone of *A. nigroaenea* consists of a small set of low-volatile, long-chain saturated and unsaturated hydrocarbons which are located in the cuticle of the insects. Although also detected in the solvent extracts of female bodies and *Ophrys* labella, these low-volatile, long-chain hydrocarbons and other cuticular components had been neglected in previous sensory ecological experiments, perhaps due to the assumption that the sex pheromones of bees and wasps were highly volatile odoriferous compounds synthesized in the mandibular glands of insects (Tengö 1979; Ågren and Borg-Karlson 1984; Kullenberg et al. 1984; Borg-Karlson 1990). The findings that the same set of compounds was also detected in the labella solvent extracts of *O. sphegodes* flowers in similar relative proportions, and that they are capable of triggering in males as many copulation attempts as those of *A. nigroaenea* females, provided the first evidence for the occurrence of a chemical mimicry between *Ophrys* flowers and the females of their hymenopteran pollinators (Schiestl et al. 1999, 2000). By mimicking the extremely specific mating signal used by insects in their intraspecific communication (Ayasse et al. 2001), sexually deceptive *Ophrys* orchids could evolve a pollination strategy of high specificity, each species being pollinated by males of only one or a few insect species (Kullenberg 1961; Paulus and Gack 1990; Paulus 2006; Vereecken and McNeil 2010).

Over the last decade, the female sex pheromones have been identified in a few hymenopteran species, and additional cases of chemical mimicry have been found in *Ophrys*-pollinator interactions. The solitary bees *Andrena nigroaenea*, *A. morio*, and *Colletes cunicularius* use as sex pheromone a mixture of long-chain saturated and unsaturated hydrocarbons, aldehydes, and esters. The specificity of the mating signal in each bee species resides in the relative amounts of a series of alkenes, which differ in chain length and in the position of the double bond (Schiestl et al. 1999, 2000; Mant et al. 2005a; Stökl et al. 2008b). The very same behaviorally active compounds were also found in the *Ophrys* species that attract specifically males of the corresponding bee species for pollination, occurring mostly in similar relative proportions in the cuticular waxes of the labellum (Schiestl et al. 1999, 2000; Schiestl and Ayasse 2002; Mant et al. 2005a; Stökl et al. 2005, 2008a,b, 2009; Vereecken and Schiestl 2008; Cortis et al. 2009; Vereecken et al. 2010a; Schlüter et al. 2011; Xu et al. 2011). Moreover, in at least 2 *Ophrys*-pollinator systems, polar compounds instead of hydrocarbons were demonstrated to be the key stimulators of male mating behavior. In *O. speculum*, the unique species pollinated by the scoliid wasp *Dasyscolia ciliata*, the active compounds were proved to be a combination of minute amounts of the unusual saturated (ω-1)-hydroxy and (ω-1)-oxo-acids (Ayasse et al. 2003, 2011), whereas in *O. normanii* and *O. chestermanii*, both pollinated by the parasitic bumblebee *Bombus vestalis*, the polar fraction of the biologically active compounds, containing fatty acids and their corresponding esters, aldehydes, and alcohols, was identified as the most attractive in eliciting the pseudocopulatory behavior in male bees (Gögler et al. 2009, 2011).

Considering the evidence from chemical analyses and field bioassays performed over the last years, it has become clear that *Ophrys* flowers are effective mimics of the females of their pollinators. Contrary to the mainstream idea of the end of the last century (Borg-Karlson 1990), it is now widely accepted that *Ophrys* flowers are capable of eliciting the same sexual excitation degree in male insects as that triggered by their own legitimate sexual partners. Surprisingly, the flowers of *O. speculum* and *O. exaltata* are found to be even more attractive than their female models (Ayasse et al. 2003; Vereecken and Schiestl 2008). In *O. speculum*, the supernormal stimulation (see Chapter 10) is due to the higher absolute amounts of the semiochemicals involved in the pollinator attraction (Ayasse et al. 2003), similar to what seems to occur in the sexually deceptive Australian orchid *Chiloglottis trapeziformis* (Schiestl 2004 and see Chapter 5). The higher attractiveness of the *O. exaltata* flowers originates in the different relative proportions of the 3 key alkenes of the odor bouquet, which are sufficient to form an exotic scent found to be preferred by the *Colletes cunicularius* males (Vereecken et al. 2007; Vereecken and Schiestl 2008).

Ophrys species attracting the same pollinator, regardless of their phylogenetic relatedness, emit a floral odor bouquet containing a similar set of behaviorally active compounds in identical relative proportions (Stökl et al. 2005; Cortis et al. 2009; Gögler et al. 2009), and often present an allopatric distribution (Ayasse et al. 2011). When 2 or more of these species occur in sympatry, the reproductive isolation is mostly achieved by a mechanical barrier provided by a different micromorphological pattern of the labellum (Schiestl and Schlüter 2009; but see Cortis et al. 2009). This assures the placement of the pollinaria in different parts of the insect body—either in the abdomen tip, as in species belonging to the monophyletic section *Pseudophrys*, or in the head, as in those of the paraphyletic section *Ophrys* (syn. *Euophrys*; Godfery 1928 nom. nud.; Godfery 1928; Devey et al. 2008; compare and contrast with Chapter 9). The reproductive isolation throughout the species-rich genus *Ophrys* is mostly effective, and predominantly achieved by the specific set of key odor compounds of the floral bouquet, as demonstrated in sympatric populations of the closely related species *O. sphegodes*, *O. exaltata*, and

O. garganica (Mant et al. 2005b; Xu et al. 2011). Nevertheless, the floral isolation barriers that hinder interspecific gene flow in sympatry may occasionally be broken down, leading to hybridization (Stökl et al. 2008a; Cortis et al. 2009; Vereecken et al. 2010a), a source of phenotypic rearrangements and novelties that has the potential to result in the formation of new species (Stökl et al. 2008a, 2009; Vereecken et al. 2010a).

The low-volatile, long-chain cuticular hydrocarbons or other fatty acid derivatives found in the cuticular waxes covering the labellum surface are key components in the attraction of specific pollinator(s) in most *Ophrys* species investigated so far. However, compounds with higher volatility may be part of the set of bioactive attractive components of the floral odor bouquet. This is the case of the monoterpene alcohol linalool, a long-range male attractant emitted by virgin females of the plasterer bee *Colletes cunicularius*, which was also detected (albeit in significantly lower absolute amounts) in the floral scent of *O. exaltata*, a species pollinated by patrolling males of this bee species (Mant et al. 2005a). The *Ophrys* floral bouquet often contains more than 100 organic compounds of different volatility (Kullenberg et al. 1984; Borg-Karlson 1990; Ayasse et al. 2000, 2003), which include mostly aliphatic compounds such as short- and long-chain hydrocarbons, primary and secondary alcohols, aldehydes, ketones, carboxylic acids, and corresponding esters, but also mono- and sesquiterpenes together with a few aromatic compounds (Ågren and Borg-Karlson 1984; Kullenberg et al. 1984; Borg-Karlson and Tengö 1986; Borg-Karlson 1990; Schiestl et al. 1997).

Even though the chemical composition of the floral odor bouquet has already been documented for several *Ophrys* species, the precise site in the labellum of biosynthesis and emission of the fragrance has received less attention. The German floral biologist Stefan Vogel first described osmophores as newly observed fragrance-producing glandular tissue in plants, specialized in the emission of highly volatile attractants for pollinators (Vogel 1961). Vogel pioneered the anatomical description of these specialized secretory structures in a large number of species, including a few *Ophrys* species, *O. fusca* and *O. lutea* in particular (Vogel 1963b; English translation in Vogel 1990). Following this early study, a detailed histological characterization of the osmophores of these 2 *Andrena*-pollinated species was provided by Ascensão et al. (2005), who showed their location both in the margins and, mostly, in the abaxial surface of the apical region of the labellum (compare and contrast with Chapter 9). A recent investigation (Francisco and Ascensão 2013) showed that in *O. bombyliflora* and *O. tenthredinifera*, osmophores are also located in the apical region of the labellum, specifically in the appendix but encompassing the margins and the abaxial surface of the adjacent area of the labellum in the latter species. These osmophores were found to synthesize a terpene-rich lipophilic secretion that could constitute a reservoir for the highly volatile long-range attractants (Francisco and Ascensão 2013), previously reported in the odor bouquets of both species (Kullenberg et al. 1984; Borg-Karlson 1990). Whether these compounds play a role in eliciting mating behavior in male Eucerini bees is currently not known.

Compared with the number of investigations on the key role of floral scent in pollinator attraction and on its implications to the ecology and evolution of the genus, only very few studies have addressed the relative role of visual and tactile signals in *Ophrys* pollination since the early research by Kullenberg and colleagues (Kullenberg 1961; Kullenberg and Bergström 1976b; Ågren et al. 1984; Borg-Karlson 1990; Borg-Karlson et al. 1993). Our current knowledge of the *Ophrys* labellum micromorphology is based on the studies performed in some species belonging to the *O. sphegodes* and *O. fusca* groups (Servettaz et al. 1994; Ascensão et al. 2005; Cortis et al. 2009) and on the comprehensive comparative survey of 32 taxa of the genus by Bradshaw and collaborators (2010), which revealed a great diversity of epidermal cell types in the labellum surface, ranging from flat cells or dome-shaped papillae to unicellular trichomes varying in length, shape, and texture along the different regions of the labellum. Moreover, the recent investigation by Francisco and Ascensão (2013) showed a set of newly observed micromorphological characters on the labellum of *O. bombyliflora* and *O. tenthredinifera*. These authors observed that in *O. bombyliflora*, the labellum ends with a distinctive fleshy apical appendix, visible only when the flower is seen in bottom view (Figure 3.4), which presumably provides a tactile stimulation to the male insects during pseudocopulation.

Another interesting finding of this study is the occurrence of a shallow groove in the apical region of the labellum of *O. bombyliflora* (Figure 3.4), likely to have a tactile guiding role similar to that of the basal groove found in the labellum of *O. fusca* and *O. lutea* (Ascensão et al. 2005). The contrasting direction and arrangement of trichomes in the labellar surface (Figure 3.4) encourage the *Eucera* male pollinators of *O. bombyliflora* and *O. tenthredinifera* to perform a cephalic pseudocopulation while stimulating the *Andrena* male pollinators of *O. fusca* and *O. lutea* (species included in the *Pseudophrys* section) to adopt a reversed position on the labellum, guiding the abdomen tip toward the stigmatic cavity so that an abdominal pseudocopulation can take place (Kullenberg 1961; Ascensão et al. 2005; Francisco and Ascensão 2013). As mentioned above, the labellum's micromorphological features constitute the unique (mechani-

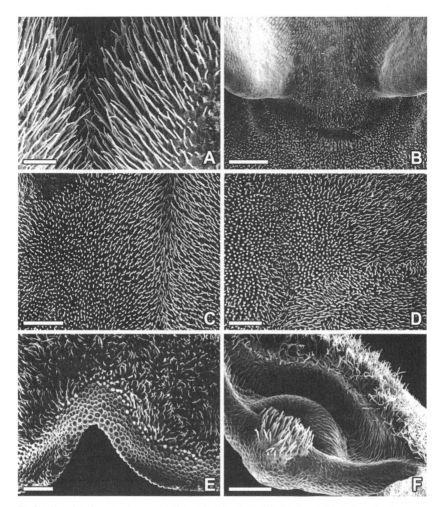

Fig. 3.4 Scanning electron micrographs of the adaxial surface of the labellum in (*A, C, E*) *Ophrys fusca* and (*B, D, F*) *Ophrys bombyliflora*: (*A, B*) basal region near the stigmatic cavity; (*C*) median-basal region comprising the speculum, showing the basal groove (*on the right*); (*D*) apical region above the appendix where trichomes are oriented toward a shallow longitudinal groove; (*E*) central notch in the apical margin composed of secretory papillae; (*F*) apical appendix forming a distinctive concavity with a protuberance and a tuft of trichomes at the tip. Scale bars: (*A, D*) = 250 μm; (*B, C, E, F*) = 500 μm. All images are part of the PhD project of A. Francisco, which has been carried out under the supervision of Prof. L. Ascensão (University of Lisbon, Portugal). Used by permission.

cal) isolation barrier between co-occurring species that attract the same pollinator, even though it could occasionally fail to prevent interspecific gene flow (Cortis et al. 2009; Vereecken et al. 2010b).

Besides its relevance in predicting the tactile stimulating role of the epidermal cells of the labellum's surface for male insect pollinators, the study of the micromorphological pattern of the *Ophrys* labellum also provides information on the optical properties of its surface, which could constitute a visual stimulus for male insects approaching the flower. The shape, size, and cuticular ornamentation of the epidermal cells of the labellum, combined with their pigment content, seem to be the determining factors in the appearance and, particularly, in the reflectivity of the diverse areas of the labellum (Kullenberg 1961; Ascensão et al. 2005; Bradshaw et al. 2010; Vignolini et al. 2012; Francisco and Ascensão 2013). The central region of the *Ophrys* labellum, the speculum (a Latin-derived word that means "mirror"), whose designation stems precisely from its typically high reflectivity, has long been considered a mimic of the insect's wings (Kullenberg 1961; Ascensão et al. 2005; Paulus 2006; Bradshaw et al. 2010). In addition to this high reflectivity, the speculum of a few species is believed to also exhibit iridescence (Bradshaw et al. 2010; Glover and Whitney 2010). The occurrence and origin of the iridescence assumed to occur in the highly reflective bright-blue speculum of the flowers of certain *Ophrys* species, such as the mirror orchid, *O. speculum*, have begun to be explored recently (Bradshaw et al. 2010; Bateman et al. 2011; Vereecken et al. 2011; Vignolini et al. 2012).

Behavioral field experiments were performed in order to assess the visual importance of both the brightly colored perianth and the labellum pattern of some *Ophrys* flowers in the attraction of the pollinators (Spaethe et al. 2007, 2010; Streinzer et al. 2009, 2010; Vereecken and Schiestl 2009). The bright-pink perianth of *O. heldreichii* was found to increase attractiveness and detectability of the flower at short distances for its male pollinator *Eucera berlandi* (Spaethe et al. 2007, 2010; Streinzer et al. 2009). A plausible explanation for this phenomenon is the resemblance of the spectral reflection of the perianth of these flowers to the flower color of other co-occurring plants, where females of pollinator bees appear to feed after emerging and where mating usually takes place (Spaethe et al. 2007, 2010; Streinzer et al. 2009). On the other hand, the white perianth of some individuals of *O. arachnitiformis* appears to play no role in the attraction of its pollinator *Colletes cunicularius* male bees (Vereecken and Schiestl 2009). In strong contrast with the attractive role of the sepals and lateral petals of the flower, a recent investigation with *O. heldreichii* has shown that the visual pattern of the labellum does not significantly influence the approach flights to the flowers, indicating that it seems to be not involved in the pollinator attraction (Streinzer et al. 2010). However, caution should be taken not to generalize this finding, as the genus *Ophrys* contains so many different species.

FLORAL SCENT EVOLUTION ABOVE THE SPECIES LEVEL
IN *OPHRYS* ORCHIDS

In their review on the diversity and distribution of floral scents, Knudsen et al. (2006; p. 2) conclude that

> Floral scent chemistry is of little use for phylogenetic estimates above the genus level, whereas the distribution and combinations of floral scent compounds at species and subspecific levels is a promising field of investigation for the understanding of adaptations and evolutionary processes in angiosperms.

Their conclusions were based on the observation that most plant families dispose of the major biosynthetic pathways for the production of most classes of floral scent compounds. Consequently, typical floral scents are usually evolutionarily labile and regarded as relatively weakly patterned, at the level of compound classes and at higher taxonomic levels in plants (but see Barkman 2001; Steiner et al. 2011; Levin et al. 2003, at lower taxonomic levels). There are of course exceptions to this tenet, some floral volatiles being known only from certain groups of flowering plants (see e.g. Kaiser and Nussbaumer 1990; Schlumpberger et al. 2004, on dehydrogeosmin in cacti; Ayasse et al. 2003, on [ω-1]-hydroxy and [ω-1]-oxo-acids in *Ophrys* orchids; Schiestl et al. 2003; Franke et al. 2009; Peakall et al. 2010, on chiloglottones in *Chiloglottis* orchids).

Sexually deceptive orchids of the genus *Ophrys* have evolved the use of cuticular hydrocarbons, particularly alkenes and their derivatives, as primary pollinator attractants (but see Ayasse et al. 2003; Gögler et al. 2009). Although a thorough analysis of the female sex pheromones of wild bees that pollinate *Ophrys* orchids is still lacking at this stage, ongoing studies on the female sex pheromones of their associated pollinators indicate that at least some phylogenetic patterning of these compounds is perceptible at the species group, subgenus, and perhaps genus levels in solitary bees (Vereecken et al., unpublished data). This phylogenetic conservatism of cuticular hydrocarbons in wild bees has the potential to drive pollinator shifts and can therefore promote the multiplication of new *Ophrys* species, especially when (1) the male bees attracted belong to diverse subgenera or genera (e.g. in the genera *Andrena* or *Eucera*); (2) the pollinators attracted and their closely related species are characterized by a partial overlap of their flight activity and distribution; and (3) the female sex pheromone signals of closely related bee species represent variations on a "chemical theme" (identical compounds, different ratios of the compounds in the total blend) that nevertheless act as strong prezygotic isolation barriers.

By contrast, *Ophrys* species that have evolved toward the attraction of male bees with a restricted distribution range that belong to less diverse subgenera/genera or to a group characterized by an original female sex pheromone chemistry are less likely to experience a spectacular diversification, unless followed by more drastic changes in the floral scent chemistry that would then be required to allow a taxonomic jump onto another group of wild bees. This could explain why certain groups of *Ophrys* species have diversified spectacularly, with now up to 15 representatives (e.g. the *O. mammosa* group), all pollinated by different solitary bee species of the genera *Andrena* and *Eucera*, whereas others have not, such as the *O. speculum* group (pollinated by scoliid wasps through the emission of hydroxydecanoic acids—not alkenes—as primary pollinator attractants) or indeed the *O. insectifera* group, which exploits primarily male insects that belong to smaller taxonomic groups of hymenopterans.

CONCLUSION

Darwin's contribution to *Ophrys* pollination may seem quite small in the light of the recent dramatic advances made in this field over the past few decades, but most contemporary orchid pollination biologists acknowledge his profound impact on the way we look at flowers, orchid flowers in particular, as being tailored to the most eccentric preferences of their pollinators (see Chapters 6 and 10). Darwin was never an armchair naturalist, and his orchid book nicely illustrates how he rigorously and thoroughly investigated the adaptive significance of even the minutest details of the orchid flower morphology. This experimental approach, the Darwinian Method, has proved successful in the field of orchid pollination and in countless other disciplines ever since. The discoveries made by Pouyanne, Coleman, Godfery, and Kullenberg among others have built an edifice whose foundations were laid by Darwin himself, and his contributions have inspired a great many modern researchers in this field.

Recent experimental studies have provided remarkable instances of convergence and divergence of floral traits, particularly in the pollinator-attracting chemical signals in the genus *Ophrys* (Stökl et al. 2005, 2008a, 2008b, 2009; Cortis et al. 2009; Gögler et al. 2009; Vereecken and Schiestl 2009; Ayasse et al. 2010, 2011; Vereecken et al. 2010a,b), and some of the details of what is now considered a textbook example of mimicry have been revealed (Schiestl et al. 1999; Ayasse et al. 2003; Spaethe et al. 2007;

Vereecken 2009; Streinzer et al. 2010; Schlüter et al. 2011). The genus *Ophrys*, with its formidable diversity in floral shapes, colors, and scents, offers a fertile loam for ecological and evolutionary studies. We have just started scratching the surface of these unique insect-plant interactions, which seem to offer countless avenues for future research. Studies on hybridization, reproductive isolation, and macro-evolutionary perspectives on this mimicry system and its origin will help describe the details of the complex and intertwined ecology of *Ophrys* orchids and their pollinators across the spectrum of selection levels, from genes to populations and species.

ACKNOWLEDGMENTS

We are grateful to Peter Bernhardt and Retha Meier for their invitation to take part in the Darwin Orchid Symposium during the International Botanical Congress 2011 held in Melbourne, Australia, as well as for their dedication and help in the publication of this volume celebrating the sesquicentennial of Darwin's opus on orchid pollination. We also thank Salvatore Cozzolino for his comments, which helped improve the manuscript. N.J.V. was financially supported by the Belgian National Fund for Scientific Research (FRS-FNRS) through a postdoctoral grant during the preparation of this chapter. A.F. is a PhD candidate under the supervision of Prof. Lia Ascensão (University of Lisbon, Portugal); the results presented in the section "Modern Studies in *Ophrys* Pollination," including the images in Figure 3.4, are part of this PhD project. A.F. is deeply indebted to Prof. Lia Ascensão for all the support, dedication, and guidance throughout her research, and is also grateful to the Portuguese Fundação para a Ciência e a Tecnologia (FCT) for financial support through a doctoral grant (SFRH/BD/18823/2004).

PART III

Darwin and His Colleagues:
Orchid Evolution in the Southern Hemisphere

Pollination of South African Orchids in the Context of Ecological Guilds and Evolutionary Syndromes

Steven D. Johnson

In a geographic context, the spatial structure of variation in pollinator abundance and [plant] community composition can also have important implications for floral evolution.
—D. A. Moeller (2005)

INTRODUCTION

Orchids have a reputation for being plant aristocrats that follow a different set of rules compared with other plants. This may be due to the natural rarity of most species, their showy and complex flowers, and the fact that many species in cultivation must be pampered by wealthy humans. Orchidology has even developed as a specialized branch of botany devoted to the study of orchids. All this has had the unfortunate consequence that orchids have tended to be studied in isolation from the other biota in natural ecosystems. In this chapter, I use the orchid flora of South Africa to illustrate my view that the ecology and evolution of orchids are best understood in the geographic context of the plant communities to which orchids belong. In particular, orchids share pollinators with other plants and thus form part of ecological guilds. Furthermore, their floral traits are also usually similar to those of other guild members in local plant communities, through either convergent evolution or, in the case of some deceptive orchids, adaptive mimicry.

Charles Darwin himself contributed to the aura of orchid nobility by devoting a book to the subject of their pollination (1862). This book, the first edition of which was published 3 years after the "Origin," constituted a major defense of his idea of adaptation through natural selection (Harder and Johnson 2009). It focused heavily on the role that floral traits play in promoting cross-pollination between flowers (Peter and Johnson 2006), a theme he later developed further for plants in general (Darwin 1876). The observations and ideas in the orchid book enthralled naturalists at the time and inspired many to conduct their own observations. Darwin corresponded extensively with leading naturalists such as Hermann Müller (1829–1883) in continental Europe and Fritz Müller (1821–1897) in South America (see Chapter 9), and incorporated many of their findings in the book's second edition. Despite being isolated from the scientific mainstream, many settler naturalists in South Africa corresponded via shipborne mail with Darwin, who in turn personally helped facilitate publication of their findings in the *Journal of the Linnean Society*. Notable among these were descriptions of the floral structures and surmised pollination mechanisms of various South African orchids (Trimen 1864; Barber 1869; Weale 1869, 1873a,b,c).

The second edition of the orchid book, published in 1877, included many of the findings by South African naturalists that had been published in the intervening 15 years (see Chapter 1). In it, for example, Darwin discusses the morphology of the flowers of *Bonatea speciosa* (see Color Plate 4), which he describes (1877; p. 264) as being the most profoundly modified of any orchid. He correctly surmised that the toothlike process at the entrance to the spur would compel moths to enter the flower from either side, but actual observations of this phenomenon did not take place until the end of the 20th century, when it was discovered that hawkmoths pollinate the orchid and carry its pollinaria on their eyes (Johnson and Liltved 1997; see Color Plate 4).

It is notable that few of the publications in Darwin's time included pollinator observations, and that most simply speculated about the functions of various flower parts under the general rubric of "functional morphology" (see Chapter 1). Darwin himself was faced by similar challenges and conceded to having made few direct observations of orchid pollinators, although this situation had improved considerably by the publication of the second edition (see Chapter 1). In South Africa, a succession of naturalists in the 19th century attempted to find the pollinator of South Africa's most spectacular orchid, the large, red-flowered species *Disa uniflora*, which remains common on Table Mountain above Cape Town. There were speculations about the function of the various floral modifications in this species (Trimen 1864), but the actual pollinator, a large satyriine butterfly, was only discovered at the end of the 19th century (Marloth 1895).

Natural history underwent a decline in many parts of the world in the first part of the 20th century, and the only notable publication on orchid pollination in South Africa from this period was a report of pollination by fungus gnats in the diminutive *Satyrium bicallosum* (Garside 1922). In 1954, Stefan Vogel published a major monograph on floral "styles" in the South African flora (Vogel 1954) that also included some original observations, such as hawkmoth pollination in *Habenaria epipactoides* (Peter et al. 2009). This publication is of particular relevance today, because it was the first major exposition of Delpino's original and much-contested concept of floral syndromes (Delpino 1868–1875). It is of historical interest to note that Vogel's monograph was much more detailed than the later work on floral syndromes by Faegri and van der Pijl (1979), yet on account of being written in German remained largely unknown in the English-speaking world. Vogel's (1954) monograph was followed by an even more detailed account of the floral morphology of South African orchids (Vogel 1959).

Detailed studies of pollination systems in South African orchids commenced in the 1980s (Steiner 1989), and the past 20 years have seen an explosion in new knowledge. Today, the orchid flora of South Africa is one of the best studied worldwide with respect to pollination systems and has provided some outstanding examples of evolutionary processes, such as pollinator-driven divergence of populations (Johnson and Steiner 1997), convergent evolution (Pauw 2006), and floral mimicry (Johnson 1994a,b). In this chapter, I start by highlighting some of the ecological guilds to which South African orchids belong, then discuss the extent to which these guilds are characterized by convergent floral syndromes, as well as the special case of advergent evolution (sensu Hines et al. 2011) in orchids that mimic flowers of rewarding sympatric plants. I then consider the evidence for the idea that pollinators have been a major driver of diversification in orchid lineages in the region, and conclude with some ideas for future research programs.

ECOLOGICAL GUILDS

Plants that become ecologically specialized for pollination by a particular pollinator species or functional pollinator group shared by other plants become members of ecological guilds. One of the early applications of the guild concept was to plants specialized for various hummingbirds in the Americas (Feinsinger 1978). The pollination guild concept has been a strong theme in South African pollination biology since the early 1990s, when it was applied to a plant guild specialized for pollination by a satyriine butterfly (Johnson and Bond 1992), and then to various guilds of plants pollinated by long-proboscid nemestrinid and tabanid fly species (Manning and Goldblatt 1996, 1997; Goldblatt and Manning 2000).

Johnson (2010) defined a pollination guild as the smallest possible unit of a pollination system. For example, in South Africa there is a long-proboscid fly pollination system that consists of several distinct pollination guilds. To date, more than 20 different pollination guilds varying from 5 to over 100 plant species have been identified in the South African flora (Table 4.1). Most of these guilds include orchid members. The orchid *Disa uniflora*, for example, is part of a guild of approximately 25 South African plant species specialized for pollination by the mountain pride butterfly *Aeropetes* (*Meneris*) *tulbaghia*.

Pollination guilds have a number of important properties, including asymmetric specialization (plants are usually more specialized than their pollinators), restricted geographic distributions, short phenophases in the case of many insect pollinators, local competition and facilitation, diffuse coevolution, convergent evolution, and advergent floral mimicry.

SPECIALIZATION

Pollination guilds arise when plants become specialized for particular pollinators. Broad surveys have shown unusually high levels of pollination system specialization in the South African flora. Johnson and Steiner (2003) found that most South African orchid species studied to date have just one known pollinator species, while their survey of related terrestrial orchid species in Europe and North America showed that these typically have several different pollinator species. The levels of specialization in South African temperate orchids match those found in the tropical epidendroid orchids (Tremblay 1992 and see Chapters 8 and 9). In their survey, Johnson and Steiner (2003) found that the trend to specialization also applies to South African species of Iridaceae, which are typically pollinated by just one insect species while being more generalist elsewhere in the world. Similarly, Ollerton et al. (2006) found that pollination systems of milkweeds (Apocynaceae) in South Africa are more specialized than those studied elsewhere. It is not yet well understood why floral specialization to particular pollinator species is well developed in the South

Table 4.1 Specialized South African pollination guilds containing orchid members

Pollinator(s) of guild	Plant species in guild	Plant families	Orchid genera represented	Habitat, Distribution	Convergent floral traits	References
Tanglewing flies (Nemestrinidae)						
Moegistorhynchus longirostris	20	Iridaceae, Orchidaceae, Geraniaceae	*Disa*	Sand plains, western Cape lowlands	Cream or pink flower color, nectar guides, very long corolla tube; early summer (Sept.-Nov.) flowering time	Manning & Goldblatt 1997; Johnson & Steiner (Johnson and Steiner 1997); Pauw (Pauw et al. 2009)
Stenobasipteron wiedmannii	19	Iridaceae, Orchidaceae, Acanthaceae, Balsaminaceae, Gesneriaceae, Lamiaceae	*Brownleea*	Forest patches in eastern region	Pale blue-pink flower color, nectar guides, long corolla tubes; autumn (Feb.-April) flowering time	Goldblatt & Manning 2000; Potgieter & Edwards 2005
Prosoeca ganglbaueri	20	Amaryliidaceae, Iridaceae, Orchidaceae, Scrophulariaceae	*Disa, Brownleea*	Drakensberg Mountains	Pink or cream flower color, nectar guides, long corolla tube; autumn (Jan.-April) flowering time	Goldblatt & Manning 2000; Anderson et al. 2005; Anderson & Johnson 2009

continued

Table 4.1 *continued*

Pollinator(s) of guild	Plant species in guild	Plant families	Orchid genera represented	Habitat, Distribution	Convergent floral traits	References
Horseflies (Tabanidae)						
Philoliche aethiopica horsefly	15	Amaryllidaceae, Iridaceae, Orchidaceae, Scrophulariaceae	*Disa, Satyrium*	Eastern region	Pink flower color, long corolla tube; summer (Nov.-March) flowering time	Johnson 2000, Johnson & Morita 2006, Johnson & Morita unpublished
Philoliche rostrata	20	Iridaceae, Orchidaceae, Geraniaceae	*Disa*	South-western Cape mountains	Pink or cream flower color, nectar guides, long corolla tube; summer (Oct.-Jan.) flowering time	Goldblatt & Manning 2000
Philoliche gulosa	13	Iridaceae, Orchidaceae, Geraniaceae	*Disa, Satyrium*	Western Cape mountains	Pink or cream flower color, nectar guides, long corolla tube; spring (Sept.- Nov.) flowering time	Goldblatt & Manning 2000
Oil bees (Melittidae)						
Rediviva peringueyi	20	Orchidaceae, Scrophulariaceae	*Ceratandra, Corycium, Disperis, Evotella, Pterygodium*	Clay flats, south-western Cape	Grayish green flower color, open or bi-saccate, oil production, no scent or soapy scent; Aug./Sept. flowering time	Steiner unpublished, Steiner 1989, Pauw 2006, Pauw & Bond 2011, Whitehead & Steiner 2001, Steiner & Cruz 2009

Rediviva gigas	19	Iridaceae, Orchidaceae, Scrophulariaceae	*Ceratandra, Corycium, Pterygodium, Satyrium*	Montane, south-western Cape	Yellow to greenish yellow, magenta to pink and white flower color, open or pouched, oil production, rarely nectar; Oct.-Nov. flowering time	Whitehead & Steiner 2001, Steiner & Cruz 2009, Steiner unpublished
Rediviva longimanus	8	Orchidaceae, Scrophulariaceae	*Pterygodium*	Western Cape	Grayish magenta, white, or cream flower color, soapy or no scent; Aug./Sept. flowering time	Whitehead & Steiner 2001, Steiner unpublished
Rediviva neliana, R. brunnea & R. pallidula	29	Orchidaceae, Scrophulariaceae	*Disperis, Corycium, Disperis, Pterygodium,*	Drakensberg mountains	Deep floral spurs, oil production; Dec.-March flowering pink to red, white	Steiner 2010, Steiner & Whitehead 1988, Steiner & Whitehead 1990, Whitehead et al. 2008
Rediviva colorata	6	Orchidaceae	*Disperis, Pterygodium*	Forest patches, Drakensberg mountains	Floral sacs or spurs, flower color white, Jan.-March flowering time	Steiner 1989, 2010, Whitehead et al. 2008
Rediviva macgregori	18	Orchidaceae, Scrophulariaceae	*Corycium, Disperis, Pterygodium*	Western/ northern Cape	Flowers grayish magenta, yellow, grayish blue; Aug.-Sept. flowering time	Whitehead & Steiner 2001, Steiner unpublished, Waterman et al. 2009
Spider-hunting wasps (Pompilidae)						
Hemipepsis spp.	60	Apocynaceae, Orchidaceae, Hyacinthaceae	*Disa, Satyrium*	Summer rainfall	Dull flower color, open shape, bitter nectar, pungent scent	Ollerton et al. 2003, Shuttleworth & Johnson 2012

continued

Table 4.1 *continued*

Pollinator(s) of guild	Plant species in guild	Plant families	Orchid genera represented	Habitat, Distribution	Convergent floral traits	References
Butterflies and moths (Lepidoptera)						
Mountain pride butterfly (*Aeropetes tulbaghia*)	21	Amaryllidaceae, Iridaceae, Orchidaceae, Crassulaceae,	*Disa*	Cape and Drakensberg mountains	Red flower color, large flowers, narrow corolla tube; late-summer flowering time	Johnson & Bond 1994
Convolvulus hawkmoth *Agrius convolvuli*	25	Amaryllidaceae, Iridaceae, Orchidaceae, Capparaceae, Rubiaceae	*Bonatea*	Eastern region	Pale flowers, very long corolla tube, evening scent	Alexandersson & Johnson 2002, Johnson & Raguso unpublished
Birds						
Malachite sunbird (*Nectarinia famosa*)	>44	Iridaceae, Amaryllidaceae,	*Disa*, *Satyrium*	Cape mountains	Very long, curved corolla tube, unscented, copious dilute nectar	Geerts & Pauw 2009

Source: Adapted and expanded from Johnson 2010.

African flora. One possible explanation is that the diversity of pollinator species within each pollinator functional group is low.

Plants in these guilds are ecologically interdependent on each other to sustain their pollinators energetically. These plants are also interdependent on their pollinators. Therefore, these pollinators can be considered keystone species.

RESTRICTED GEOGRAPHIC DISTRIBUTION

The geographic distribution of guilds is determined by the distribution of their pollinator. Although some guilds are widely distributed, as is the case in the large guild of plants specialized for pollination by the long-billed malachite sunbirds (Geerts and Pauw 2007), most tend to be geographically localized. As examples, the guild of plants for pollination by the long-proboscid fly *Moegistorhynchus longirostris* (Nemestrinidae) is confined to a narrow strip of the Cape west coast (Manning and Goldblatt 1997), and the guild of plants pollinated by the oil bee *Rediviva colorata* is confined to few small forest patches close to the Drakensberg range in the eastern part of South Africa (Whitehead et al. 2008; Steiner 2010; see Figure 4.1). It is not yet known which factors account for the narrow distributions of many long-proboscid fly species, as their larval biology is for the most part unknown. In the case of oil bees, it has been suggested that soil preferences for nesting sites may control distributions (Waterman et al. 2011).

The association between the distribution of a guild of oil-producing orchids and their oil bee pollinator *Rediviva peringueyi* has been explored by Pauw and Bond (2011). They found that the diversity of local communities of oil-producing orchids declines in accordance with the rate of pollination. Because the orchids provide only a small fraction of the oil needed by bees (most of it is obtained from species of Scrophulariaceae), it is likely that the loss in guild composition is caused by the rarity of the bees rather than the converse. Those guild members that persist in areas with very low rates of pollination manage to do so only by virtue of their capacity for clonal reproduction (Pauw and Hawkins 2011). These clonal orchids thus even manage to extend their ranges beyond that of their pollinator (Pauw and Bond 2011).

In a series of experiments, it was shown that oil-producing orchids translocated beyond the range of their oil-collecting bee pollinator into the range of another species of oil-collecting bee suffer from highly reduced rates of pollination (Waterman et al. 2011). This reflects the high

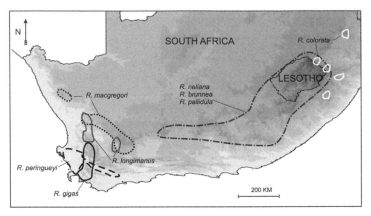

Fig. 4.1 Distribution of oil-collecting *Rediviva* bees associated with various pollination guilds in South Africa (see Table 1). Modified in part from Waterman et al. 2011 by Steve Johnson.

levels of specialization in these guilds and implies that oil-collecting bees are not functionally equivalent, despite being similar morphologically. One possible explanation is that specialization in the interactions between orchids and oil bees is mediated by specific blends of chemical volatiles, which elicit innate behavioral responses (Waterman et al. 2011). Interestingly, when seeds of these orchids were translocated outside their natural distribution, they still germinated freely, suggesting that interactions with pollinators are more specific than those with mycorrhizal fungal symbionts, and thus may play a more important role in determining the geographic distributions of terrestrial orchids (Waterman et al. 2011).

One of the features of the distribution of pollination guilds in South Africa is that they tend to be confined to either the winter rainfall biotic zone in the Cape or the summer rainfall biotic zone in the eastern part of South Africa. For example, almost all the oil-bee and long-proboscid-fly pollination guilds are restricted to either the winter or the summer rainfall region. The recently described pompilid wasp pollination guild (Shuttleworth and Johnson 2012) is confined almost entirely to the summer rainfall grassland region.

Another feature of pollinator distributions is that they can be specific to particular habitats. Plants in the guild pollinated by the mountain pride butterfly are confined to the montane habitats of this insect. In a study of the orchid *Disa uniflora*, it was found that plants growing on streamsides in a steep rocky habitat are far more likely to be pollinated than those that flower on streamsides in flat, open landscapes (Johnson and Bond 1992).

SHORT PHENOPHASES

The adult flight periods of insects largely determine the flowering times of plants in pollination guilds. An example of the effects of pollinator flight periods on orchid flowering times is the evolution of early- and late-flowering forms of the oil-producing orchid *Pterygodium catholicum* (Steiner et al. 2011). Early-flowering forms on the lowlands are pollinated by *Rediviva peringueyi*, which is active in spring, while the late-flowering montane form is pollinated by *R. gigas*, which is active in mid- to late summer. In addition to flowering time, these forms differ in floral scent chemistry (Steiner et al. 2011).

The effect of fly phenophase on the flowering time of orchids is also evident in the *Disa draconis* complex (Johnson and Steiner 1997). *Disa draconis* flowers in late spring when its pollinator, the nemestrinid fly *Moegistorhynchus longirostris*, is on the wing, while its close relative *Disa harveiana* subsp. *harveiana* flowers in summer (December–January) when its pollinator, the tabanid fly *Philoliche rostrata*, is on the wing (Johnson and Steiner 1997).

COMPETITION AND FACILITATION

Although guilds can have 20 or more plant species, there are typically 3–6 guild members in any particular plant community. These species often flower sequentially, but substantial overlap in flowering times can nevertheless occur, meaning that plants could compete in terms of reproductive interference. In early studies of guilds of oil-producing plants in the Drakensberg, it was recognized that sympatric species of *Diascia* as well as *Disperis* often place pollen and pollinaria on different parts of the pollinator's body (Steiner and Whitehead 1988; Steiner 1989). This has been confirmed in studies showing that placement of pollinaria on different parts of the body of *Rediviva* bees is a general feature of guilds of oil-producing orchids (Steiner 1989, 2010; Pauw 2006; Waterman et al. 2011). Crossing experiments show that there are also strong crossing barriers between sympatric orchids, implying that coexistence is not only a consequence of mechanical isolation arising from differential placement of pollinaria (Waterman et al. 2011).

Studies of guilds of plants pollinated by long-proboscid flies also show that sympatric species belonging to the same guild tend to place pollen on different parts of the body of flies (Manning and Goldblatt 1996, 1997; Goldblatt and Manning 2000). However, irids make up the majority of species in these guilds, and reproductive interference seems unlikely to explain placement of orchid pollinaria on these flies. This is because orchids can easily lose pollen to stigmas of other orchid species, but not to stigmas of non-orchids.

While sympatric members of the same guild can potentially compete, they also cross-subsidize the energetic requirements of their pollinator and may thus facilitate each other's pollination (Kevan and Baker 1983; Waser 1983). Furthermore, abundant members of the guild may act as magnet species that increase local pollinator abundance to the benefit of rarer members of the guild (Johnson et al. 2003b). In guilds of oil-producing plants in which species of Scrophulariaceae contribute far more oil to bees than do orchids, there is some evidence that availability of Scrophulariaceae is a key determinant of the pollination success of orchids in local communities (Pauw and Bond 2011). Facilitation is most evident in species that lack nectar and rely on mimicry of rewarding guild members for attraction of pollinators (see below; and compare with Chapters 2, 7, 10, 11). In the deceptive orchid *Eulophia zeyheriana*, pollination success increases in proportion to proximity to flowers of *Wahlenbergia cuspidata* (Campanulaceae), which is the main food plant for the *Lipotriches* bees that pollinate both species (Peter and Johnson 2008). Similarly, in the deceptive species *Disa ferruginea*, rates of pollination are increased severalfold when the rewarding species *Tritoniopsis triticea* is present in the same community (Johnson 1994a,b). In these deceptive species, facilitation is probably due to both magnet effects and local conditioning of pollinator behavior.

FLORAL SYNDROMES

As first noted by Vogel (1954), striking patterns of convergent floral evolution are evident in South African plants. These "pollination syndromes" are associated with specialized pollination systems and guilds, many of which include orchid members (see Color Plate 5), and are assumed to reflect the specific sensory preferences and morphologies of different pollinator groups. Floral syndromes can occur at several hierarchical levels in the flora. For example, plants pollinated by long-proboscid flies share a number of general features, including long, narrow floral tubes, while those belonging to the *Prosoeca ganglbaueri* guild tend to flower in late summer/autumn and have cream or pink corollas with tubes 20–50 mm in length.

It was recently realized that floral convergence can even be evident at the local population level. Different plant species in the *Prosoeca ganglbaueri* guild vary extensively across their geographic range, but are highly convergent at particular sites (Anderson and Johnson 2009). This was attributed to diffuse coevolution between the fly and the plant guild. Deceptive orchids in the guild do not participate in coevolution, as their empty flowers cannot influence the evolution of fly proboscis length. Instead, it was proposed that these orchids track the evolutionary changes in the rest of the guild. A similar pattern of diffuse coevolution and trait tracking by orchids has been described in the *Moegistorhynchus longirostris* pollination guild on the Cape west coast (Pauw et al. 2009).

Although the mechanism for the evolution of corolla tube length is now well understood (Alexandersson and Johnson 2002; Anderson et al. 2008), we are still far from understanding convergence in traits such as flower color (see Chapter 11) and floral scent. Methods for studying convergent evolution are still in their infancy. Where phylogenies are available, it is possible to study convergent evolution between closely related lineages even within genera (Johnson et al. 1998; Waterman et al. 2009; van der Niet et al. 2011). In a recent study of the guild of South African plants pollinated by pompilid wasps, Shuttleworth and Johnson (2012) used congeners of guild members as a control to test for the extent of floral convergence. Using a bee vision model as a proxy for how wasps might perceive flowers, they found that spectra of flowers in the pompilid guild were more similar to those of background vegetation than were the spectra of congeners. They attributed this to selection for crypsis, as evidence suggests that pompilid wasps find flowers mainly using scent cues. Pauw (2006) reported a striking pattern of convergence in the spectral reflectance of the yellow flowers of orchids in the *Rediviva peringueyi* oil-bee guild. However, he did not test whether related species pollinated by other oil bees differ in spectral reflectance. An interesting general feature of oil-bee pollination guilds in South Africa is that while the orchids often share the same yellow coloration, species of *Diascia* (Scrophulariaceae) that are also part of these guilds usually have flowers that are pink or reddish-purple (Steiner and Whitehead 1988).

In a recent study, Steiner et al. (2011) analyzed the floral scent chemistry of oil-producing orchids in South Africa to determine the extent to which convergence has occurred at the level of pollination guilds. They found limited evidence for convergence and instead found that scent chemistry tends to be strongly conserved within lineages. Shuttleworth and Johnson (2012) reached a similar conclusion in the study of the floral scent of dozens of species in the pompilid wasp pollination guild. They found no clear pattern of convergence in floral scent among the wasp-pollinated species. These studies raise interesting questions about the functional importance of floral scent blends in these highly specialized systems. They suggest that selection may act on a few compounds, perhaps not even the dominant ones in the blend. This appears to be the case in sexually deceptive orchids where the active compounds that elicit behavioral responses in pollinators are often of low volatility or constitute minor peaks in the floral headspace (Schiestl et al. 1999).

In contrast to flowers specialized for hymenopteran pollination, patterns of convergence in floral volatiles among plants pollinated by carrion flies are now well established. For example, van der Niet et al. (2011) showed that the South African carrion fly–pollinated orchid *Satyrium pumilum* produces a scent dominated by oligosulfide compounds. These same compounds dominate the scent of many other South African plants pollinated by carrion flies (Jurgens et al. 2006; Johnson and Jurgens 2010) and have been shown to play a key role in mediating the attraction of carrion flies (Shuttleworth and Johnson 2010). In this case, at least, a clear pattern of convergence in floral scent chemistry is evident at the guild level.

FLORAL MIMICRY

Many orchids (approximately one-third of all species worldwide, by one estimate) do not produce floral rewards. In the South African orchid flora, the most important genera in which pollination by deception occurs are *Disa* (about 50% of species are nonrewarding) and *Eulophia* (all species appear to be nonrewarding). For species that do not offer rewards, pollinators must be somehow duped into visiting the flowers. Sexual deception is the most notorious of systems of deception, but has been reported in just 2 South African species (Steiner et al. 1994). Most nonrewarding orchid species bear a generalized resemblance to rewarding flowers and exploit the instincts of food-seeking insects (compare with Chapters 2, 7, 10).

A smaller proportion of orchids mimics the flowers of specific plants (see Color Plate 5) and thereby exploits their pollinators. These "Batesian mimics" are of special interest here, as they are often members of specialized ecological guilds and exemplify the complex interdependency of orchids and other guild members in a community context (see Chapter 5). Most of the proposed examples of Batesian mimicry involve orchids that are members of guilds of plants pollinated by long-proboscid flies or butterflies. One possible explanation is that these pollinators rely mainly on visual cues to select flowers, and the orchids therefore do not require the additional complexity of mimicking the scent blends of the flowers of model rewarding plants. Behavioral studies in which pollinators are offered a choice between flowers of orchids that are proposed Batesian mimics and flowers of their putative model show that in most cases, pollinators are unable to distinguish between the nonrewarding orchids and their models (Table 4.2). This raises interesting questions about which traits have become modified to accomplish this effective ruse.

The role of color in Batesian floral mimicry is now well established. Generally, flowers of Batesian orchid mimics closely match the spectral reflectance of flowers of their proposed models (Table 4.2; Figure 4.2). Furthermore, experimental modification of the mimics' flower color has an adverse effect on pollination. Peter and Johnson (2008) removed ultraviolet reflectance from the perianth of the orchid mimic *Eulophia zeyheriana* with sunscreen and thereby demonstrated that its pollinator, a halictid bee, selects flowers on the basis of their color. This effect of color manipulation was subsequently also demonstrated for the orchid's model, *Wahlenbergia cuspidata* (Welsford and Johnson 2012).

The strongest evidence for the adaptive significance of color in a Batesian mimicry system was recently obtained in a study of the functional significance of color in the orchid *Disa ferruginea* (Newman et al. 2012). Red morphs of this orchid co-occur with the red flowers of an iris, *Tritoniopsis triticea*, in the western Cape mountains, while an orange morph of the orchid co-occurs with the orange flowers of an asphodel, *Kniphofia uvaria* (Xanthorrhoeaceae), in the southern Cape mountains (Johnson 1994b). Newman et al. (2012) reciprocally translocated inflorescences of these orchids between the 2 regions, so that their pollinator, the butterfly *Aeropetes tulbaghia*, was offered a choice between orchids of contrasting flower colors. In accordance with the mimicry hypothesis, that pollinators should select flower colors according to their conditioning experience with local rewarding food plants, the butterflies preferred red morphs of the orchids in the western Cape and orange morphs in the southern Cape (Figure 4.2).

Although color is clearly a target of pollinator-mediated selection on the flowers of deceptive orchid mimics, the roles of floral morphology and inflorescence structure are less well understood. In a study of the orchids *Brownleea galpinii* and *Disa cephalotes*, which appear to mimic the inflorescences of *Scabiosa columbaria* (Dipsacaceae), Johnson et al. (2003a) found that experimental modification of the inflorescence structure of *B. galpinii* from a capitulum (which resembles that of *Scabiosa*) to a raceme (inferred to occur in ancestors of this orchid) had a strong negative

Table 4.2 Orchids proposed to be Batesian mimics of rewarding plant species, and the level of evidence for mimicry

Proposed orchid mimic	Model	Pollinator	Evidence	References
Disa amoena	*Watsonia wilmsii* (Iridaceae)	*Prosoeca ganglbaueri* (Nemestrinidae)	1	Goldblatt & Manning 2000
Disa nervosa	*Watsonia densiflora* (Iridaceae)	*Philoliche aethiopica* (Tabanidae)	1, 2	Johnson & Morita 2006
Disa pulchra	*Watsonia lepida* (Iridaceae)	*Philoliche aethiopica* (Tabanidae)	1, 2, 3	Johnson 2000
Disa nivea	*Zaluzianskya microsiphon* (Scrophulariaceae)	*Prosoeca ganglbaueri* (Nemestrinidae)	1, 2, 3	Anderson & Johnson 2006, 2009; Anderson et al. 2005
Disa ferruginea (red morph)	*Tritoniopsis triticea* (Iridaceae)	*Aeropetes tulbaghia* (Nymphalidae)	1, 2, 3, 4	Johnson 1994a; Newman et al. 2012
Disa ferruginea (orange morph)	*Kniphofia uvaria* (Asphodelaceae)	*Aeropetes tulbaghia* (Nymphalidae)	1, 2, 3, 4	Johnson 1994a; Newman et al. 2012
Disa karooica	*Pelargonium*	*Philoliche gulosa* (Tabanidae)	1, 2	Combs and Pauw 2009
Disa cephalotes, Brownleea galpinii	*Scabiosa columbaria* (Dipsacaceae)	*Prosoeca sp.* (Nemestrinidae)	1, 2, 3	Johnson et al. 2003a
Eulophia parviflora	*Wahlenbergia cuspidata* (Campanulaceae)	*Lipotriches sp.* (Halictidae)	1, 2, 3, 4.	Peter & Johnson 2008

Key for evidence: 1. matching distributions; 2. matching spectra; 3. behavioral tests with pollinators; 4. enhanced fitness in presence of the model.

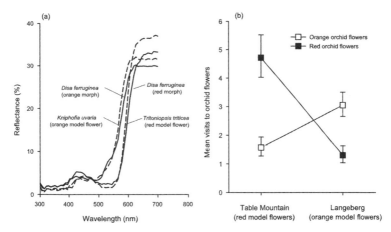

Fig. 4.2 Spectral reflectance patterns of flowers of the Batesian mimic *Disa ferruginea* and their implications for pollination success. (*a*) Spectra of *D. ferruginea* flowers (*solid lines*) match those of co-occuring rewarding flowers (*dashed lines*). (*b*) The implications of orchid flower color for rate of visitation in an experiment in which *D. ferruginea* inflorescences were reciprocally translocated between Table Mountain, where the model plants have red flowers, and the Langeberg Mountains, where the models have orange flowers. Modified from Newman et al. 2012. Photos: Steve Johnson.

effect on pollinator visitation. Anderson et al. (2005) obtained indirect evidence for selection on flower size in the deceptive orchid *Disa nivea*. In this orchid, flower size is correlated with flower size of *Zaluzianskya microsiphon* (Scrophulariaceae), its proposed model, across the distribution range of both species.

THE GUILD CONTEXT OF ORCHID DIVERSIFICATION

Pollination guilds can be considered to be constituents of the ecological niches for the reproductive biology of plants (Johnson 2010). The parameters of these niches are defined not only by the pollinator involved but also by the local plant community, which affects the abundance of the pollinator as well as its conditioned foraging preferences (Johnson 2010). Diversification of angiosperms through evolutionary shifts between different pollinators (Stebbins 1970) can also be thought of as diversification through shifts in pollination guild membership. This model of guild shifts has proved highly applicable to diversification of South African orchids (Johnson 1997; Johnson and Steiner 1997; Waterman et al. 2009).

Guild shifts are best understood in the context of the restricted geographic distributions of the pollinators of most guilds (Figure 4.3). For plants that extend their range beyond the geographic confines of one guild, there is strong selection for traits that attract a different pollinator (Johnson and Steiner 1997) or that provide reproductive assurance through selfing or clonality (Pauw and Bond 2011). Plant species whose distribution extends across the distribution of 2 different pollination guilds can develop local ecotypes. Some of the best-documented examples involve irids in the genus *Lapeirouisa* and orchids in the *Disa draconis* complex that have ecotypes pollinated by the very long-proboscid nemestrinid fly *Moegistorhynchus longirostis* on the Cape sand plain and the horsefly *Philoliche rostrata* in the surrounding mountains (Johnson and Steiner 1997; Pauw et al. 2009). Other examples include ecotypes of *Pterygodium catholicum* adapted to different oil-collecting bees (Steiner et al. 2011) or that have shifted from pollination by oil bees to a largely clonal habit (Pauw and Bond 2011; Pauw and Hawkins 2011). Steiner (1998) attributed the shift from pollination by oil-collecting bees to pollination by hopliine beetles in the genus *Ceratandra* to the expansion of the range of this genus into the southern Cape, where oil bees appear to be very rare or absent.

At a macroevolutionary scale, evidence for the role of shifts between pollination guilds in the speciation of South African orchids is provided from phylogenetic studies showing that a high proportion of divergence events are associated with shifts between different pollination guilds (Johnson et al. 1998; Waterman et al. 2009). In a study of sister species in the Cape flora, van der Niet and Johnson (2009) found that shifts between different pollinators are as frequent as shifts between different soils, and often coincide with soil shifts (van der Niet et al. 2006). One possible explanation is that the distributions of pollinators are often controlled by interactions between soils and insect larvae (Waterman et al. 2009).

Because pollinators are often associated with particular types of vegetation, species that make transitions between habitats often adapt to different pollination guilds. A good example of this phenomenon was recently provided in a phylogenetic study of the oil-producing orchid genus *Huttonaea* in which a shift from grassland to forest habitat resulted in a shift from membership in the *Rediviva neliana* guild to membership in a guild pollinated by the forest oil-collecting bee *R. colorata* (Steiner 2010).

CONCLUSIONS

Although Darwin and his contemporaries established the adaptationist framework for pollination biology, they studied orchids in isolation

Fig. 4.3 Orchids and other plants belonging to wasp, bee, and beetle pollination guilds in South Africa. The orchid *Disa sankeyi* (*a*) and milkweed *Asclepias macropus* (*b*) are adapted for pollination by *Hemipepsis* (pompilid wasps). The orchid *Corycium dracomontanum* (*c*) and scroph *Diascia purpurea* (*d*) are part of a guild of plants pollinated by the oil-collecting bees *Rediviva neliana* (here visiting *C. dracomontanum*), *R. brunnea* (here visiting *D. purpurea*), and *R. pallidula*. The cetoniine beetle, *Atrichelaphinus tigrina*, pollinates flowers of the orchid *Satyrium microrrynchum* (*e*) and the milkweed *Xysmalobium involucratum* (*f*). Photos: Steve Johnson.

from the geographic context of the various pollination guilds to which they belong and thus made little progress in identifying the factors that drove orchid diversification. We now realize that the landscape in which orchids evolve is a complex mosaic of different selective factors, including a complex spatial geographic mosaic of different pollination guilds. Evolutionary modification of orchids to become members of these guilds occurs in response not just to the sensory modalities and morphology of pollinators but also to other plants in the guilds. We now realize that co-occurring plants can influence the evolution of orchid flowering times (compare with Chapter 12), site of placement of pollinaria on the pollinator's body (compare with Chapters 8 and 9), and the color and shape of flowers, as exemplified by Batesian mimicry in deceptive orchids (see Chapter 2). However, we still lack a proper understanding of the role that other aspects of the floral phenotype, such as scent emissions, play in shifts between different pollination guilds. It is clear that progress will require a broad set of analytical tools, along with a combination of microevolutionary studies of selection and macroevolutionary studies of diversification patterns in orchid lineages.

ACKNOWLEDGMENTS

Kim Steiner, Bruce Anderson, Craig Peter, and Anton Pauw have shared my passion for orchid biology and helped develop some of the ideas in this chapter. Peter Weston, Timo van der Niet, Kim Steiner, and Kathy Johnson made many useful suggestions for improving the manuscript.

FIVE

Phylogeny of Orchidaceae Tribe Diurideae and Its Implications for the Evolution of Pollination Systems

Peter H. Weston, Andrew J. Perkins, James O. Indsto, and Mark A. Clements

INTRODUCTION

Charles Darwin wrote almost nothing about Australian plants or animals in his first book about the pollination of orchids (Darwin 1862),
because very little was then known about the biology of Australian orchids beyond their descriptive morphology. This gap in knowledge would
be rectified handsomely over the next 150 years. Starting with Robert
Fitzgerald (Fitzgerald 1875–1895), numerous observers, most of whom
were amateur orchid enthusiasts, subsequently gave us a rich natural history of the amazingly diverse, sometimes bizarre, relationships between
Australian orchids, especially the terrestrial species, and their pollinators.
They also provided detailed accounts of the mechanics of self-pollination,
a few of which Darwin (1877) was able to discuss in the second edition of
his orchid book (see Chapters 1, 6, and 7). Recently, Adams and Lawson
(1993), van der Cingel (2001), and Bower (2001) reviewed most of this
literature, showing in the process that Australia had the potential to provide numerous examples of the kinds of phenomena Darwin theorized
about in his orchid book (and see Chapter 6 and 7).

The sophistication of tools for interpreting the evolutionary relationships between flowering plants and their pollinators and for testing
hypotheses that seek to explain those patterns has grown dramatically
since Darwin's time, especially over the last 50 years. Central to this analytical tool kit are methods for reconstructing phylogeny and associated

hypotheses of character evolution (e.g. Schuh and Brower 2009; Lemey et al. 2009). The results of those analyses can now be further processed using analytical tools for testing a wide range of descriptive and explanatory hypotheses (e.g. papers in Eggleton and Vane-Wright 1994 and Harvey et al. 1996).

The History of Phylogenetic Research on the Tribe Diurideae

Darwin wrote little about the phylogeny of orchids or any other groups of organisms, despite his recommendation (Darwin 1859; p. 420) that biological classification "must be strictly genealogical in order to be natural." Although he explicitly proposed a treelike model of evolving, divergently branching, ancestral-descendant lineages (Darwin 1859; plate facing p. 117) that forms the basis of the standard techniques of phylogeny reconstruction employed in this chapter, he did not propose any method of phylogenetic analysis (Stevens 1984). Instead, Darwin interpreted available hierarchical classifications as estimates of phylogeny on which he superimposed hypotheses of adaptation. In the case of the orchids, John Lindley's intuitive classification of the family (Lindley 1846) provided the phylogenetic basis for Darwin's evolutionary inferences. It would be another 130 years before the first attempt was made to reconstruct the phylogeny of any group of Australian orchids using explicit analytical methods (Lavarack 1976). Even then, Lavarack employed a phenetic rather than an overtly phylogenetic approach to try to analyze relationships among Australian terrestrial orchids using morphological characters.

The first explicitly phylogenetic analysis of Australian orchids would not appear for another 10 years, when Burns-Balogh and Funk (1986) published a morphology-based cladistic analysis of the Orchidaceae, represented by a selection of terminal taxa that mostly corresponded to tribal and subtribal units recognized by Dressler (1981). Their trees, however, grouped these units into more inclusive putative clades that differed radically from the higher taxa that Dressler (1981) had recognized. Burns-Balogh and Funk assigned members of Dressler's tribe Diurideae to 2 different subfamilies, the Spiranthoideae and Neottioideae, suggesting that his diurids were grossly polyphyletic. Dressler (1993b) responded with his own intuitively derived cladogram of the Orchidaceae in which the tribe Diurideae, with the addition of the subtribes Cryptostylidinae and Rhizanthellinae, was treated as a clade.

The results of both Burns-Balogh and Funk's and Dressler's analyses were soon shown to be inaccurate by the first molecular systematic analyses of Australian terrestrial orchids (Cameron et al. 1999; Kores et al. 2000, 2001), based on parsimony analyses of alignments of chloroplast DNA sequences (cpDNA); these were followed soon after by a comparable analysis of an alignment of sequences of the Internal Transcribed Spacers (ITS) of the nuclear ribosomal DNA (nrDNA) repeat unit (Clements et al. 2002).

Those first molecular systematic analyses represented a huge progressive step in our understanding of the membership of, and phylogeny within, the tribe Diurideae. The chloroplast and ITS trees were largely congruent and decisively resolved several long-standing problems in the systematics of this group. First, although they showed that the Diurideae, as circumscribed by Dressler (1993b), were polyphyletic, a clade that included most of the genera that had been assigned to this tribe could be recognized as the Diurideae. These analyses also brought some surprises. The Diurideae were resolved as the sister group of the tribe Cranichideae, a group composed mostly of taxa that had until recently been treated as part of a distinct subfamily, the Spiranthoideae. Perhaps most surprising was the discovery that none of the South American taxa that had previously been included in the Diurideae as the subtribe Chloreinae turned out to belong within the diurids. They did not even form a clade, with most grouping together in the Cranichideae, while South American *Codonorchis* found softly incongruent placements in the chloroplast and ITS trees outside both the Diurideae and the Cranichideae. Similarly surprising was the placement of *Pterostylis* in the Cranichideae, not the Diurideae. On the other hand, *Cryptostylis* and *Rhizanthella*, the positions of which had been debated for years, grouped with the diurids.

The Need for a Phylogenetic Analysis of the Diurideae Based on All Available Evidence

A good case can be made for the assembly and simultaneous phylogenetic analysis of a supermatrix of all available characters for the Diurideae. The chloroplast and nuclear DNA data sets are complementary in that some clades that are strongly supported by published trees from one are either weakly supported or not found at all in trees from the other. Moreover, data for a diverse array of "traditional" characters have accumulated for the diurids over the past 30 years as a result of systematic surveys of floral development (Rasmussen 1982; Kurzweil 1987, 1988; Kurzweil et al. 2005), anatomy (Stern et al. 1993; Pridgeon 1993, 1994; Kurzweil et al. 1995; Pridgeon and Chase 1995), embryology (Clements 1999), and seed coat micromorphology (Molvray and Kores 1995), and these can usefully augment data for macromorphological characters as well as indicate the phylogenetic positions of the great majority of species that have not been sequenced for any DNA loci. All these characters represent independent tests of phylogenetic relationships and thus have the potential to increase tree resolution and levels of support for clades.

The primary aim of this chapter, then, is to present a phylogenetic analysis of chloroplast, ITS, and morphological data sets representing all currently recognized genera of the tribe Diurideae, to provide as fully resolved and strongly corroborated a hypothesis of phylogenetic relationships as possible.

TESTING HYPOTHESES ABOUT THE EVOLUTION OF POLLINATION SYSTEMS IN THE DIURIDEAE USING METHODS OF ANCESTRAL CHARACTER-STATE RECONSTRUCTION

Our second aim was to use our best estimate of diurid phylogeny to reconstruct key features of the evolution of pollination systems in the Diurideae and thus to test several evolutionary hypotheses (see e.g. Donoghue 1989). Speculation on the evolutionary origins of deceptive pollination systems in the Orchidaceae (e.g. Dressler 1981; Dafni and Bernhardt 1989) has often assumed that such systems have evolved repeatedly from rewarding, nectar-producing ancestors, but molecular phylogenetic analyses of higher-level relationships within the Orchidaceae (e.g. Freudenstein et al. 2004) suggest that the family may be primitively nectarless (Cozzolino and Widmer 2005). We aimed to determine the most likely ancestral state for this character in the Diurideae and to reconstruct transitions between the presence and the absence of nectar production in the tribe.

We also wanted to go further and reconstruct the phylogenetic relationships between different kinds of pollination systems in the Diurideae: pollination by various functional classes of insects lured by nectar reward, food mimicry, and sexual deception, and autogamy or apomixis. The definition of those classes warrants some discussion. Because many orchids are highly specialized in being pollinated by one or a few closely related pollinator species (Schiestl and Schlüter 2009 and references therein), we need a basis for establishing relationships between specialist pollinators in order to understand the evolution of pollinator shifts (see Chapters 8 and 9). Without such a basis, all we have is a collection of disconnected 1:1 pairings of plants and pollinators. In some cases, the most appropriate connections for this purpose are the phylogenetic relationships between the pollinators (e.g. Mant et al. 2002, 2005b).

Unfortunately, knowledge of the detailed phylogenetic relationships between closely related species of insects is rarely available. Moreover, many species of orchids are pollinated by classes of functionally similar animals, which are not necessarily closely related. For example, *Thelymitra ixioides* is pollinated by bees in the genera *Exoneura* and *Lasioglossum* (Sydes and Calder 1993), which belong to different families, the long-tongued Apidae and the short-tongued Halictidae (Danforth et al. 2006 and references therein). Although *Thelymitra ixioides* is not adapted for pollination by just one species, it is still a bee specialist (see Chapter 7). It now seems clear that diversification in pollination systems has come about overwhelmingly by a process of evolutionary shifts between pollinators or pollinator classes rather than cospeciation between plants and animals (e.g. Mant et al. 2002; Tripp and Manos 2008; van der Niet and Johnson 2012). We aimed to reconstruct the history of diversification of pollination systems in the Diurideae by reconstructing the history of such shifts on our best estimate of phylogeny.

In doing so, we were interested in determining if similar pollination systems have evolved repeatedly in different lineages. Such repeated patterns would suggest adaptive change (Fenster et al. 2004). In particular, we wanted to test an explanatory hypothesis for the evolution of deceptive pollination systems: that deceptive systems are selectively advantageous by minimizing geitonogamous pollination, the self-pollination process of a flower receiving pollen from another flower on the same plant, and also pollinator-mediated intrafloral selfing arising through repeated probings by the same insect on the same flower (Johnson and Nilsson 1999).

A prediction of that hypothesis is that the evolution of a rewarding system from an ancestrally deceptive, multiflowered one is likely to be accompanied by a decrease in flower number per plant to one, because the combination of multiple flowers with nectar rewards, pollinated by an optimally foraging pollinator, would render a plant vulnerable to inbreeding through geitonogamy. Conversely, one would expect the evolution of a deceptive system from an ancestrally rewarding, single-flowered one to be accompanied by an increase in flower number per plant to more than one (compare with Chapters 2–4). Because many diurid orchids multiply vegetatively by producing multiple daughter tubers at the ends of stolon-like roots instead of, or in addition to, producing replacement tubers (Pridgeon and Chase 1995; Jones 2006), care must be taken in categorizing plants as single- or multiflowered. Species of *Corybas*, for example, the annual shoots of which are consistently single flowered, are in fact multiflowered plants, because they form clonal patches of genetically identical individuals.

Various authors have discussed the evolution of floral diversity in the Diurideae based on phylogenetic frameworks that were subsequently shown to be misleading. Burns-Balogh and Bernhardt (1988), Dafni and Bernhardt (1989), and Bernhardt (1995a), for instance, accepted the phylogenetic hypothesis produced by Burns-Balogh and Funk (1986), and proceeded to build an evolutionary scenario for the tribes Thelymitreae and Geoblasteae sensu Burns-Balogh and Funk. The Thelymitreae have since been shown to be monophyletic and were treated as the subtribe Thelymitrinae by Jones (in Pridgeon et al. 2001), but the Geoblasteae are a polyphyletic amalgam of superficially similar clades now placed in the tribes Diurideae and Cranichideae. Nevertheless, falsification of a phylogenetic hypothesis does not necessarily falsify hypotheses of recurrent processes framed around that phylogeny. Our third aim was thus to re-evaluate earlier ideas (based on different phylogenetic hypotheses) about the evolution of traits and pollination systems in the tribe.

METHODS

Taxa Used as Terminal Entities in Our Phylogenetic Analyses

The terminals for which we sampled character data for phylogenetic analysis were the genera recognized by Jones (2006). This should not be taken as an endorsement of his generic-level taxonomy—some of us do not see the need for most of the extensive taxonomic changes in the Diurideae that have been proposed over the past 10 years. However, using these taxa as terminals allowed a more informative analysis of the Diurideae than if we had used the more broadly circumscribed genera of Jones's taxonomic treatment of the Diurideae in Pridgeon et al. (2001). DNA sequence data are available for almost all Jones's (2006) genera but only sporadically within them, and Jones's book provides a useful framework and reference work within which morphological and pollination data could be collated.

Taxon sampling of molecular data was designed to include at least one representative of each genus of Diurideae while maximizing the number of characters sampled for each terminal taxon. For 25 genera it was necessary to combine sequence data from different species of the same genus, producing terminals that were composites of sequences from 2 or 3 species. Morphological data represent the totality of known variation within each genus. Terminals were coded as polymorphic where more than one state of a character was recorded for a genus. Thus, all genera with more than one species were represented by terminals that were composites of data drawn from multiple species. This approach to coding terminals is unproblematic if the genera are monophyletic, but can produce confounded results if some of the species contributing to a terminal are most closely related to those belonging to another terminal. Previously published phylogenetic analyses (e.g. Clements et al. 2002; Indsto et al. 2009; Mant et al. 2002, 2005b) have tested and corroborated the monophyly of most of the terminals in our analysis. However, as a secondary check of the monophyly of terminals, all morphological characters were traced on our best phylogenetic tree in Mesquite (Maddison and Maddison 2011), using the parsimony criterion, to compile a list of unequivocal autapomorphies of genera.

Sources of DNA Sequences Used in Our Phylogenetic Analyses

DNA sequences for the nuclear locus ITS1-5.8S-ITS2 and chloroplast loci *rbcL*, *matK* (which appears to be a nonfunctional pseudogene in Orchidaceae; Kores et al. 2000) and the *trnL-trnF* intergenic spacer were downloaded from GenBank or retrieved from our own sequence archives (Table 5.1). Various procedures had been used to acquire tissue samples, extract DNA from them, and amplify and sequence the targeted genes by the original authors of those sequences (Cameron et al. 1999; Kores et al. 2000, 2001; Clements et al. 2002). Tissue samples were mostly collected from wild populations and usually consisted of a single whole leaf, or in the case of leafless species, part of an inflorescence, which was usually enclosed in a snap-lock polythene bag with indicator silica gel for rapid dehydration (Chase and Hills 1991). A voucher specimen, usually consisting of a mature inflorescence or whole flower, was also collected from the same plant as the tissue sample. The vouchers were either pressed and dried or preserved in 70% ethanol and accessioned in one of the following registered herbaria: CANB, CHR, K, NCU, OKL. Total genomic DNA was extracted from 10- to 100-mg portions of the tissue samples; targeted genes were amplified using the polymerase chain reaction (PCR), and sequenced using variants of the Sanger DNA sequencing technique, as specified in the original publications (Cameron et al. 1999; Kores et al. 2000, 2001; Clements et al. 2002).

Characters Used in Our Phylogenetic Analyses

Alignment of the *rbcL* sequences was trivial, requiring the insertion of no indels. ClustalX with default parameters was used to align the *matK*,

Table 5.1 Genbank accessions used for molecular phylogenetic analyses

Terminal taxon	trnL-F Genbank accession	matK Genbank accession	rbcL Genbank accession	ITS Genbank accession
Disperis	D. capensis AJ409402	D. capensis AJ310022	D. capensis AY381120	D. capensis AJ000128
Disa	D. glandulosa AJ409401	D. glandulosa AJ310021	D. glandulosa AF274006	D. uniflora AJ000112
Satyrium	S. nepalense AJ409450	S. nepalense AJ310070	—	S. odorum EF601513
Habenaria	H. repens AJ409418	H. repens AJ310036	H. repens AF074177	H. propinquior AF348035
Zeuxine	Z. strateumatica AJ409458	Z. strateumatica AJ310080	—	Z. oblonga AF348073
Goodyera	G. viridiflora AJ409417	G. viridiflora AJ310035	G. repens FJ571330	Goodyera repens AF366896
Pachyplectron	P. arifolium AJ409434	P. arifolium AJ310051	—	P. arifolium AJ539522
Cranichis	C. fertilis AJ409392	C. fertilis AJ310013	C. fertilis AF074137	C. cililabia AJ539506
Ponthieva	P. racemosa AJ409439	P. racemosa AJ310056	P. racemosa AF074223	P. racemosa AJ539508
Chloraea	fonkii AJ409383	C. fonkii AJ310004	C. magellanica AJ542403	C. magellanica AJ539523
Oligochaetochilus	O. pictus AJ409446	O. pictus AJ310063	—	O. rufus AF348056
Cryptostylis	C. subulata AJ409395	C. subulata AJ310015	C. subulata AF074140	C. subulata AF074140
Coilochilus	C. neocaledonicus AJ409388	C. neocaledonicus AJ310009	C. neocaledonicus AY381114	—
Acianthella	A. confusa AJ409370	A. confusa AJ309999	—	A. amplexicaulis AF347973
Stigmatodactylus	S. sikokianus AJ409453	S. sikokianus AJ310075	—	—
Corysanthes	C. diemenica AJ409389	C. diemenica AJ310010	C. diemenica AF074135	C. fimbriata AF348009
Nematoceras	—	—	—	N. macranthum AF348010
Corybas	C. neocaledonicus AJ409390	C. neocaledonicus AJ310011	—	C. aconitiflorus AF348008
Singularybas	—	—	—	S. oblongus AF391774
Anzybas	—	—	—	A. unguiculatus AF348013
Cyrtostylis	C. huegelii AJ409399	C. huegelii AJ310019	—	C. robusta AF348021
Nemacianthus	N. caudatus AJ409369	N. caudatus AJ309990	—	N. caudatus AF347976
Acianthus	A. fornicatus AJ409374	A. fornicatus AJ309994	A. exsertus AF074101	A. fornicatus AF347977
Spurianthus	—	—	—	S. atepalus AF347974

Genus				
Townsonia	—	—	—	*T. deflexa* AY135375
Corunastylis	*C. fimbriata* AJ409413	*C. fimbriata* AJ310031	*C. fimbriata* AY368345	*C. apostasioides* AF391783
Genoplesium	—	—	—	*G. baueri* AF391784
Prasophyllum	*P. brevilabre* AJ409441	*P. brevilabre* AJ310058	—	*P. australe* AF348053
Mecopodum	—	—	—	—
Chiloterus	—	—	—	—
Microtis	*M. unifolia* AJ409429	*M. unifolia* AJ310046	*M. parviflora* AF074194	*M. unifolia* AF348046
Hydrorchis	—	—	—	*H. orbicularis* AF348044
Microtidium	—	—	—	*M. atratum* AF348043
Rhizanthella	—	—	—	*R. slateri* AF348059
Adenochilus	*A. nortonii* AJ409375	*A. nortonii* AJ309995	*A. nortonii* AY381108	*A. nortonii* AF347979
Eriochilus	*E. cucullatus* AJ409410	*E. cucullatus* AJ310028	*E. cucullatus* AF074166	*E. cucullatus* AF348030
Leptoceras	*L. menziesii* AJ409421	*L. menziesii* AJ310039	—	*L. menziesii* AF348039
Praecoxanthus	*P. aphyllus* AJ409440	*P. aphyllus* AJ310057	—	*P. aphyllus* AF348052
Cyanicula subgenus *Trilobatae*	*C. sericea* AJ409398	*C. sericea* AJ310018	*C. caerulea* AF074116	*C. amplexans* AF348016
Cyanicula subgenus *Cyanicula*	*C. gemmata* GQ866817	*C. gemmata* GQ866553	*C. gemmata* AY381116	*C. gemmata* AF348018
Elythranthera	*E. emarginata* AJ409407	*E. emarginata* AJ310027	—	*E. brunonis* AF348028
Ericksonella	—	*E. saccharata* GQ866558	—	*E. saccharata* AF347995
Glossodia	*G. major* AJ409415	*G. major* AJ310033	*G. major* AF074173	*G. major* AF348034
Drakonorchis	*D. barbarossa* AJ409406	*D. barbarossa* AJ310026	—	*D. barbarossa* AF347986
Arachnorchis	*A. falcata* AJ409377	*A. falcata* AJ309998	—	*A. longicauda* AF347992
Jonesiopsis	—	*J. cairnsiana* AJ309996	—	*J. filifera* AF347990
Petalochilus	*P. catenatus* AJ409376	*P. catenatus* AJ309997	—	*P. carneus* AF347988
Stegostyla	—	—	—	*S. congesta* AF347989

continued

Table 5.1 *continued*

Terminal taxon	trnL-F Genbank accession	matK Genbank accession	rbcL Genbank accession	ITS Genbank accession
Caladenia	*C. latifolia* AJ409378	*C. latifolia* AJ310000	—	*C. marginata* AF347993
Pheladenia	—	*P. deformis* GQ866574	—	*P. deformis* GQ866524
Phoringopsis	—	—	—	*P. dockrillii* AF347982
Thynninorchis	—	—	—	*T. huntiana* AF321582
Arthrochilus	—	—	—	*A. oreophilus* AF347984
Simpliglottis	—	—	—	*S. valida* AY042161
Chiloglottis	—	—	—	*C. diphylla* AF348001
Myrmechila	*M. trapeziformis* AJ409382	*M. trapeziformis* EF065585	*M. trapeziformis* AF074124	*M. trapeziformis* AY042153
Drakaea	*D. concolor* AJ409405	*D. concolor* AJ310025	—	*D. glyptodon* AF348027
Caleana	*C. major* AJ409379	*C. major* AJ310001	—	*C. major* AF321585
Sullivania	*S. nigrita* AJ409436	*S. nigrita* AJ310053	—	*S. nigrita* AF348051
Spiculaea	*S. ciliata* AJ409452	*S. ciliata* AJ310072	—	*S. ciliata* AF348063
Burnettia	—	—		*B. cuneata* AF347985
Aporostylis	—	—	*A. bifolia* AY381109	*A. bifolia* AF347980
Rimacola	*R. elliptica* AJ409449	*R. elliptica* AJ310066	*R. elliptica* AY381133	*R. elliptica* AF348060
Waireia				*W. stenopetala* AF348072
Leporella	*L. fimbriata* AJ409420	*L. fimbriata* AJ310038	*L. fimbriata* AY381124	*L. fimbriata* AF348038
Lyperanthus	*L. suaveolens* AJ409424	*L. suaveolens* AJ310041	—	*L. suaveolens* AF348040
Megastylis	*M. rara* AJ409427	*M. rara* AJ310044	*M. rara* AY381126	—
Pyrorchis	*P. nigricans* AJ409448	*P. nigricans* AJ310065	*P. nigricans* AF074187	*P. nigricans* AF348057
Calochilus	*C. campestris* AJ409380	*C. campestris* AJ310002	*C. robertsonii* AF074118	*C. campestris* AF321589
Epiblema	*E. grandiflorum* JQ712999			*E. grandiflorum* AF348029
Thelymitra	*T. ixioides* AJ409455	*T. ixioides* AJ310077	*T. sp.* AF074232	*T. longifolia* AF348070
Diuris	*D. laxiflora* AJ409403	*D. laxiflora* AJ310023	*D. sulphurea* AF074152	*D. drummondii* AF348023
Orthoceras	*O. strictum* AJ409433	*O. strictum* AJ310050	*O. strictum* AF074204	*O. strictum* AF348048

trnL-trnF, and nuclear ITS1-5.8S-ITS2 sequences, followed by some manual adjustment.

Morphological, anatomical, and embryological characters (hereafter simply referred to as "morphology") were compiled from publications as well as from our own observations of living and preserved plant specimens. Our sources of published data included Clements (1999), Clements et al. (2002), Dixon and Tremblay (2009), Dockrill (1969), Hallé (1977), Hopper and Brown (2004, 2006a, 2006b, 2007), Jones (2006), Kurzweil (1987, 1988), Kurzweil et al. (1995, 2005), Linder and Kurzweil (1999), Molvray and Kores (1995), Nicholls (1969), Pridgeon (1993, 1994), Pridgeon and Chase (1995), Pridgeon et al. (2001, 2003), Rasmussen (1982), and Stern et al. (1993). The morphological characters and their states are listed in Appendix 5.1.

PHYLOGENETIC METHODS

Our phylogenetic data matrix consisted of 5 partitions: morphology, ITS nrDNA, trnL-trnF, matK, and rbcL. All characters were treated as unordered. Indels in the sequence alignments were treated as missing data. All partitions were analyzed separately as well as in the following combinations: molecular (ITS nrDNA + *trnL-trnF* + *matK* + *rbcL*) and all data (morphology + ITS nrDNA + *trnL-trnF* + *matK* + *rbcL*) under the maximum parsimony (MP) criterion, using PAUP version 4.0b10. Heuristic searches were performed starting with 200 replicate searches, each starting from a randomized topology that was subjected to branch swapping using the tree bisection-reconnection algorithm, to conduct a thorough search of tree space. Three trees were sampled per iteration and fed into an additional round of branch swapping. Each of these partitions and combinations was also subjected to parsimony bootstrapping by PAUP, in most cases using default heuristic search settings and 1000 bootstrap pseudoreplicates, with the following exceptions: for the analysis of the complete data set, 5000 pseudoreplicates were used rather than 1000; for morphology and trnL-F partitions, bootstrapping was conducted using the faststep search option instead of default settings, because searches using any form of branch swapping bogged down in huge numbers of equally parsimonious trees in both cases.

The molecular partitions were also analyzed separately and in combination under the Bayesian probability criterion as implemented in MrBayes 3.1.2, after first selecting appropriate models of molecular evolution for each of the 4 molecular partitions using MrModelTest 2.3 (available from J. A. A. Nylander's website at http:/www.abc.se~nylandermrmo deltest2mrmodeltest2.html). In the combined analysis, we unlinked the sampling of state frequencies and substitution rates for the 4 DNA data partitions. We ran each Markov Chain Monte Carlo (MCMC) analysis for 4–6 million generations, sampling every 1000th generation. The average standard deviation of split frequencies reduced to below 0.01 after 720 thousand to 1.5 million generations, thus providing a burn-in of less than 25% of sampled generations for all partitions. Majority-rule consensus trees were constructed from the trees sampled after the burn-in period in each of the 5 MCMC runs.

Bootstrapped parsimony and Bayesian probability trees were compared to find instances of strong incongruence between the partitions. Strong incongruence between 2 trees was defined as the presence of at least one pair of noncombinable components (sensu Nelson and Platnick 1981), both with maximum parsimony bootstrap percentages (MPBS) higher than 88% (equivalent to 2 uncontradicted synapomorphies; see Weston and Barker 2006) and/or conflicting Bayesian posterior probabilities (BPP) higher than 0.95.

DATA SOURCES AND METHODS FOR ANCESTRAL TRAIT RECONSTRUCTIONS

For each genus of Diurideae, data were extracted from the literature and collated for the number of recognized species per genus; the number of species for which insect pollinators have been identified; the genera, families, superfamilies, and orders of pollinators; the gender of pollinators and the putative kind of pollinator attraction; the number of species known to produce nectar; the location of nectaries; and the number of species known to be obligately autogamous or agamospermous.

The number of accepted species in each genus was calculated by adding the number of species treated by Jones (2006) to the number of species subsequently named.

Our survey of the pollination literature of the Diurideae relies heavily on Bower (2001) and references cited by him, augmented by subsequently published studies, including Alcock (2000), Faast et al. (2009), Farrington et al. (2009), Hopper and Brown (2004, 2006a, 2006b, 2007), Indsto et al. (2006, 2007), Jones (2006), Peakall et al. (2010), and Phillips et al. (2009b, 2013), and see Chapter 7. A genus of insects was counted as a pollinator of a genus of orchids if individuals of the insect had been observed or inferred to pollinate flowers of the orchid. Evidence of pollination was interpreted as including the removal of pollinaria from flowers, the deposition of pollinaria onto the stigma of a flower, or the carrying of rigorously identified pollinaria. Our records of pollinators therefore included categories of confirmed and probable pollinators in the sense of Adams and Lawson (1993), both of which can reasonably be justified as classes of known pollinators for our purposes (Weston et al. 2005).

Only 2 Australian orchid taxa, *Sullivania minor* and *Corunastylis apostasioides*, have been corroborated as agamospermous as a result of detailed investigation (Clements, unpublished data), so this trait was not mapped on the phylogeny. The number of species suggested to be self-pollinating on the basis of observation or inference (e.g. by fruit set ratios) was tallied from Bower (2001) and Jones (2006). Classes of putative pollinator attractions and/or rewards (labellum nectar, sepaline nectar, food deception, sexual deception) were inferred from a combination of pollinator information (where known) and floral traits such as nectar provision, floral scent, and floral display. Data for functional numbers of flowers per plant were extracted from the taxonomic treatments cited above and from our own observations of herbarium specimens.

Data on the reproductive biology of the outgroup terminals *Disperis*, *Brownleea*, *Disa*, "other Diseae-Orchideae" (Diseae-Orchideae; see Chapter 4), Pterostylidinae, Chloraeinae, and "other Cranichideae" (Cranichideae) were extracted from Benitez-Vieyra et al. (2009), Johnson et al. (1998), Jones and Clements (2002), Larsen et al. (2008), Lehnebach and Riveros (2003), Steiner (2001), and Valdivia et al. (2011). Carlos Lehnebach kindly shared the unpublished results of his searches for nectar production in 10 species of Chloraeinae, all of which he had found to be nectarless. "Other Cranichideae" and "other Diseae-Orchideae" were not scored for pollinator families because of the very high pollinator diversity in Diseae-Orchideae (Pridgeon et al. 2001) and poor knowledge of pollinators in most basal Cranichideae (Pridgeon et al. 2003).

Traits were mapped onto a modified version of our best phylogeny using maximum parsimony and, where possible, maximum likelihood, implemented in Mesquite (Maddison and Maddison 2011). Our best tree was modified by resolving the basal-most relationships within the outgroup clades Cranichideae and Diseae-Orchideae, according to the results of Douzery et al. (1999) and Salazar et al. (2003), to allow us to resolve ancestral character-state assignments for the stem group node of the Diurideae.

RESULTS

Phylogenetic Analyses

Our phylogenetic data matrix, which is deposited in TreeBASE (http:/purl .orgphylotreebasephylowsstudyTB2:S14678), consisted of 5854 characters, of which 3265 were invariant and 1733 were parsimony informative. They were partitioned as follows (Table 5.2): ({morphological characters: 1–102}, {molecular partition: characters 103–5854 [ITS nrDNA: 103–1017], [chloroplast DNA partition: characters 1018–5854 (*trnL-trnF* characters 1018–2705), (*matK* characters 2706–4561), (*rbcL* characters 4562–5854)]}).

Key features of the data partitions and the matrix as a whole are listed in Table 5.2, along with summary statistics of the results of the parsimony analyses. A Bayesian probability phylogram for the combined molecular data set, annotated with posterior probability values for components (sensu Nelson and Platnick 1981), is shown in Figure 5.1. Figure 5.2 shows a phylogram of one of 4242 equally parsimonious trees from the morphology partition, optimized using the ACCTRAN option and annotated with bootstrap support indices. A phylogram of 1 of the 9 equally parsimonious trees from the full matrix, optimized using the ACCTRAN option and annotated with bootstrap support indices, is shown in Figure 5.3. Table 5.3 compares the bootstrapped parsimony trees and Bayesian probability trees for the various data partitions with the bootstrapped parsimony tree for the full data set. It lists all the components found in the strict consensus of 9 equally parsimonious trees for the full data set, along with all conflicting components that received greater than 50% bootstrap support (MPBS) in any of the parsimony trees or greater than 0.95 posterior probability (BPP) in any of the Bayesian probability trees.

Trees constructed using maximum parsimony and Bayesian probability criteria showed strong congruence when used to analyze the same data set, consistent with the results of a meta-analysis of 504 published studies by Rindal and Brower (2011). Moreover, parsimony bootstrap percentages proved to be useful predictors of posterior probabilities, with every bootstrap value of 73% or more corresponding with a posterior probability of greater than 0.95.

The trees from the different data partitions were largely congruent with one another. However, although the chloroplast is usually uniparentally inherited as a single unit without recombination and is therefore expected to show congruence between its partitions, we did find 2 instances of strong incongruence between the *trnL-F* and *matK* trees. One involved incongruence among the outgroups in the subtribe Goodyeri-

Table 5.2 Statistics of trees produced by parsimony analyses of different combinations of partitions of the data set

Data set	All data	Molecular	cpDNA	trnL-trnF	matK	rbcL	ITS	Morphology
Number of taxa	73	71	53	49	49	29	68	73
Number of characters	5854	5752	4837	1688	1856	1293	915	102
Number of invariant characters	3265	3265	3070	996	1012	1070	195	0
Parsimony informative characters	1733	1637	1039	394	526	107	598	96
Number of equally parsimonious trees	9	52	36	72	26	8975	12	4242
Tree length	10473	9847	4219	1548	2214	384	5570	556
Consistency index	0.4021	0.4092	0.5757	0.6447	0.533	0.6276	0.2873	0.3273
Retention index	0.5796	0.5746	0.6228	0.6537	0.6235	0.568	0.5555	0.6929

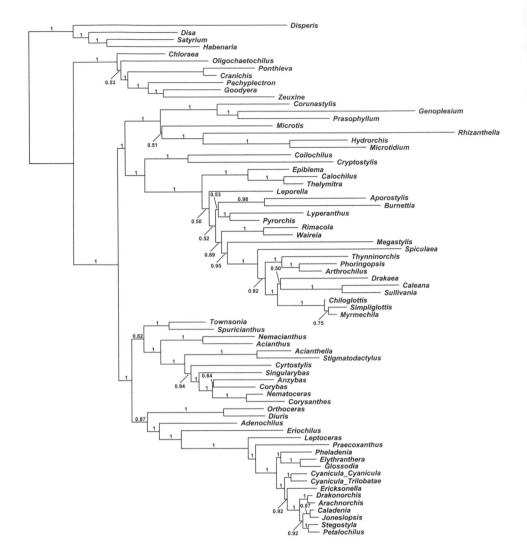

Fig. 5.1 Bayesian probability phylogram for the combined molecular data set, with branches annotated with their posterior probability values.

nae, with *matK* finding *Zeuxine* and *Goodyera* to be more closely related to each other than to *Pachyplectron* with 98% bootstrap support, in contrast to *Goodyera*, and *Pachyplectron* forming a clade to the exclusion of *Zeuxine* with 91% bootstrap support with *trnL-F*. However, both trees supported the monophyly of the Goodyerinae as a whole. Incongruence nested so deeply within the outgroups seems unlikely to have affected

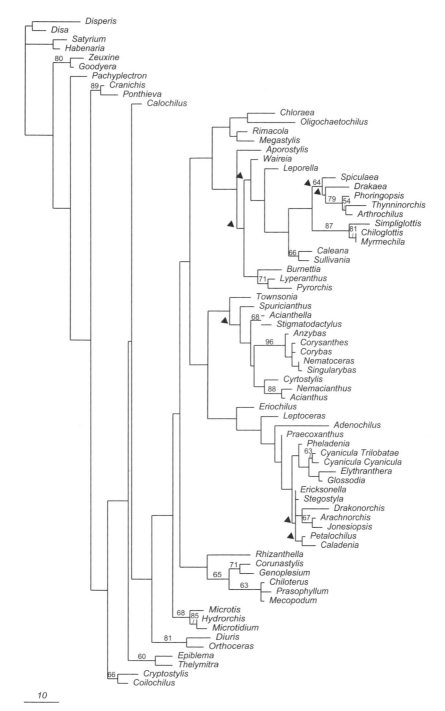

Fig. 5.2 Phylogram for tree number 1 of 4242 equally parsimonious trees produced from the morphological data set, optimized using the ACCTRAN option in PAUP*, with branches annotated with their
bootstrap support indices. Arrows indicate the nodes that collapse in the strict consensus tree.

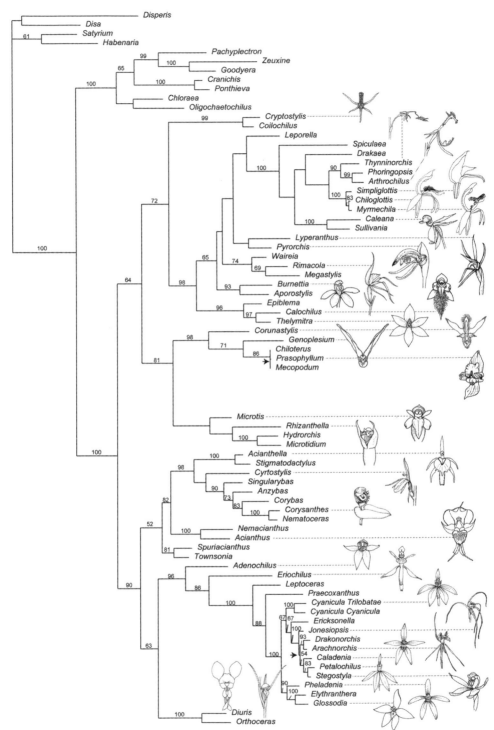

the topology reconstructed within the Diurideae. The other incongruity between cpDNA partitions involved the position of *Eriochilus*. This genus was nested within the subtribe Caladeniinae by *matK* with 98% bootstrap support and by *rbcL* with 60% bootstrap support and posterior probability of 0.77. In contrast, *trnL-F* placed *Eriochilus* as the sister group of the subtribe Diuridinae with 70% bootstrap support and posterior probability of 1.0. These incongruities within the chloroplast DNA partition are probably best explained as the result of stochastic variation in the evolution of *trnL-F*, the consensus tree for which also found a number of other components that were more weakly incongruent with both the *matK* and the *rbcL* trees.

Of 69 putative clades in the parsimony tree constructed from the full data set, only 5 showed strong incongruence between it and trees derived from the morphological, cpDNA, or ITS data partitions (components bolded in Table 5.3). The component with the highest bootstrap support that was incongruent with a more highly supported clade from another partition was (*Thelymitra* + *Epiblema*) in the ITS tree, with a bootstrap value of 97%. This was incongruent with the (*Thelymitra* + *Calochilus*) component in the cpDNA tree, which received 100% bootstrap support. Interestingly, the *Thelymitra* + *Epiblema* + *Calochilus* component in the ITS tree (subtribe Thelymitrinae) received a bootstrap value of 100%, but was neither supported nor contradicted by the bootstrap consensus tree for the cpDNA partition. The only other partition in which *Epiblema* was sampled was morphology, which found 60% bootstrap support in favor of the (*Thelymitra* + *Epiblema*) grouping but weakly found the Thelymitrinae to be paraphyletic. Given the apparent closeness of relationship between these 3 genera, either incomplete lineage sorting or horizontal chloroplast transfer by hybridization (see e.g. Knowles 2009) would be a plausible explanation for strong incongruence within this clade.

Fig. 5.3 Phylogram for the first of 9 equally parsimonious trees from the full data matrix of morphological and molecular characters, optimized using the ACCTRAN option in PAUP*, and with branches annotated with their bootstrap support indices. Arrows indicate the nodes that collapse in the strict consensus tree. Drawings of the following orchid species are connected to their generic names by dashed lines (*from the top*): *Cryptostylis leptochila, Thynninorchis huntianus, Arthrochilus prolixus, Simpliglottis valida, Chiloglottis reflexa, Myrmechila formicifera, Caleana major, Lyperanthus suaveolens, Pyrorchis nigricans, Rimacola elliptica, Burnettia cuneata, Calochilus campestris, Thelymitra ixioides, Corunastylis fimbriata, Genoplesium baueri, Prasophyllum australe, Microtis parviflora, Rhizanthella slateri, Acianthella amplexicaulis, Cyrtostylis reniformis, Corysanthes fimbriata, Acianthus fornicatus, Adenochilus nortonii, Eriochilus cucullatus, Cyanicula caerulea, Jonesiopsis filamentosa, Arachnorchis tentaculata, Caladenia latifolia, Petalochilus catenatus, Stegostyla cucullata, Pheladenia deformis, Glossodia major, Diuris aurea,* and *Orthoceras strictum.* All orchid drawings are reproduced by permission of the Royal Botanic Gardens and Domain Trust, Sydney. By courtesy of Maurizio Rossetto.

Table 5.3 Support indices for tree components in bootstrapped parsimony and Bayesian probability trees for all data partitions and several combinations of data partitions

Putative clade	all data MPBS	molecular MPBS	molecular BPP	cpDNA MPBS	cpDNA BPP	trnL-F MPBS	trnL-F BPP	matK MPBS	matK BPP	rbcL MPBS	rbcL BPP	ITS MPBS	ITS BPP	morphology MPBS
Disa + Satyrium + Habenaria	58	74	1.0	97	1.)	70	1.0	89	1.0	78	1.0	–65	0.59	–<50
Disa + Disperis	–58	–74	–1.0	–97	–1.0	–70	–1.0	–89	–1.0	–78	–1.0	65	–0.59	<50
Satyrium + Habenaria	61	54	1.0	55	1.0	–<50	0	84	1.0	nt	nt	<50	0.63	<50
Zeuxine + Goodyera	100	100	1.0	91	1.0	–91	–1.0	98	1.0	nt	nt	100	1.0	80
Goodyera + Pachyplectron	–100	–100	–1.0	–91	–1.0	91	1.0	–98	–1.0	nt	nt	–100	–1.0	–80
Zeuxine + Goodyera + Pachyplectron (Goodyerinae)	**99**	**99**	**1.0**	**98**	**1.0**	**75**	**1.0**	**96**	**1.0**	0	0.76	–62	**–1.0**	–<50
Ponthieva + Cranichis (Cranichidinae)	100	100	1.0	99	1.0	nt	nt	nt	nt	100	1.0	100	1.0	89
Zeuxine + Goodyera + Cranichidinae	–99	–99	–1.0	–98	–1.0	–75	–1.0	–96	–1.0	0	–0.76	62	1.0	–<50
Goodyerinae + Cranichidinae	65	59	1.0	78	1.0	–<50	0.76	62	0.97	0	0	–<50	–0.79	–<50
Chloraea + Oligochaetochilus	<50	–<50	–0.53	<50	0.85	<50	0.95	0	–0.90	nt	nt	–<50	–0.79	<50
Goodyerinae + Cranichidinae + Chloraea	–<50	–<50	–0.53	–<50	–0.85	–<50	–0.95	0	–0.9	nt	nt	–<50	–0.79	–<50
Goodyerinae + Cranichidinae + Oligochaetochilus	–<50	<50	0.53	–<50	–0.85	–<50	–0.95	0	0.9	nt	nt	–<50	0.79	–<50

Taxon													
Goodyerinae + Cranichidinae + Chloraea + Oligochaetochilus (Cranichideae)	100	100	1.0	93	-<50	0.83	62	1.0	51	0.95	100	1.0	-<50
Cryptostylis + Coilochilus (Cryptostylidinae)	99	80	1.0	100	<50	1.0	96	1.0	52	0.90	nt	nt	66
Phoringopsis + Arthrochilus	99	98	1.0	nt	nt	nt	nt	nt	nt	nt	100	1.0	-54
Arthrochilus + Thynninorchis	-99	-98	-1.0	nt	nt	nt	nt	nt	nt	nt	-100	-1.0	54
Phoringopsis + Arthrochilus + Thynninorchis (Arthrochilus s.l.)	90	73	1.0	nt	nt	nt	nt	nt	nt	nt	79	1.0	79
Chiloglottis + Myrmechila	83	62	-0.75	nt	nt	nt	nt	nt	nt	nt	69	-0.77	81
Chiloglottis + Myrmechila + Simpliglottis (Chiloglottis s.l.)	100	100	1.0	nt	nt	nt	nt	nt	nt	nt	100	1.0	87
Arthrochilus s.l. + Chiloglottis s.l.	<50	0	-1.0	nt	nt	nt	nt	nt	nt	nt	0	-1.0	-64
Arthrochilus s.l. + Chiloglottis s.l. + Drakaea	<50	0	-1.0	54	-<50	-0.53	<50	-0.67	nt	nt	0	-1.0	-64
Caleana + Sullivania (Caleana s.l.)	100	100	1.0	99	99	1.0	100	1.0	nt	nt	100	1.0	66

continued

Table 5.3 continued

Putative clade	all data MPBS	molecular MPBS	molecular BPP	cpDNA MPBS	cpDNA BPP	trnL-F MPBS	trnL-F BPP	matK MPBS	matK BPP	rbcL MPBS	rbcL BPP	ITS MPBS	ITS BPP	morphology MPBS
Chiloglottis s.l. + *Drakaea* + *Caleana s.l.*	–<50	0	1.0	76	1.0	65	0.94	65	1.0	nt	nt	0	1.0	–<50
Arthrochilus s.l. + *Chiloglottis s.l.* + *Drakaea* + *Caleana s.l.*	<50	57	0.92	nt	nt	nt	nt	nt	nt	nt	nt	0	0.94	–64
Arthrochilus s.l. + *Drakaea* + *Spiculaea*	–<50	–57	–0.92	–76	–1.0	–65	–0.94	–68	–1.0	nt	nt	0	–1.0	64
Arthrochilus s.l. + *Chiloglottis s.l.* + *Drakaea* + *Caleana s.l.* + *Spiculaea* (Drakaeinae)	100	94	1.0	89	0.98	<50	0.98	97	1.0	nt	nt	88	1.0	<50
Drakaeinae + *Leporella*	<50	–<50	–0.95	–58	–0.87	–<50	–0.77	–79	–1.0	0	0.59	–<50	0	<50
Lyperanthus + *Pyrorchis* (*Lyperanthus s.l.*)	92	88	1.0	57	0.57	–<50	–0.77	52	0.94	nt	nt	83	1.0	71
Drakaeinae + *Leporella* + *Lyperanthus s.l.*	<50	0	–0.95	–58	–0.87	–<50	–0.59	–79	–1.0	0	0	0	0	–<50
Rimacola + *Megastylis*	69	–82	–1.0	0	–0.71	–<50	–0.59	–63	–0.84	0	0	nt	nt	<50
Rimacola + *Waireia*	-69	82	1.0	nt	nt	nt	nt	nt	nt	nt	nt	97	1.0	–<50
Rimacola + *Megastylis* + *Waireia*	74	0	–0.95	nt	nt	nt	nt	nt	nt	nt	nt	nt	nt	–<50

Drakaeinae + Leporella + Lyperanthus s.l. + Rimacola + Megastylis + Waireia	<50	–<50	–0.52	–54	0	–<50	–0.59	56	0.94	0	0	0	0	–<50
Burnettia + Aporostylis	93	94	0.98	nt	nt	nt	nt	nt	nt	nt	nt	95	1.0	–<50
Drakaeinae + Leporella + Lyperanthus s.l. + Megastylis + Waireia + Burnettia + Aporostylis (Drakaeinae + Megastyliidinae)	65	–51	0.50	0	0.73	nt	nt	nt	nt	0	0	–<50	0	<50
Thelymitra + Calochilus	**97**	**100**	**1.0**	**100**	**1.0**	**100**	**1.0**	**100**	**1.0**	nt	nt	**-97**	**-1.0**	**-60**
Thelymitra + Calochilus + Epiblema	–97	–100	–1.0	–100	–1.0	–100	–1.0	nt	nt	nt	nt	97	1.0	60
Thelymitra + Calochilus + Epiblema (Thelymitrinae)	96	83	1.0	0	–0.73	–<50	–0.52	nt	nt	93	1.0	100	1.0	–<50
Drakaeinae + Megastyliidinae + Thelymitrinae	98	97	1.0	94	1.0	–<50	–0.52	100	1.0	51	1.0	96	1.0	–<50
Cryptostylidinae + Drakaeinae + Megastyliidinae + Thelymitrinae	72	67	1.0	52	0.76	–<50	0	–<50	0.82	<50	0	<50	0	–<50

continued

Table 5.3 continued

Putative clade	all data MPBS	molecular MPBS	molecular BPP	cpDNA MPBS	cpDNA BPP	trnL-F MPBS	trnL-F BPP	matK MPBS	matK BPP	rbcL MPBS	rbcL BPP	ITS MPBS	ITS BPP	morphology MPBS
Prasophyllum + *Mecopodum* + *Chiloterus* (*Prasophyllum s.l.*)	86	nt	nt	nt	nt	nt	nt	nt	nt	nt	nt	nt	nt	63
Prasophyllum s.l. + *Genoplesium*	71	81	1.0	nt	nt	nt	nt	nt	nt	nt	nt	84	1.0	−71
Corunastylis + *Genoplesium*	−71	−81	−1.0	nt	nt	nt	nt	nt	nt	nt	nt	−84	−1.0	71
Prasophyllum s.l. + *Genoplesium* + *Corunastylis*	98	93	1.0	100	1.0	97	1.0	90	1.0	nt	nt	88	1.0	65
Hydrorchis + *Microtidium*	100	94	1.0	nt	nt	nt	nt	nt	nt	nt	nt	100	1.0	85
Hydrorchis + *Microtidium* + *Rhizanthella*	81	81	1.0	nt	nt	nt	nt	nt	nt	nt	nt	90	1.0	−68
Hydrorchis + *Microtidium* + *Microtis*	−81	−81	−1.0	nt	nt	nt	nt	nt	nt	nt	nt	−90	−1.0	68
Hydrorchis + *Microtidium* + *Rhizanthella* + *Microtis*	<50	0	0.51	nt	nt	nt	nt	nt	nt	nt	nt	0	−0.84	−<50

Taxon														
Prasophyllum s.l. + *Genoplesium* + *Corunastylis* + *Hydrorchis* + *Microtidium* + *Rhizanthella* + *Microtis* (*Prasophyllinae*)	81	70	1.0	99	1.0	<50	0.92	91	1.0	76	0.99	<50	1.0	~<50
Cryptostylidinae + Drakaeinae + Megastylidinae + Thelymitrinae + Prasophyllinae	**64**	**70**	**1.0**	**-88**	**-0.99**	**-<50**	**0**	**-62**	**-1.0**	**0**	**-0.76**	**73**	**1.0**	**~<50**
Acianthella + *Stigmatodactylus*	100	77	1.0	100	1.0	87	1.0	100	1.0	nt	nt	nt	nt	68
Corysanthes + *Nematoceras*	100	100	1.0	nt	nt	nt	nt	nt	nt	nt	nt	100	1.0	~<50
Corysanthes + *Nematoceras* + *Corybas*	83	75	-0.64	87	1.0	57	0.86	67	1.0	nt	nt	67	0.78	~<50
Corysanthes + *Nematoceras* + *Corybas* + *Anzybas*	73	70	1.0	nt	nt	nt	nt	nt	nt	nt	nt	70	0.99	~<50
Corysanthes + *Nematoceras* + *Corybas* + *Anzybas* + *Singularybas* (*Corybas s.l.*)	90	55	1.0	nt	nt	nt	nt	nt	nt	nt	nt	54	1.0	96

continued

Table 5.3 *continued*

Putative clade	all data MPBS	molecular MPBS	molecular BPP	cpDNA MPBS	cpDNA BPP	trnL-F MPBS	trnL-F BPP	matK MPBS	matK BPP	rbcL MPBS	rbcL BPP	ITS MPBS	ITS BPP	morphology MPBS
Corybas s.l. + *Cyrtostylis*	98	<50	0.94	65	0.99	0	0.86	<50	0.94	nt	nt	~<50	–0.95	~<50
Acianthella + *Stigmatodactylus* + *Corybas s.l.* + *Cyrtostylis*	98	98	1.0	100	1.0	<50	1.0	99	1.0	nt	nt	71	1.0	~<50
Acianthus + *Nemacianthus*	100	100	1.0	100	1.0	98	1.0	100	1.0	nt	nt	100	1.0	88
Acianthus + *Nemacianthus* + *Acianthella* + *Stigmatodactylus* + *Corybas s.l.* + *Cyrtostylis*	82	91	1.0	100	1.0	<50	0.99	99	1.0	76	1.0	68	1.0	<50
Townsonia + *Spuricianthus*	81	74	1.0	nt	nt	nt	nt	nt	nt	nt	nt	78	1.0	0
Acianthus + *Nemacianthus* + *Acianthella* + *Stigmatodactylus* + *Corybas s.l.* + *Cyrtostylis* + *Townsonia* + *Spuricianthus* (Acianthinae)	52	~<50	0.82	nt	nt	nt	nt	nt	nt	nt	nt	~<50	–0.87	<50
Acianthinae + Prasophyllinae	–90	–94	–1.0	88	0.95	–<50	0	60	0.95	0	0	–82	–1.0	~<50

Diuris + Orthoceras (Diuridinae)	100	100	1.0	100	1.0	100	1.0	100	1.0	0	0.99	100	1.0	81
Diuridinae + Eriochilus	-96	-97	-1.0	-93	-1.0	70	1.0	-99	-1.0	-60	-0.77	-69	-1.0	<-50
Arachnorchis + Drakonorchis	93	99	1.0	97	1.0	0	0	94	1.0	nt	nt	95	1.0	-67
Arachnorchis + Jonesiopsis	-93	-99	-1.0	nt	nt	nt	nt	nt	nt	nt	nt	-95	-1.0	67
Jonesiopsis + Caladenia	0	52	0.87	nt	nt	nt	nt	nt	nt	nt	nt	54	0.78	-67
Petalochilus + Caladenia	**-83**	**-86**	**-1.0**	**100**	**1.0**	**72**	**0.95**	**99**	**1.0**	nt	nt	**-92**	**-1.0**	**0**
Petalochilus + Stegostyla	83	86	1.0	nt	nt	nt	nt	nt	nt	nt	nt	80	1.0	0
Petalochilus + Stegostyla + Caladenia	54	-52	-0.92	nt	nt	nt	nt	nt	nt	nt	nt	-92	-1.0	0
Arachnorchis + Drakonorchis + Jonesiopsis + Caladenia	-54	0	-0.92	nt	nt	nt	nt	nt	nt	nt	nt	92	1.0	0
Jonesiopsis + Caladenia + Petalochilus + Stegostyla	-54	0	0.92	nt	nt	nt	nt	nt	nt	nt	nt	-92	-1.0	<-50
Arachnorchis + Drakonorchis + Jonesiopsis + Caladenia + Petalochilus + Stegostyla														
(Caladenia s.l.)	100	99	1.0	100	1.0	82	1.0	100	1.0	nt	nt	96	1.0	0

continued

Table 5.3 continued

Putative clade	all data MPBS	molecular MPBS	molecular BPP	cpDNA MPBS	cpDNA BPP	trnL-F MPBS	trnL-F BPP	matK MPBS	matK BPP	rbcL MPBS	rbcL BPP	ITS MPBS	ITS BPP	morphology MPBS
Caladenia s.l. + *Ericksonella*	67	50	0.92	0	0.84	nt	nt	nt	nt	nt	nt	–75	–0.85	0
Cyanicula subgenus *Cyanicula* + *Cyanicula* subgenus *Trilobatae* (*Cyanicula*)	100	100	1.0	99	1.0	54	1.0	98	1.0	64	0.76	92	1.0	63
Caladenia s.l. + *Ericksonella* + *Cyanicula*	67	75	1.0	0	–0.85	0	0	0	–0.83	nt	nt	81	1.0	–<50
Glossodia + *Elythranthera*	100	100	1.0	0	0.93	0	0.9	0	0.82	nt	nt	100	1.0	<50
Glossodia + *Elythranthera* + *Pheladenia*	90	89	1.0	0	–0.85	nt	nt	0	–0.83	nt	nt	84	0.99	–<50
Caladenia s.l. + *Ericksonella* + *Cyanicula* + *Glossodia* + *Elythranthera* + *Pheladenia*	100	100	1.0	93	1.0	92	1.0	92	1.0	100	1.0	100	1.0	<50
Caladenia s.l. + *Ericksonella* + *Cyanicula* + *Glossodia* + *Elythranthera* + *Pheladenia* + *Praecoxanthus*	88	84	1.0	53	0.89	0	–0.53	55	0.93	nt	nt	77	0.97	<50

Caladenia s.l. + Ericksonella + Cyanicula + Glossodia + Elythranthera + Pheladenia + Praecoxanthus + Leptoceras	100	100	1.0	100	1.0	98	1.0	100	1.0	nt	nt	100	1.0	~<50
Caladenia s.l. + Ericksonella + Cyanicula + Glossodia + Elythranthera + Pheladenia + Praecoxanthus + Leptoceras + Eriochilus	86	92	1.0	93	1.0	-70	-1.0	98	1.0	60	0.77	<50	0.92	~<50
Caladenia s.l. + Ericksonella + Cyanicula + Glossodia + Elythranthera + Pheladenia + Praecoxanthus + Leptoceras + Eriochilus + Adenochilus (Caladeniinae)	96	97	1.0	78	1.0	-70	-1.0	0	0.99	<50	0	69	1.0	<50
Diuridinae + Caladeniinae	63	51	0.87	60	-0.99	-<50	0	0	0	-<50	-0.76	-<50	-0.87	-<50

continued

Table 5.3 *continued*

Putative clade	all data MPBS	molecular MPBS	molecular BPP	cpDNA MPBS	cpDNA BPP	trnL-F MPBS	trnL-F BPP	matK MPBS	matK BPP	rbcL MPBS	rbcL BPP	ITS MPBS	ITS BPP	morphology MPBS
Acianthinae + Diuridinae + Caladeniinae	**90**	**94**	**1.0**	**−88**	**−0.99**	−<50	0	−62	**−1.0**	−<50	−0.76	**82**	**1.0**	−<50
Cryptostylidinae + Drakaeinae + Megastylidinae + Thelymitrinae + Prasophyllinae + Acianthinae + Diuridinae + Caladeniinae (Diurideae)	100	100	1.0	99	1.0	−<50	0	99	1.0	<50	0.98	99	1.0	−<50
Cranichideae + Diurideae	100	100	1.0	100	1.0	85	1.0	100	1.0	77	1.0	98	1.0	<50

Notes: A negative number in a cell represents the highest support index for any component that is incongruent with the one in this cell.

Tree components in the parsimony tree constructed from the full data set that are strongly incongruent with trees derived from the morphological, cpDNA, or ITS data sets are bolded. By strong incongruence we mean incongruence between components with BPP > 0.95 and/or MPBS > 88%.

MPBS = maximum parsimony bootstrap support; BPP = Bayesian posterior probability; nt = component not tested.

Other instances of strong incongruence between data sets involved the subtribe Goodyerinae among the outgroups, the unstable position of *Caladenia* sensu Jones (2006) within *Caladenia* s.l., and the (Cryptostylidinae + Drakaeinae + Megastylidinae + Thelymitrinae + Prasophyllinae) component of the tree derived from the full data set. The first of these involved instability of the position of *Pachyplectron*, with cpDNA very strongly placing it within the Goodyerinae (MPBS = 98%) but with ITS much less strongly placing it as sister to (Cranichidinae + *Zeuxine* + *Goodyera*) (MPBS = 62%, BPP = 1.0). Morphology was unable to arbitrate between these alternatives, because it weakly resolved the Cranichideae to be paraphyletic, with *Pachyplectron* closer to the Cranichidinae and Diurideae than to the Goodyerinae. The ITS and *trnL-F* sequences of *Pachyplectron* are quite divergent from those of other Cranichideae, and the instability in the placement of this genus by those partitions may be due to a combination of sequence misalignment and a saturated level of sequence divergence. The (Cryptostylidinae + Drakaeinae + Megastylidinae + Thelymitrinae + Prasophyllinae) component of the trees derived from both the ITS partition and the full data set was strongly supported by ITS (MPBS = 73%, BPP = 1.0) but strongly contradicted by the cpDNA tree, which clustered Prasophyllinae with the Acianthinae with 88% bootstrap support. Morphology very weakly favored the cpDNA tree by grouping the Prasophyllinae with the Acianthinae and Caladeniinae with <50% bootstrap support, but it seems most prudent to regard the position of the Prasophyllinae as unresolved within the Diurideae at this stage. Lastly, the position of *Caladenia* sensu Jones (2006) varied significantly between the cpDNA tree, in which it strongly clustered with *Petalochilus* (MPBS = 100%), and the ITS tree, in which it clustered weakly with *Jonesiopsis* (MPBS = 54%), nested strongly within a larger clade that also included *Arachnorchis* and *Drakonorchis* (MPBS = 92%). Morphology weakly favored the cpDNA tree by clustering the spider orchids, *Jonesiopsis* and *Arachnorchis* (MPBS = 67%).

Our full data set yielded 9 equally parsimonious trees, which differed from one another only in the resolution of relationships within the *Prasophyllum-Chiloterus-Mecopodum* clade and *Caladenia* s.l. One of those trees was arbitrarily chosen for illustration and annotated with bootstrap percentages (Figure 5.1). We treated a conservatively modified version of the strict consensus of these trees as our "best tree" for further analysis and discussion. It was modified by collapsing those branches that were strongly incongruent with trees constructed from the cpDNA, ITS, and morphology data partitions and by using the results of Douzery et al. (1999) and Salazar et al. (2003) to resolve basal relationships within Diseae-Orchideae and Cranichideae.

All but 9 terminals were found to be characterized by unequivocal autapomorphies when morphological characters were traced on our best tree (results not shown). Five of these terminals were monotypic genera that are monophyletic by definition. Two of the terminals lacking morphological autapomorphies, *Chiloglottis* and *Myrmechila*, have been strongly corroborated as monophyletic by more detailed molecular phylogenetic analyses (Clements et al. 2002; Mant et al. 2002, 2005b). The other 2 genera lacking autapomorphies were *Hydrorchis* and *Chiloterus*, both of which include 2 very similar species and are thus unlikely to be grossly polyphyletic. All terminals that included DNA sequences from 2 or 3 species were found to be characterized by morphological autapomorphies and were thus corroborated as monophyletic.

Pollinators, Reproductive Traits, and Other Features of the Genera of Tribe Diurideae

Data on the number of species, selected reproductive traits, and pollinators of the genera of Diurideae are tabulated in Table 5.4. Variation in several of these traits warrants comment here. The number of species in each genus observed or inferred to be insect pollinated is mostly low relative to the total number of species. This should not be taken to imply that few species are insect pollinated, only that few species have been investigated by pollination biologists. The taxa in which pollinators are known for the majority of species are mostly sexually deceptive genera that have been the subjects of intensive study. Food-deceptive taxa have been less popular subjects of research, partly for the practical reason that they are rarely visited by pollinators (Neiland and Wilcock 1998; see Chapters 2, 4, 7, 10).

The number of species in each genus known to secrete floral nectar is also generally low relative to total species numbers (Table 5.4) because no systematic studies have been made of the occurrence or structure of nectaries in the Diurideae. *Prasophyllum*, which is pollinated by insects in at least 7 different families, is thought to be predominantly if not exclusively nectariferous (Bower 2001). Most records have been the result of observations of the presence of liquid exudate on the labella of flowers as part of studies of pollination biology (e.g. Bernhardt and Burns-Balogh 1986b), while some (e.g. Indsto et al. 2007) have relied on refractometer-based readings of the sugar content of water used to irrigate labella. For some genera (e.g. *Prasophyllum*), the number of species in which nectar has been recorded is probably a gross underestimate of the true total.

Of the 30 genera of diurids for which pollinators have been identified

Table 5.4 Data on the number of species, selected reproductive traits, pollinators, and references to pollination studies of the genera of Diurideae

Genus	Number of species	Number of species observed or inferred to be insect-pollinated	Number of species known to secrete nectar	Genera (and families) of insect pollinators	Gender of insect pollinators	Putative pollinator attraction (Bower 2001, Jones 2006)	Reference(s) to pollinators	Number of species known to be autonomously autogamous (Bower 2001, Jones 2006)
Acianthella	7	0	0	—	—	—	—	2
Stigmatodactylus	10	0	0	—	—	—	—	2
Corysanthes	17	1	0	unidentified Mycetophilidae	?	food deception	Jones (1970)	2
Nematoceras	21	0	0	?	?	?	—	3
Corybas	12	0	0	?	?	?	—	0
Singularybas	1	1	0	*Exechia* (Mycetophilidae)	unknown	food deception	Miller in Bower (2001)	1
Anzybas	6	1	0	?	?	?	—	1
Cyrtostylis	6	1	6	*Atrichopleura* (Empididae)	unknown	nectar reward	Bower (2001)	0
Nemacianthus	1	1	1	unidentified Diptera	unknown	nectar reward	Bower (2001)	1
Acianthus	9	1	9	unidentified Diptera	unknown	nectar reward	Bower (2001)	1
Spuricianthus	1	0	?	?	?	?	—	?
Townsonia	2	0	0	—	—	—	—	2

continued

Table 5.4 continued

Genus	Number of species	Number of species observed or inferred to be insect-pollinated	Number of species known to secrete nectar	Genera (and families) of insect pollinators	Gender of insect pollinators	Putative pollinator attraction (Bower 2001, Jones 2006)	Reference(s) to pollinators	Number of species known to be autonomously autogamous (Bower 2001, Jones 2006)
Diuris	71	4	2	*Eurys* (Pergidae)	unknown	nectar reward	Indsto *et al.* (2007)	0
				Exoneura (Apidae)	female	nectar reward	Indsto *et al.* (2007)	
				Gasteruption (Gasteruptiidae)	male	food deception	Beardsall *et al.* (1986)	
				Lasioglossum (Halictidae)	unknown	food deception	Bower (2001)	
				Leioproctus (Colletidae)	male, female	food deception	Beardsall *et al.* (1986)	
				Trichocolletes (Colletidae)	male, female	food deception	Beardsall et al. (1986), Indsto *et al.* (2006)	
Orthoceras	2	0	0	—	—	—	—	2
Adenochilus	2	0	0	?	?	food deception	—	1
Eriochilus	8	1	3	*Halictus* (Halictidae)	unknown	nectar reward	Bates (1987)	0

Genus				Pollinator	Sex	Pollination strategy	Reference	
Leptoceras	1	1	1	*Lasioglossum* (Halictidae)	unknown	nectar reward	Erickson (1965), Bower (2001), Hopper and Brown (2006a)	0
Praecoxanthus	1	0	0	unidentified Apoidea	unknown	nectar reward	Jones (2006)	0
Cyanicula subgenus *Trilobatae*	4	0	0	?	?	food deception	Bower (2001)	0
Cyanicula subgenus *Cyanicula*	6	1	0	?	?	food deception	—	0
				Neophyllotocus (Scarabaeidae)	male and female	food deception	Peakall (1987)	
Elythranthera	2	0	0	?	?	food deception	—	0
Ericksonella	1	0	0	?	?	food deception	—	0
Glossodia	2	0	0	?	?	food deception	—	0
Drakonorchis	4	4	0	*Thynnoides* (Tiphiidae)	male	sexual deception	Phillips et al. (2009)	0
Arachnorchis	156	46	2	*Aeolothynnus* (Tiphiidae)	male	sexual deception	Bower in Phillips et al. (2009)	2
				Exoneura (Apidae)	female	nectar reward	Faast et al. (2009)	

continued

Table 5.4 *continued*

Genus	Number of species	Number of species observed or inferred to be insect-pollinated	Number of species known to secrete nectar	Genera (and families) of insect pollinators	Gender of insect pollinators	Putative pollinator attraction (Bower 2001, Jones 2006)	Reference(s) to pollinators	Number of species known to be autonomously autogamous (Bower 2001, Jones 2006)
				Lasioglossum (Halictidae)	female	nectar reward	Faast *et al.* (2009)	
				Lestricothynnus (Tiphiidae)	male	sexual deception	Phillips *et al.* (2009)	
				Lophocheilus (Tiphiidae,)	male	sexual deception	Bower in Phillips *et al.* (2009)	
				Macrothynnus (Tiphiidae)	male	sexual deception	Phillips *et al.* (2009)	
				Neozeleboria (Tiphiidae)	male	sexual deception	Bower in Phillips *et al.* (2009)	
				Phymatothynnus (Tiphiidae)	male	sexual deception	Phillips *et al.* (2009)	
				Phyllotocus (Scarabaeidae)	unknown	food deception	Phillips *et al.* (2009)	
				Simosyrphus (Syrphidae)	unknown	nectar reward	Faast *et al.* (2009)	
				Simosyrphus (Syrphidae)	unknown	food deception	Farrington *et al.* (2009)	

Genus				Pollinator (Family)	Sex/caste	Strategy	Reference	
Jonesiopsis	50	9	0	Tachynomyia (Tiphiidae)	male	sexual deception	Stoutamire (1983)	
				Thynnoides (Tiphiidae)	male	sexual deception	Phillips et al. (2009)	
				Zaspilothynnus (Tiphiidae)	male	sexual deception	Phillips et al. (2009)	
				Chilothynnus (Tiphiidae)	male	sexual deception	Phillips et al. (2009)	1
				Phymatothynnus (Tiphiidae)	male	sexual deception	Phillips et al. (2009)	
Petalochilus	34	2	0	Trigona (Apidae)	female (workers)	food deception	Adams et al. (1992)	10
				Lasioglossum (Halictidae)	unknown	food deception	Farrington et al. (2009)	
				Syrphus (Syrphidae)	unknown	food deception	Ulherr (1967)	
Stegostyla	18	1	0	Hylaeus (Colletidae)	unknown	food deception	Bates (1982)	1
Caladenia	6	1	1	Leioproctus (Colletidae)	unknown	food deception	Erickson (1965)	0
Pheladenia	1	1	0	Lasioglossum (Halictidae)	unknown	food deception	Rogers in Bower (2001)	0
Corunastylis	50	7	7	Caviceps (Chloropidae)	unknown	nectar reward	Garnet (1940), Bower (2001)	3
				Oscinosoma (Chloropidae)	unknown	nectar reward	Garnet (1940), Bower (2001)	

continued

Table 5.4 *continued*

Genus	Number of species	Number of species observed or inferred to be insect-pollinated	Number of species known to secrete nectar	Genera (and families) of insect pollinators	Gender of insect pollinators	Putative pollinator attraction (Bower 2001, Jones 2006)	Reference(s) to pollinators	Number of species known to be autonomously autogamous (Bower 2001, Jones 2006)
Genoplesium	1	0	0	unknown	unknown	nectar reward	Garnet (1940), Bower (2001)	1
Prasophyllum	116	13	13	Milichiidae	—	—	—	6
				Campsomeris (Scoliidae)	unknown	nectar reward	Peakall (1987)	
				Helphilus (Syrphidae)			Bernhardt and Burns-Balogh (1986b)	
				Lasioglossum (Halictidae)	unknown	nectar reward	Jones in Bower (2001)	
				Leioproctus (Colletidae)	male	nectar reward	Bernhardt and Burns-Balogh (1986b), Peakall (1987)	
				Melanostoma (Syrphidae)	unknown	nectar reward	Jones in Bower (2001)	
				Melangyna (Syrphidae)	unknown	nectar reward	Bernhardt and Burns-Balogh (1986b)	
				Microdon (Syrphidae)	unknown	nectar reward	Bernhardt and Burns-Balogh (1986b)	

Genus					Pollinator genus (Family)	Sex/caste	Reward	Reference
Mecopodum	4	0	?	0	Syrphus (Syrphidae)	unknown	nectar reward	Bernhardt and Burns-Balogh (1986b)
Chiloterus	3	0	3	0	unidentified Sphecidae	unknown	nectar reward	Bates (1984)
Microtis	18	1	1	3	unidentified Tiphiidae	male	nectar reward	Jones in Bower (2001)
					Pterocormus (Ichneumonidae)	unknown	nectar reward	Jones (1972)
					?	?	?	—
					?	?	nectar reward	—
					Iridomyrmex (Formicidae)	female (workers)	nectar reward	Peakall and Beatty (1989), Bower (2001)
Hydrorchis	2	0	0	2	—	—	—	—
Microtidium	1	0	0	1	—	—	—	—
Rhizanthella	2	1	0	0	Megaselia (Phoridae)	unknown	food deception	George and Cooke (1981)
Phoringopsis	3	0	0	0	?	?	sexual deception	—
Thynninorchis	2	1	0	1	Arthrothynnus (Tiphiidae)	male	sexual deception	Brown (1996)
Arthrochilus	10	1	0	0	Arthrothynnus (Tiphiidae)	male	sexual deception	Brown (1996)
Simpliglottis	12	11	0	1	Neozeleboria (Tiphiidae)	male	sexual deception	Bower and Brown (2009)
					Eirone (Tiphiidae)	male	sexual deception	Bower and Brown (2009)

continued

Table 5.4 *continued*

Genus	Number of species	Number of species observed or inferred to be insect-pollinated	Number of species known to secrete nectar	Genera (and families) of insect pollinators	Gender of insect pollinators	Putative pollinator attraction (Bower 2001, Jones 2006)	Reference(s) to pollinators	Number of species known to be autonomously autogamous (Bower 2001, Jones 2006)
Chiloglottis	10	8	0	*Neozeleboria* (Tiphiidae)	male	sexual deception	Mant *et al.* (2005)	0
				Chilothynnus (Tiphiidae)	male	sexual deception	Mant *et al.* (2005)	
Myrmechila	5	4	0	*Neozeleboria* (Tiphiidae)	male	sexual deception	Mant *et al.* (2005)	0
Drakaea	10	7	0	*Zaspilothynnus* (Tiphiidae)	male	sexual deception	Hopper and Brown (2007)	0
Caleana	1	1	0	*Lophyrotoma* (Pergidae)	male	sexual deception	Cady (1965)	0
				Pterygophorus (Pergidae)	male	sexual deception	Bates (1989)	
Sullivania	13	2	0	*Eirone* (Tiphiidae)	male	sexual deception	Hopper and Brown (2006b)	0
				Thynnoturneria (Tiphiidae)	male	sexual deception	Bower (2001)	
Spiculaea	1	1	0	*Thynnoturneria* (Tiphiidae)	male	sexual deception	Alcock (2000)	0

Burnettia	1	0	?	?	food deception	Jones (2006)	0
Aporostylis	1	0	?	?	food deception	Molloy (1990)	0
Rimacola	1	0	unidentified Hymenoptera	?	?	Jones (2006)	0
Waireia	1	0	—	—	—	—	1
Leporella	1	1	*Myrmecia* (Formicidae)	male	sexual deception	Peakall (1989)	0
Lyperanthus	2	0	?	?	food deception	Bower (2001)	0
Megastylis	6	0	?	?	?	Bower (2001)	0
Pyrorchis	2	2	?	?	nectar reward	Bower (2001)	0
Calochilus	27	0	*Campsomeris* (Scoliidae)	male	sexual deception	Bower (2001)	27
Epiblema	1	0	?	?	food deception	Bower (2001)	0
Thelymitra	110	6	*Amegilla* (Apidae)	unknown	food deception	Bernhardt and Meier (this volume)	60
			Eurys (Pergidae)	unknown	food deception	Dafni and Calder (1987)	
			Exoneura (Apidae)	unknown	food deception	Bernhardt and Meier (this volume)	

continued

Table 5.4 *continued*

Genus	Number of species	Number of species observed or inferred to be insect-pollinated	Number of species known to secrete nectar	Genera (and families) of insect pollinators	Gender of insect pollinators	Putative pollinator attraction (Bower 2001, Jones 2006)	Reference(s) to pollinators	Number of species known to be autonomously autogamous (Bower 2001, Jones 2006)
				Lasioglossum (Halictidae)	female	food deception	Bernhardt and Burns-Balogh (1986a), Dafni and Calder (1987), Sydes and Calder (1993), Bernhardt and Meier (this volume)	
				Leioproctus (Colletidae)	unknown	food deception	Bernhardt and Meier (this volume)	
				Nomia (Halictidae)	female	food deception	Cropper and Calder (1990)	
				Syrphus (Syrphidae)	unknown	food deception	Dafni and Calder (1987)	
Cryptostylis	25	5	0	*Lissopimpla* (Ichneumonidae)	male	sexual deception	Coleman (1928), Gaskett (2012)	0
Coilochilus	1	0	0	—	—	—	Bower (2001)	1

at least to family, 24 are reported as pollinated by species of only one family of insects (Table 5.4). Jones's (2006) genera mostly seem to have been circumscribed using floral morphological character states associated with particular families of pollinators. Of the 6 genera pollinated by insects belonging to more than one family, 4 include nectariferous species that account for a disproportionate share of the observed pollinator diversity. In *Arachnorchis*, for instance, all 44 of the species that have been shown to be sexually deceptive are pollinated exclusively by tiphiid wasps (Phillips et al. 2009b), but 1 nectariferous species, *Arachnorchis rigida*, is pollinated by at least 3 different families of insects (Faast et al. 2009).

The gender of pollinating insects has frequently been recorded for hymenopteran pollinators but less frequently for other orders (Table 5.4). As one would expect, putatively sexually deceptive orchids are all pollinated exclusively by males. However, some well-corroborated food-deceptive orchids are also exclusively pollinated by male members of the Hymenoptera, at least in parts of their range. For instance, Indsto et al. (2006) found that all 14 individuals of *Trichocolletes venustus* that they caught carrying pollinaria of *Diuris maculata* in the Sydney region were males. In contrast, further south, in Victoria, Beardsell et al. (1986) collected a wider range of male and female pollinators, including 6 individuals of 2 *Trichocolletes* species, all of which were female. Pollination exclusively by male insects has sometimes been taken as strong evidence in itself for sexual deception, and this is how Coleman (1932) and Bates (1977) interpreted their collections of exclusively male bees carrying pollinaria of *Diuris behrii* and *D. sulphurea*. Bower (2001) was skeptical of this conclusion on the grounds that the coloration and morphology of *Diuris* flowers do not mimic the putative insect models, and that the insects did not display mating behavior. Moreover, Dafni and Bernhardt (1989) noted that these orchids often commence blooming just as naïve male bees emerge from their pupae, well before the emergence of females, which could explain variation in gender ratios of pollinators at different times and locations (see also Chapter 3). Our own observations of *Diuris sulphurea* (unpublished data) have also shown this species to be nectariferous and thus not deceptive but rewarding, so gender ratios of pollinators should be interpreted carefully.

Ancestral Character-State Reconstructions

Data for presence or absence of nectaries, when mapped onto our best tree, unequivocally supported a history of multiple origins of labellum nectaries within the Diurideae from a nectarless ancestor, under both parsimony and maximum likelihood criteria. Although the likelihood criterion implemented in Mesquite does not allow polymorphic terminals, we were able to sidestep this difficulty by splitting the 9 polymorphic terminals and coding the 2 members of each pair unequivocally to the alternative states. The proportional likelihood of the nectarless state at the crown node of Diurideae was 0.996.

When data for pollination systems were mapped onto our best tree, coded as 2 kinds of nectar reward, food deception, sexual deception, and autonomous autogamy (coded as missing data), the most parsimonious ancestral state for the tribe was resolved unequivocally as food deception (Figure 5.4). Sexual deception was mapped as having arisen at least 4 times and possibly as many as 6 times in the Diurideae (Figure 5.4). Pollination by sexual deception of male tiphiid wasps has evolved at least twice, in the Drakaeinae and Caladeniinae, with secondary losses in *Arachnorchis* and *Jonesiopsis*, or possibly as many as 4 times, in Drakaeinae, *Arachnorchis*, *Jonesiopsis*, and *Drakonorchis*, with no secondary losses. One species of tiphiid wasp, *Zaspilothynnis nigripes*, has been convergently exploited by orchid species in different subtribes, the Drakaeinae (*Drakaea livida*) and the Caladeniinae (*Arachnorchis pectinata*, as *Caladenia pectinata*) (Phillips et al. 2013). Pollinator shifts between different tiphiid pollinators have previously been documented in the sexually deceptive Drakaeinae (Mant et al. 2002, 2005b), but the switch from the most recent common ancestor of *Caleana* plus *Sullivania* to *Caleana* involves a shift between different hymenopteran families, from Tiphiidae to Pergidae (see Chapter 6).

Data for pollinator families, when mapped onto our best tree under the parsimony criterion, unequivocally supported an ancestral state of Halictidae for the stem node of Diurideae (Figure 5.5). Data for pollinator orders and superfamilies, when similarly mapped (trees not shown), were equivocal for the stem of Diurideae, attributing either order Hymenoptera or Diptera and either superfamily Apoidea or Vespoidea to it. Ancestral states could not be estimated for the trait "pollinator families" using maximum likelihood because of the high degree of polymorphism within several terminals.

Autonomous autogamy was inferred to have arisen at least 23 times in the Diurideae, mostly within genera from xenogamous or geitonogamous ancestors, with no evidence of reversal (Figure 5.6; see Chapter 7). Species with a modal number of one flower per plant were inferred to have evolved twice in the Diurideae, both times in the subtribe Caladeniinae (Figure 5.7), with multiple reversals to multiflowered plants. None of the species in this subtribe that is known to produce nectar are consistently one flowered. Of the 9 to 12 inferred origins of nectar production in the

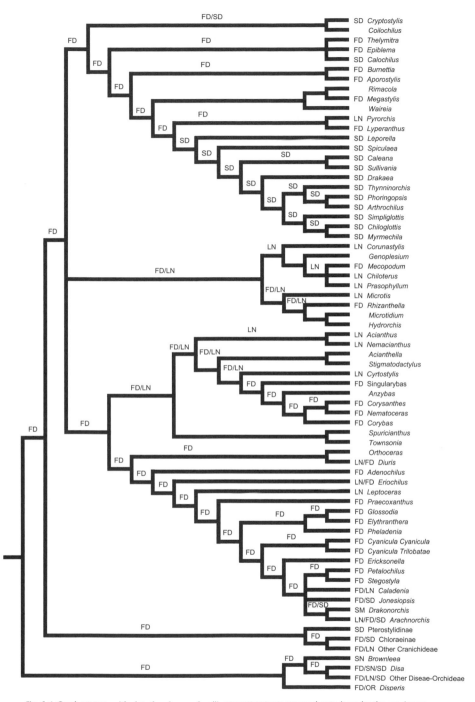

Fig. 5.4 Our best tree, with data for classes of pollinator attractants mapped onto it under the maximum parsimony criterion. This unequivocally supported a history of multiple origins of nectar production within the Diurideae from a nectarless, food-deceptive ancestor. *FD*: food deception; *LN*: nectar secreted from the labellum; *OR*: oil reward; *SD*: sexual deception; *SN*: nectar secreted in a spur of the dorsal sepal.

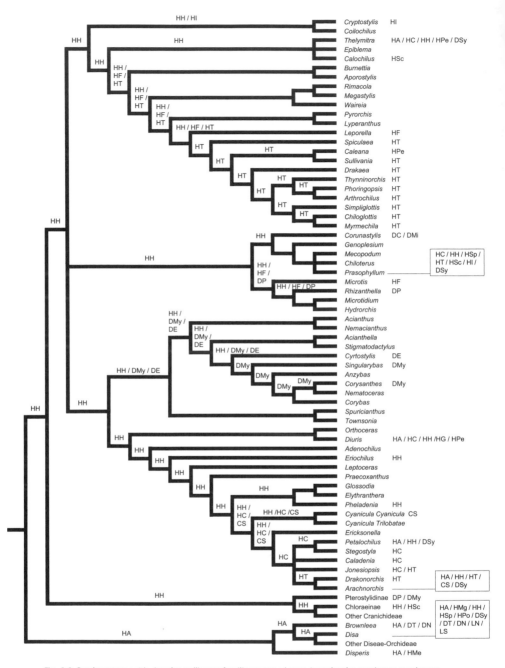

Fig. 5.5 Our best tree, with data for pollinator families mapped onto it under the maximum parsimony criterion. Key to family codes: COLEOPTERA: CS—Scarabaeidae; DIPTERA: DC—Chloropidae, DE—Empididae, DMi—Milichiidae, DMy—Mycetophilidae, DN—Nemestrinidae, DP—Phoridae, DSy—Syrphidae, DT—Tabanidae; HYMENOPTERA: HA—Apidae, HC—Colletidae, HF—Formicidae, HH—Halictidae, HI—Ichneumonidae, HMg—Megachilidae, HMe—Melittidae, HPe—Pergidae, HPo—Pompilidae, HSc—Scoliidae, HSp—Sphecidae, HT—Tiphiidae; LEPIDOPTERA: LN—Nymphalidae, LS—Sphingidae.

Diurideae, at least 7 occurred without associated reductions in flower number to one per plant.

DISCUSSION

Phylogeny of the Tribe Diurideae

The cladogram presented in Figures 5.4–5.7, based on the most comprehensive sample of diurid genera and character data yet assembled, is the most highly resolved and strongly supported estimate of the phylogeny of the tribe that has been produced so far. It was derived from a data set with 61% more characters than the largest matrix for the Diurideae previously analyzed (Kores et al. 2001). At the same time, this tree also takes into account the small number of putative clades that were strongly incongruent between trees derived from different partitions in the data set. It does this by collapsing the stems of disputed clades, rendering the tree a little less fully resolved than the optimal trees from parsimony and Bayesian analyses of the full and molecular data sets, but consistent with conflicting patterns in the data that are likely to have been generated by incomplete lineage sorting or hybridization events between closely related lineages.

The Ancestral Pollinators of the Tribe Diurideae Were Bees

Our character-mapping analysis unequivocally identified the Halictidae as the pollinator family of the most recent common ancestor of the Diurideae. However, it can be argued that this result is likely to be unrealistically precise, perhaps reflecting the higher probability of collecting halictids than other bee families on generalist, bee-pollinated flowers in Australia. According to the *Atlas of Living Australia* (2012), the number of records of halictids in Australian natural history collections is almost twice that for colletids, the next most commonly collected bee family, and almost 8 times that of the third most commonly collected family, the Apidae. Recent surveys of bee communities (Gollan et al. 2011; Lentini et al. 2012) suggest an even greater bias in the abundance of Halictidae, with individuals of this family outnumbering all other families 5- to 10-fold at 2 sites in southeastern Australia. This supports the idea that halictids might be disproportionately represented among records of pollinators of generalist, bee-pollinated orchids. Perhaps the character state "Halictidae" in our analysis is really a surrogate for a functional pollinator category "generalist bee pollination." A more cautious interpretation of

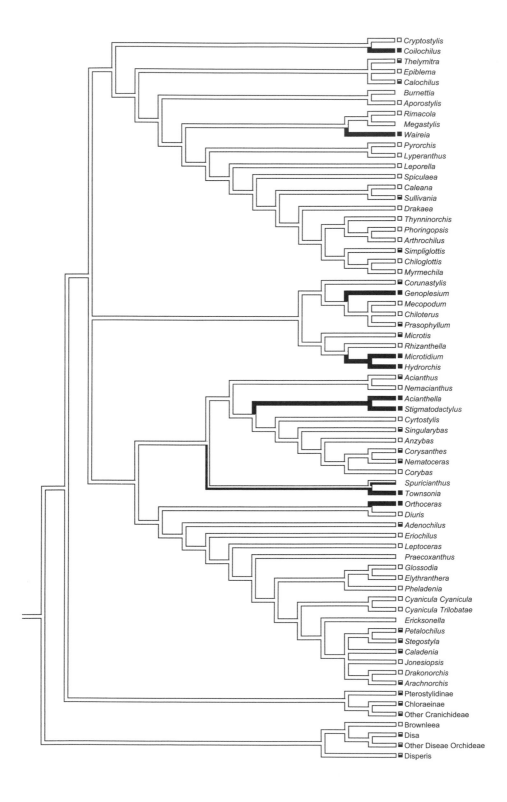

Cryptostylis
Coilochilus
Thelymitra
Epiblema
Calochilus
Burnettia
Aporostylis
Rimacola
Megastylis
Waireia
Pyrorchis
Lyperanthus
Leporella
Spiculaea
Caleana
Sullivania
Drakaea
Thynninorchis
Phoringopsis
Arthrochilus
Simpliglottis
Chiloglottis
Myrmechila
Corunastylis
Genoplesium
Mecopodum
Chiloterus
Prasophyllum
Microtis
Rhizanthella
Microtidium
Hydrorchis
Acianthus
Nemacianthus
Acianthella
Stigmatodactylus
Cyrtostylis
Singularybas
Anzybas
Corysanthes
Nematoceras
Corybas
Spuricianthus
Townsonia
Orthoceras
Diuris
Adenochilus
Eriochilus
Leptoceras
Praecoxanthus
Glossodia
Elythranthera
Pheladenia
Cyanicula Cyanicula
Cyanicula Trilobatae
Ericksonella
Petalochilus
Stegostyla
Caladenia
Jonesiopsis
Drakonorchis
Arachnorchis
Pterostylidinae
Chloraeinae
Other Cranichideae
Brownleea
Disa
Other Diseae Orchideae
Disperis

our result would be to conclude simply that the most recent common ancestor of the Diurideae was bee pollinated.

Nectar Production Has Evolved Repeatedly from Food-Mimicking, Nectarless Ancestors

One of our main aims was to determine the most likely ancestral state for nectar production in the Diurideae, to test whether the tribe was primitively nectariferous, as assumed in previous discussions of the evolution of its pollination systems. Our answer to this question was unequivocal: the most parsimonious character-state assignment for the most recent common ancestor of the Diurideae was found to be absence of a nectary, as also shown in recent analyses of the tribe Orchideae (e.g. Inda et al. 2012 and see Chapter 2). The absence of floral nectar secretions was also found to be the maximum likelihood estimate of the ancestral state of this character in the Diurideae, with a statistically significant proportional likelihood. As nectar is the only tangible reward known to be produced by any taxon in the Diurideae, it follows that pollination systems in the tribe would have evolved from a primitively rewardless ancestral system.

Consequently, hypotheses concerning selective pressures favoring the evolution of deceptive pollination in the Diurideae have been predicated on a false assumption. We should really have been asking what selective pressures would have favored the evolution of nectar production from nectarless ancestors in this tribe. An intuitively appealing hypothesis is that nectar production evolved as an additional incentive to induce more foraging animals to visit flowers in which fecundity is limited by scarcity of pollinators (Neiland and Wilcock 1998). However, provision of nectar would encourage optimally foraging insects to work a multiflowered inflorescence from one end to the other, resulting in geitonogamous pollination. Under these circumstances, selection pressures favoring morphologies or processes that prevent geitonogamy would be high. An evolutionary change that would function as such an adaptation would be the evolution, from multiflowered, rewardless ancestors, of single-flowered plants that offer a floral nectar reward.

Yet we could find no phylogenetic evidence for such an association in the Diurideae, with most lineages that acquired nectar production

Fig. 5.6 Our best tree, with data for autogamy/apomixis versus xenogamy/geitonogamy mapped onto it under the maximum parsimony criterion. *White*: xenogamy/geitonogamy; *black*: autogamy/apomixis; *white/black*: polymorphism or equivocal ancestral reconstruction.

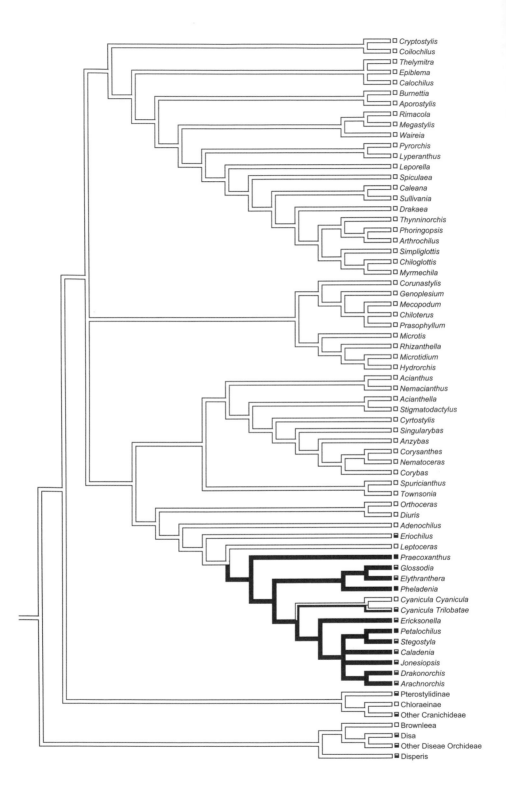

Cryptostylis
Coilochilus
Thelymitra
Epiblema
Calochilus
Burnettia
Aporostylis
Rimacola
Megastylis
Waireia
Pyrorchis
Lyperanthus
Leporella
Spiculaea
Caleana
Sullivania
Drakaea
Thynninorchis
Phoringopsis
Arthrochilus
Simpliglottis
Chiloglottis
Myrmechila
Corunastylis
Genoplesium
Mecopodum
Chiloterus
Prasophyllum
Microtis
Rhizanthella
Microtidium
Hydrorchis
Acianthus
Nemacianthus
Acianthella
Stigmatodactylus
Cyrtostylis
Singularybas
Anzybas
Corysanthes
Nematoceras
Corybas
Spuricianthus
Townsonia
Orthoceras
Diuris
Adenochilus
Eriochilus
Leptoceras
Praecoxanthus
Glossodia
Elythranthera
Pheladenia
Cyanicula Cyanicula
Cyanicula Trilobatae
Ericksonella
Petalochilus
Stegostyla
Caladenia
Jonesiopsis
Drakonorchis
Arachnorchis
Pterostylidinae
Chloraeinae
Other Cranichideae
Brownleea
Disa
Other Diseae Orchideae
Disperis

remaining multiflowered. Another adaptation that would fulfill this function would be the reconfiguration of pollinaria into a position where the pollinia could contact a receptive stigma only a set time after their removal from the anther, rendering them incapable of pollinating another flower immediately (Peter and Johnson 2006). However, most species of Diurideae have liquid viscidia, and their pollinaria seem physically incapable of markedly changing position after removal (see Chapters 1 and 2). Of the multiflowered, nectar-producing taxa with solid viscidia, only *Corunastylis* and *Microtis* have been reported as exhibiting bending of the pollinarium after its removal from the anther (Bower 2001), but in both cases this movement occurs immediately upon removal and so would not prevent geitonogamy.

Multiflowered, self-compatible, nectariferous species are likely to have mixed mating systems, an intermediate area in the continuum between xenogamy and obligate autogamy. Such systems have presented a challenging problem for evolutionary theorists (Goodwillie et al. 2005), but a number of examples of facultatively autogamous plants with attractive flowers that are also visited by animal pollinators have been well documented (e.g. Zhang et al. 2005; Perez et al. 2009). In the Diurideae, the selective advantage of the combination of increased pollination rates with multiple flowers has apparently outweighed the selective disadvantage of lower-quality sex. A model to explain this result would presumably have to invoke rapid purging of the genetic load of deleterious recessive alleles soon after the evolution of nectar production.

The Evolution of Breeding Systems

The rewardless systems that have been documented in the Diurideae comprise autonomous autogamy, agamospermy, food-source deception, and sexual deception (Dafni and Bernhardt 1989; Bower 2001). We found that autonomous autogamy is most parsimoniously reconstructed as having evolved repeatedly in different lineages of the tribe, as first suggested by Darwin (1877) for the Orchidaceae as a whole, with no evidence of reversal, thus confirming conventional wisdom that the transition from

Fig. 5.7 Our best tree, with data for modal number of flowers per plant (one versus more than one) mapped onto it. One flower per plant was inferred to have evolved twice in the Diurideae, both times in the subtribe Caladeniinae, with multiple reversals to multiflowered plants. On comparing this figure with Figure 5.4, showing the pattern of evolution of nectar production, it is clear that the 2 characters are not associated. *White*: multiflowered plants; *black*: single-flowered plants; *white/black*: polymorphism or equivocal ancestral reconstruction.

xenogamy-geitonogamy to obligate autogamy is a predominantly uni-directional evolutionary change (Lande and Schemske 1985). Once a lineage has lost the morphological features that prevent self-pollination as well as those that attract suitable pollinators, increased fecundity resulting from autogamy is likely to maintain microevolutionary selective processes in the direction of autogamy (Lloyd 1979). Macroevolutionary processes of differential speciation and extinction would then regulate the balance between xenogamous and geitonogamous systems on the one hand and autogamy on the other, with net diversification rates predicted to favor xenogamous systems. This prediction could be tested empirically, given a combination of detailed information on breeding systems and well-resolved phylogenetic trees within clades that include both xenogamous-geitonogamous and autogamous species.

Species in at least 2 genera present examples of Lloyd's (1979) model of delayed self-fertilization, a kind of mixed mating system. For example, in *Calochilus caeruleus*, some populations of *C. campestris*, and possibly other species of *Calochilus*, the flowers have not lost the ability to attract—and be cross-pollinated by—male scoliid wasps; but the morphological features preventing self-pollination break down during anthesis, and the flowers eventually self-pollinate if they have not been cross-pollinated earlier during anthesis (Bower 2001). Dafni and Bernhardt (1989; p. 221) stated that "it is undoubtedly true that the majority of Australian orchids will [be] self-pollinated mechanically in the absence of appropriate pollinators." This idea was based exclusively on earlier literature compiled by an earlier generation of Australian naturalists. However, it was never tested explicitly. Bower (2001) reported only a few other documented examples of delayed or facultative self-pollination. Given Lloyd's (1979; p. 71) conclusion that this kind of system is "always advantageous whenever it is possible," it seems surprising that the process of delayed self-pollination has not been reported more often among diurid orchids (but see Chapter 7).

Sexual Deception Has Evolved Repeatedly from Food-Mimicking, Nectarless Ancestors

Sexually deceptive systems were found to have evolved at least 4 times and possibly as many as 7 times in different lineages in the Diurideae, always from food-deceptive ancestors and involving males of a range of different hymenopteran clades in the families Tiphiidae, Scoliidae, Pergidae, Formicidae, and Ichneumonidae, as suggested by Dafni and Bernhardt (1989), based on a radically different phylogenetic framework.

Australia remains the global center of diversity for sexually deceptive pollination systems in the Orchidaceae (see Chapter 6), although recent discoveries of sexually deceptive clades in tropical South America (Blanco and Barboza 2005, and references therein) have resulted in the Neotropics rivaling Australia as a center of both specific and phylogenetic diversity of sexually deceptive lineages. Interestingly, sexual deception is thus far known only from the terrestrial tribes Diurideae and Cranichideae in Australia but only from epiphytic taxa in the tribes Epidendreae and Cymbidieae in the Neotropics, with one sexually deceptive terrestrial species, *Geoblasta pennicillata* (tribe Cranichideae), known from subtropical South America (Benitez-Vieyra et al. 2009).

A plausible explanation for the repeated evolution of sexual deception among Australian terrestrial orchids has yet to be proposed. A potential explanation that seemed promising but which has so far failed to find empirical support in the Diurideae is the suggestion that some insect pheromones are structurally similar to common chemical constituents of plant tissues and would therefore be preadapted to function as pseudopheromones in attracting pollinators (Schiestl and Cozzolino 2008; see Chapter 3). This idea was based on the discovery that the pheromones emitted by the female models of Hymenoptera of *Ophrys* orchids are complex cocktails of many organic compounds (Chapter 3) that are mostly common constituents of the leaves of *Ophrys* species (Schiestl et al. 2000; Ayasse et al. 2003). In contrast, the pseudopheromones emitted by species of a sexually deceptive diurid genus, *Chiloglottis*, and the pheromones emitted by their models (tiphiid wasps in 2 genera) are composed of a unique class of volatile compounds, 2,5-dialkylcyclohexan-1,3-diones, also known as chiloglottones (Franke et al. 2009). Although derived from precursors that are general intermediates in fatty-acid biosynthesis, chiloglottones are not close chemical relatives of any other known class of compound that has been found in *Chiloglottis*, and it is difficult to imagine how they first evolved.

The selection pressure favoring the evolution of sexual deception from food-deceptive ancestors seems obvious enough: sexual deception is a more effective way of luring common pollinators (Scopece et al. 2010). This contrast in efficacy is easily seen by comparing the rates at which observers have reported seeing individual insects visiting flowers of species representing the 2 pollination systems in the field. Bower (2006), for instance, collected a total of 1253 male individuals of 4 species of *Neozeleboria* (Tiphiidae) visiting flowers of 2 species of the sexually deceptive genus *Chiloglottis* during 612 observational "exposures" that totaled 30.6 hours. By contrast, Bernhardt and Burns-Balogh (1986a) caught only

3 female individuals of 2 native bee species and observed only a few more visiting flowers of *Thelymitra megcalyptra* (as *T. nuda*) during 2 weeks of fieldwork on a population of this food-deceptive orchid (see Chapter 7). Similarly, Indsto et al. (2006) observed only 4 individuals of the native bee species *Trichocolletes venustus* visiting flowers of *Diuris maculata*, another food mimic, over a total observational period of about 8 hours.

Specialized Food Mimics Have Evolved Repeatedly from Generalized Food-Mimicking Ancestors

Specialized food-source deception is a pollination system in which the flowers of the mimic closely resemble those of a model species, clade, or guild that offers a tangible reward. The flowers of both mimic and model are pollinated by the same animal species. Close resemblance here is taken to mean similar enough that the predominant pollinators would be incapable of reliably distinguishing mimic from model (Dafni and Bernhardt 1989; see Chapter 7). Additional criteria have been proposed as essential tests for the demonstration of this system (Cozzolino and Widmer 2005; Johnson et al. 2003a), but it is usually very difficult to apply all these together, and no Australian examples have achieved this. However, several of these tests have been passed to our satisfaction by 3 taxa in the Diurideae, *Diuris maculata*, *Thelymitra megcalyptra* (as *T. nuda*), and *T. antennifera*. Although this small number of examples probably underestimates the true number of specialized food mimics, this pollination system is probably both uncommon and phylogenetically derived within the tribe.

    D. maculata mimics a guild of papilionoid legumes commonly known as "egg and bacon peas" in the tribes Mirbelieae and Bossiaeeae, and both orchid and models are pollinated by species of the colletid bee genus *Trichocolletes* (Beardsell et al. 1986; Indsto et al. 2006) that specialize in collecting nectar from the flowers of papilionoid legumes. *Thelymitra megcalyptra* (syn. *T. nuda*) mimics a guild of blue- to mauve-flowered lilioid monocots in the family Asparagaceae s.l., all of which are pollinated by bees of the halictid genus *Lasioglossum* (Bernhardt and Burns-Balogh 1986a). *Thelymitra antennifera*, on the other hand, mimics a guild of yellow-flowered eudicots that are also pollinated by *Lasioglossum* bees but also by *Eurys*, a pergid wasp, and *Syrphus* hoverflies (Dafni and Calder 1987). The mitra of the *Thelymitra* flower is an intricately ornamented, hoodlike structure formed from the column wings, which have fused and greatly expanded above the column, partly obscuring it. Pollinating insects mistake the mitra for a cluster of poricidal anthers when foraging for pollen, and pollinate the flowers in the process of attempting to sonicate them (see Chapter 7).

Other species of *Diuris* (e.g. *D. alba*; Indsto et al. 2007) and *Thelymitra* (e.g. *T. epipactoides*; Cropper and Calder 1990) are generalized, deceptive food mimics, and this kind of pollination system seems most likely to be ancestral in the Diurideae. The existence of generalized food deception (Johnson and Nilsson 1999), also known as nonmimicry deception (Dafni and Bernhardt 1989), is difficult to test empirically (Gigord et al. 2002) because it is based to some extent on the absence of evidence: the absence of nectar or any other tangible floral reward and the lack of a close resemblance to the flowers of any other species. The flowers of generalized food mimics are typically conspicuously colorful and/or sweetly scented. In the Diurideae, this combination of characteristics is found in lineages scattered across the phylogeny (Bower 2001), and was inferred unequivocally by parsimony-based optimization as the ancestral pollination system for the tribe.

In combination, the results of our character-mapping analyses suggest that the most recent common ancestor of the Diurideae was pollinated by bees that were deceptively induced to visit its nectarless flowers by their resemblance to flowers providing nectar and/or pollen rewards. Lineages that attracted pollinators by offering nectar or the false promise of sex, or that dispensed with cross-pollination altogether, evolved multiple times from such an ancestor. This scenario is remarkably similar to the conclusions of Inda et al. (2012) concerning evolution of the tribe Orchideae.

The Origin of Food Mimicry in the Orchidaceae

Our conclusion renders explanatory hypotheses for the origin of deceptive pollination systems from nectar-producing ancestors within the Diurideae (Dafni and Bernhardt 1989, as the tribe Geoblasteae sensu Burns-Balogh and Funk 1986) redundant, because such transitions, if they ever occurred, did so late in the evolution of the tribe and involved few if any lineages. Theorizing about the origin of deceptive pollination systems needs to focus on events that preceded the origin of the Diurideae. We favor an alternative scenario first suggested by Neiland and Wilcock (1998) in which generalized food deception originated very early in the evolution of the Orchidaceae. Phylogenetic relationships at the base of the Orchidaceae (Cameron 2004; Freudenstein et al. 2004) suggest that deceptive pollination originated in the most recent common ancestor of the subfamilies Vanilloideae, Cypripedioideae (see Chapter 10), Orchidoideae, and Epidendroideae. The subfamily Apostasioideae is the sister group of this clade, and is the only group of orchids known to offer pollen as an edible reward to their pollinators (Kocyan and Endress 2001). These nectarless, bee-pollinated flowers closely resemble those of some near-basal members of the Asparagales such as the Hypoxidaceae, and the flowers of the most recent common ancestor of the orchids might also have shared these features.

According to this hypothesis, deceptive pollination by generalized food deception would have evolved in association with the origin of pollinia, from an ancestor in which powdery monads or tetrads of pollen performed the dual function of being a vehicle for male gametophytes and an edible reward for pollinators. Packaging pollen grains in pollinia was an adaptation that maximized the fecundity of flowers that carried thousands of ovules, but which precluded pollen functioning as a tangible reward. Floral nectaries, which in the Orchidaceae mostly develop as part of the labellum, evolved independently of the septal nectaries of the Asparagales and originated numerous times in different orchid lineages as a class of adaptations that enhanced plant fecundity by attracting more pollinators than rewardless flowers (Neiland and Wilcock 1998).

Some predictions of this scenario are easier to test than others. A reconstruction of the evolutionary origins of nectar production in the Orchidaceae as a whole has not been published, but such an analysis would be straightforward given sufficient information about the phylogenetic distribution of nectaries and a well-resolved phylogeny. Comparing the relative fecundity of nectariferous and rewardless taxa has already been done for a subsample of orchids (Neiland and Wilcock 1998), confirming the reproductive benefit of nectar production. However, reconstructing the phylogenetic history of pollination systems is likely to yield somewhat ambiguous results given the diversity of pollinators and rewards within subclades of both the Orchidaceae and the Asparagales, and the technical difficulty of demonstrating cases of food deception.

Tests of Other Evolutionary Hypotheses

We have shown that the evolutionary scenarios of Burns-Balogh and Bernhardt (1988), Dafni and Bernhardt (1989), and Bernhardt (1995a) are probably inaccurate with respect to the phylogenetic relationships of rewarding and rewardless pollination systems in the Diurideae, but how do other aspects of those models fare in the light of our results? All these scenarios envisaged sexual deception evolving repeatedly from food-deceptive ancestors, and this prediction has been clearly corroborated by our results. Likewise, their suggestion that autogamous breeding systems had arisen repeatedly in different lineages of Australasian terrestrial orchids from diverse ancestral pollination systems has also been strongly borne out by our results. Burns-Balogh and Bernhardt (1988) suggested that speciation by hybridization might have contributed substantially to diversification within the genus *Thelymitra*. Strong phylogenetic congruence between our chloroplast, nuclear, and morphological data sets refutes the prospect that this might have been a dominant process in the evolution of the Diurideae s.s., as proposed by Fitzgerald (1875–1895), before the differentiation of generic-level clades, but the few instances of strong phylogenetic incongruence that we found could represent residual evidence of ancient hybridization events (see Edens-Meier et al. 2013).

We suggest the reader also examine the recent literature on hybridization and speciation in *Thelymitra* spp. endemic to New Zealand, as reviewed here in Chapter 7. Evidence of speciation by hybridization has not been found in the few detailed phylogenetic studies of interspecific relationships within some genera (e.g. Mant et al. 2002, 2005b; Indsto et al. 2009; Peakall et al. 2010). However, much more work is needed at this level before ideas regarding the role of hybridization as an evolutionary process in the Diurideae can be rigorously tested.

Macroevolutionary Patterns

Some patterns evident in the raw data presented in Table 5.4 warrant comment in the light of our phylogenetic results. First, comparisons of the numbers of species in some sister-group pairs show striking imbalance in diversity. For example, *Caladenia* s.l. (*Caladenia* + *Petalochilus* + *Stegostyla* + *Jonesiopsis* + *Arachnorchis* + *Drakonorchis*) includes 268 named species, but its sister group is a single species, *Ericksonella saccharata*. The progressive sister groups encountered as one climbs down the tree—*Cyanicula* (10 spp.), *Pheladenia* + *Glossodia* + *Elythranthera* (5 spp.), *Praecoxanthus* (1 sp.), *Leptoceras* (1 sp.), *Eriochilus* (spp.), *Adenochilus* (2 spp.)—all repeatedly emphasize this imbalance. This pattern can partly be attributed to the high diversity of sexually deceptive species in *Jonesiopsis*, *Arachnorchis*, and *Drakonorchis*, on the grounds that sexually deceptive taxa are likely to have higher speciation rates than food-deceptive clades (Xu et al. 2012). However, even if we arbitrarily excluded these genera from consideration, the remaining 58 nectariferous and food-deceptive species of *Caladenia*, *Petalochilus*, and *Stegostyla* would still greatly outnumber all their successive sister groups down to

Adenochilus. We are unable to explain this pattern, but note that similarly unbalanced sister-group contrasts are common in the Diurideae and warrant a detailed analysis of net diversification rates in different lineages (compare with Sauquet et al. 2009).

Descriptive Studies of Pollination and Breeding Systems Are Needed for a Broader Sample of Taxa in the Tribe Diurideae

A second obvious observation is that our knowledge of the reproductive biology of the Diurideae is still fragmentary and inadequate, despite the impressive efforts of generations of amateur and professional biologists. For 19 genera we can make only educated guesses about their likely pollinators and pollination systems, based on floral traits; several of these include common, widespread, easily accessible species. Pollinators of a further 12 multispecies genera of diurid orchids are known from only one species. A lot of simple, descriptive studies of the natural history of these taxa still need to be completed.

WHAT WOULD CHARLES DARWIN HAVE THOUGHT?

We started this chapter by observing that Charles Darwin was almost completely deprived of information on the reproductive biology of the tribe Diurideae when he wrote his first book on the pollination biology of orchids (Darwin 1862). By the time he wrote the second edition (Darwin 1877), he was able to cite Robert Fitzgerald's description of self-pollination in 2 species of *Thelymitra* as rare examples of autogamy in a family in which he saw cross-fertilization as "the rule." He also had access by then to examples of obligate insect pollination in species of *Acianthus*, *Caladenia*, *Caleana*, *Cyrtostylis*, and *Corysanthes* provided by Fitzgerald and Thomas Cheeseman from New Zealand. Sadly, the most fascinating aspects of the pollination biology of the Diurideae and the phylogenetic reconstruction of their evolutionary history only started to be revealed long after Darwin's death. What would he have made of the repeated origins of obligate autogamy in numerous lineages of Diurideae that we have documented? This result might have caused him to doubt his central proposition, that "nature abhors perpetual self-fertilisation" (see Chapter 1). How would Darwin have explained the repeated origin of sexually deceptive lineages in the Diurideae? These and other questions have presented significant challenges to evolutionary theorists, and some of them continue to taunt Darwin's intellectual descendants today.

SUMMARY

The phylogenetic relationships between genera in the tribe Diurideae (Orchidaceae subfamily Orchidoideae) were reconstructed by analyzing morphological characters and alignments of nuclear and chloroplast DNA sequences, using bootstrapped parsimony and Bayesian probability criteria. We considered our best tree to be the one constructed using all our data, assembled under the parsimony criterion, with stems of the few strongly incongruent nodes collapsed. Data on the presence or absence of floral nectaries, pollinator orders, superfamilies and families, pollination systems, and functional numbers of flowers per plant were mapped onto our best tree using maximum parsimony and, where possible, maximum likelihood criteria. The results of our character-mapping analyses suggest that the most recent common ancestor of the Diurideae was pollinated by bees that were deceptively induced to visit its nectarless flowers by their resemblance to flowers providing nectar and/or pollen rewards. Species and their descendants that attracted pollinators by offering nectar or the false promise of sex, or that dispensed with cross-pollination altogether, evolved multiple times from such ancestors.

ACKNOWLEDGMENTS

The Hermon Slade Orchid Fund, administered by the National Herbarium of New South Wales, is thanked for supporting P.H.W.'s visit to the Jodrell Laboratory, Royal Botanic Gardens, Kew, in 2005, where he commenced assembling the morphological matrix analyzed here. We are also grateful to Mark Chase, Mike Fay, and Vincent Savolainen for facilitating that visit and for useful discussions of characters and methods. The Australian Orchid Foundation is thanked for supporting J.O.I.'s visit to Melbourne in July 2011, during which he presented the results of this study in Symposium 057, "Orchid-Pollination: The Post-Darwinian Revolution," at the 18th International Botanical Congress. Carlos Lehnebach kindly provided unpublished information on the absence of nectar secretion in selected species of subtribe Chloraeinae. Michael Batley is thanked for providing references to published surveys of bee communities in Australia. We are grateful to Helen Stevenson and Lesley Elkan for assistance in producing the figures. Barbara Briggs, editor Peter Bernhardt, and referee Steven Johnson provided numerous helpful suggestions for improving our chapter; its residual flaws remain our responsibility. We are grateful to the Royal Botanic Gardens and Domain Trust,

Sydney, for giving us permission to reproduce drawings of orchids in our Figure 5.3 that were originally published in *Flora of New South Wales*, vol. 4 (1993).

APPENDIX 5.1

Morphological Character States Used to Construct a Phylogeny of the Diurideae

States 1–31 represent vegetative states (entire organs or tissues). States 32–50 refer to reproductive characters related to the inflorescence or perianth, excluding the labellum. States 51–72 refer exclusively to labellum characters (entire organ or epidermal sculpturing). States 73–89 refer to column characters (male, female, and sterile organs). States 90–98 refer to pollinarium and individual pollinia. State 99 refers only to carpel development. State 100 refers only to the composition of the viscidium. Finally, states 101 and 102 refer exclusively to seed characters.

1. Monocarpic growth: (0) absent (1) present (e.g. *Burnettia*).
2. Root tuber presence: (0) present (e.g. *Diuris*) (1) absent (e.g. *Cryptostylis*).
3. Root tuber shape: (0) elongated (ovoid to ellipsoid) (e.g. *Diuris*) (1) globose (e.g. *Caladenia*) (2) irregular (e.g. *Genoplesium*).
4. Tuber stele: (0) not dissected (e.g. *Diuris*) (1) dissected (e.g. *Eriochilus*).
5. Replacement tuber presence: (0) absent (e.g. *Drakaea*) (1) present (e.g. *Diuris*).
6. Replacement tuber connection: (0) terminating a short dropper (1) produced at the end of a short, descending dropper for a few years and then sessile (e.g. *Cyanicula*).
7. Root tuber tunica: (0) absent (1) present.
8. Tunica morphology: (0) enclosing upper half of root tuber (e.g. *Caladenia*) (1) entirely enclosing root tuber (e.g. *Cyanicula*).
9. Stolonoid roots: (0) absent (1) present (e.g. *Acianthus*).
10. Non-stolonoid roots: (0) present (e.g. *Prasophyllum*) (1) absent.
11. Spiranthosomes: (0) absent (1) present.
12. Leaf development: (0) fully developed, chlorophyllous (e.g. *Diuris*) (1) vestigial, achlorophyllous (e.g. *Thynninorchis*).
13. Papillate leaf epidermal cells: (0) absent (1) present.

14. Multicellular leaf trichome distribution: (0) absent (1) present adaxially only (2) present both surfaces (3) present on margins only.
15. Leaf trichome type: (0) nonglandular (1) both glandular, nonglandular (2) glandular.
16. Leaf stomata distribution: (0) abaxial only (1) both surfaces (2) adaxial only.
17. Leaf mesophyll: (0) homogeneous (1) heterogeneous.
18. Leaf longevity: (0) less than one year (1) more than one year.
19. Number of basal leaves per shoot: (0) more than 2 (1) 2 (2) 1 (3) none.
20. Number of cauline phyllomes per shoot: (0) 2 or more (1) 1 (2) none.
21. Cauline leaf morphology: (0) at least one foliose (1) bracteose.
22. Leaf lamina base: (0) cuneate (1) cordate (2) obcuneate.
23. Leaf lamina length:width ratio: (0) linear (LW ratio > 10) (1) elliptical-lanceolate (10 < LWratio > 2) (2) broad (LW ratio ~ 1).
24. Leaf pseudopetiole: (0) absent (1) present (2) basal leaf petiolate, cauline leaves sessile.
25. Leaf attitude: (0) horizontal (1) suberect to erect.
26. Anastomosing veins between longitudinal veins: (0) mostly simple (e.g. *Caladenia*) (1) mostly reticulate (e.g. *Corybas*).
27. Leaf margins: (0) entire (1) lobed.
28. Leaf transverse section: (0) flat (1) V-shaped (2) concave (3) cylindrical (hollow with connate margins) (4) terete with adaxial groove.
29. Peduncle adnation: (0) free (1) basal part adnate to adaxial leaf surface.
30. Peduncle: (0) elongated (1) vestigial.
31. Stem trichomes: (0) absent (1) present.
32. Inflorescence: flower number: (0) one (1) more than one.
33. Flower orientation: (0) resupinate (1) not resupinate.
34. Weather-controlled flower opening: (0) flowers staying open after anthesis until withering (1) flowers staying open only in warm, sunny weather.
35. Sepal length (heteromorphy): (0) dorsal sepal much longer than lateral sepals (1) dorsal and lateral sepals of similar length (2) dorsal sepal much shorter than lateral sepals.
36. Sepal width (heteromorphy): (0) dorsal sepal much wider than lateral sepals (1) dorsal and lateral sepals of similar width (2) dorsal sepal much narrower than lateral sepals.
37. Dorsal sepal tip shape: (0) obtuse to acute (1) apiculate (2) attenuate-filiform (3) swollen (4) attenuate-filiform with swollen tip.
38. Sepal trichomes (presence and type): (0) absent (1) present, nonglandular (2) present, glandular.
39. Dorsal sepal concavity: (0) flat to shallowly concave or convex in TS (1) cucullate (2) involute.

40. Dorsal sepal spur: (0) absent (1) present.
41. Lateral sepal margins: (0) flat to shallowly concave or convex in TS (1) incurved-involute.
42. Sepal connation: (0) free (1) lateral sepals basally fused along adjacent margins (2) lateral sepals completely fused along adjacent margins (3) all sepals basally fused along adjacent margins.
43. Lateral petal presence: (0) present (1) absent.
44. Lateral petal shape (basal claw): (0) not clawed (1) differentiated into narrow basal claw and much wider distal lamina (e.g. *Diuris*).
45. Lateral petal shape (limb): (0) spathulate (1) ovate-lanceolate (2) linear (3) two-lobed (4) oblong (5) oblanceolate.
46. Lateral petal attitude: (0) spreading to erect (1) reflexed, parallel to ovary.
47. Lateral petal trichomes: (0) absent (1) simple (2) glandular.
48. Relative petal length: (0) lateral petals much longer than labellum (1) lateral petals of similar length to labellum (2) lateral petals much shorter than labellum.
49. Relative petal width: (0) labellum much wider than lateral petals (1) labellum of similar width to lateral petals (2) labellum much narrower than lateral petals.
50. Galea: (0) Dorsal sepal and lateral petals free (1) dorsal sepal and lateral petals coherent, forming a galea.
51. Labellum identity: (0) labellum morphologically distinct from lateral petals (1) labellum petaloid (e.g. *Thelymitra*).
52. Labellum attachment: (0) sessile (1) hinged on a short claw (2) basally adnate to column.
53. Labellum mobility: (0) labellum immobile (1) labellum readily pushed away from column and springing back when released (2) labellum tremulous (3) labellum actively mobile.
54. Labellum lamina shape (lobing): (0) unlobed (1) 2-lobed (2) 3-lobed (3) galeate with 2 backward-pointing spurs (4) 3-lobed with 1 backward-pointing spur.
55. Backward-pointing, insectiform, saddle-like structure distal to basal claw of labellum: (0) absent (1) present.
56. Adaxial sessile, hemispherical calli (presence): (0) absent (1) present.
57. Adaxial sessile, hemispherical calli (distribution): (0) covering whole bckward-pointing, insectiform, saddle-like structure (1) covering distal tip of backward-pointing, insectiform, saddle-like structure (2) central on lamina.
58. Adaxial clavate, fingerlike calli (presence): (0) absent (1) present.
59. Adaxial clavate, fingerlike calli (distribution): (0) parallel, longitudinal rows (1) basal semicircle (2) covering lamina (3) clustered in a sub-basal

tuft (4) clustered around the base of backward-pointing, insectiform, saddle-like structure (5) clustered around the middle of backward-pointing, insectiform, saddle-like structure (6) forming a central longitudinal band of crowded calli (7) forming a loose cluster.

60. Adaxial longitudinal ridge-like calli (presence): (0) absent (1) present.
61. Adaxial longitudinal ridge-like calli (distribution): (0) single central ridge (1) 2 parallel ridges (2) 2 slightly divergent ridges (3) > 2 parallel ridges (4) 2 slightly convergent ridges.
62. Adaxial papilla-like calli (presence): (0) absent (1) present.
63. Adaxial papilla-like calli (distribution): (0) covering labellum lamina (1) covering lateral labellum lobes (2) forming 2 longitudinal lines.
64. Adaxial paired, basal, lozenge-like calli (presence): (0) absent (1) present.
65. Adaxial single, basal, moundlike callus (presence): (0) absent (1) present.
66. Adaxial stalked, dendritic calli: (0) absent (1) present.
67. Abaxial, papilla-like calli: (0) absent (1) present.
68. Labellum margin: (0) entire (1) crenate (2) dentate (3) papillose (4) simple-laciniate (5) simple- and branching-laciniate.
69. Labellum auricle presence: (0) absent (1) present (e.g. *Corybas*).
70. Labellum basal tube: (0) absent (1) present, with free margins (e.g. *Corybas*) (2) present, with connate margins (e.g. *Anzybas*).
71. Labellum adaxial trichomes: (0) absent (1) present, nonglandular (e.g. *Eriochilus*) (2) present, glandular (e.g. *Cryptostylis*) (3) present, papillose (e.g. *Rhizanthella*) (4) present, multicellular, dendritic (e.g. *Thynninorchis*).
72. Labellum nectar secretory glands: (0) absent (1) present.
73. Adnation of style and staminal filament: (0) complete (1) basal only.
74. Column foot presence: (0) absent (1) present.
75. Column foot extent: (0) vestigial (e.g. *Leporella*) (1) extending to basal margins of lateral sepals (e.g. *Paracaleana*) (2) extending beyond margins of lateral sepals (e.g. *Drakaea*).
76. Column anterior basal paired glands: (0) absent (1) present.
77. Anterior column appendage: (0) absent (1) present (e.g. *Stigmatodactylus*).
78. Column wings (presence): (0) present (1) staminodes not developing beyond primordia.
79. Column wings (direction of expansion): (0) expanded laterally only (1) expanded laterally and apically (2) expanded apically only.
80. Column wings (shape of lateral expansion): (0) expanded narrowly and evenly throughout column (e.g. *Rimacola*) (1) expanded broadly and evenly throughout column (e.g. *Epiblema*) (2) expanded throughout column but most at base, gradually narrowing to tip (e.g. *Paracaleana*) (3) expanded throughout column but most widely at base, forming curved basal auricles (e.g. *Drakaea*) (4) throughout column but expanded

gradually, most widely in the middle (e.g. *Caleana*) (5) expanded abruptly in the middle as triangular lobes, with incurved tips (e.g. *Arthrochilus*) (6) expanded laterally as smoothly curved auricles only in the distal part of the column (e.g. *Caladenia*) (7) expanded laterally and abruptly as axehead-shaped auricles only in the distal part of the column (e.g. *Pterostylis*).

81. Column wings (shape of apical expansion): (0) extended apically only as a minute tooth (1) extended substantially but shorter than anther (2) extended to anther tip (3) extended beyond anther tip.

82. Column wings connation: (0) column wings free from each other (1) column wings connate behind and above the anther to form a mitra.

83. Mitra shape: (0) unlobed (e.g. *Calochilus*) (1) bilobed (e.g. *Epiblema*) (2) trilobed (e.g. *Thelymitra*) (3) unlobed but with a dentate apical margin (e.g. *Adenochilus*).

84. Column auricles: (0) absent (1) present (e.g. *Lyperanthus*).

85. Column auricle form: (0) toothlike (e.g. *Corunastylis*) (1) papillose (e.g. *Lyperanthus*).

86. Staminal filament-style (length relative to anther): (0) short (shorter than anther) (1) long (longer than anther).

87. Anther position relative to rostellum: (0) acrotonic (1) mesotonic (2) basitonic.

88. Anther orientation: (0) erect (1) recumbent (e.g. *Disa*).

89. Anther apex: (0) rostrate (1) acute (2) truncate (3) emarginate.

90. Viscidium number: (0) 1 (1) 2.

91. Pollinarium stipe: (0) absent (1) pollinia attached to viscidium by a hamulus (2) pollinium attached to viscidium by a stipe.

92. Pollinarium caudicles: (0) absent (1) present.

93. Pollinium number: (0) 2 (1) 4 (2) 8.

94. Pollinium shape: (0) curved (1) clavate (2) bilobed (3) subdeltoid (4) reniform (5) oblong.

95. Pollen grain aggregation: (0) not aggregated (monads) (1) aggregated into tetrads.

96. Pollen apertures: (0) monoaperturate (1) inaperturate.

97. Pollen color: (0) yellow (1) cream (2) white.

98. Pollinium structure: (0) mealy-granular (1) waxy (2) sectile.

99. Lateral carpel apices at initiation: (0) separate (1) connate.

100. Viscidium structure: (0) solid (1) liquid (e.g. *Drakaea*).

101. Seed morphology class (Molvray and Kores 1995): (0) orchidoid (1) spiranthoid (2) diurid (3) goodyeroid (4) vanilloid.

102. Embryo type (Clements 1999): (0) Orchideae (1) Spiranthinae (2) Goodyerinae (3) Diurideae.

SIX

Pollination of Spider Orchids (*Caladenia* syn. *Arachnorchis*) by Wasps . . . and Others: A Lingering Post-Darwinian Mystery

Sophie Petit

INTRODUCTION

The genus *Caladenia* R.Br. refers to a group of fewer than 400 orchid species, most of which are located in southern Australia and a large proportion of which is threatened (Hopper and Brown 2004; Dixon and Hopper 2009; Phillips et al. 2009a). Taxonomic classification of *Caladenia* orchids is still in turmoil (Hopper 2009), with general agreement on the 6 subgenera described by Hopper and Brown (2004; but see also Chapter 5). Taken in its broadest sense, *Caladenia* had been represented by an alliance of genera as determined by Jones et al. (2001). This chapter focuses on the pollination of sexually deceptive spider orchids (they appear spidery to the human eye because of their long, narrow sepals and lateral petals), which were placed by Jones et al. (2001) in the genus *Arachnorchis* (>132 spp.), although undoubtedly the discussion in part applies to other related species, such as in the genera *Jonesiopsis* and *Drakonorchis* (see Chapter 5). The inclusion of spider orchids under the genus *Caladenia* has received broad support (Hopper 2009), but the name *Arachnorchis* is used in this chapter to refer specifically to spider orchids.

This extraordinary group of orchids typically displays 1 to 2 flower(s) (up to 6 in *Arachnorchis hirta*, which does not look like a "classic" *Arachnorchis* and produces nectar; Jones 2006) and 1 hairy, basal leaf. Some species occur in one color, or vary subtly or dramatically in coloration and markings, such as *A. colorata*, *A. behrii*, and *A. dinema*. Some have strange scents (from a human perspective), but most are unscented (Bower 2001). They are dormant as underground tubers in the dry season (summer) and grow during the wet season (winter). They are mycorrhizal, principally via collar infection (Dixon and Hopper 2009; Dixon and Tremblay 2009). Most species are sexually deceptive, although many are food deceptive, and some of those are probably called sexually deceptive mistakenly (C. Bower, personal communication). Nevertheless, each sexually deceptive species seems to have a specific relationship with a haplodiploid species of thynnine wasp (Hymenoptera: Tiphiidae). Attracted by a pheromone mimic, the male wasp confuses the labellum with a receptive wingless female and inadvertently picks up or deposits pollinia in pseudocopulation.

Charles Darwin would no doubt have been delighted to discover such an exquisitely wicked pollination system; but like many of us, he would also have suffered from the difficulty of observing pollination, which is relatively rare in spider orchids containing many threatened species (Petit and Dickson 2005; Phillips et al. 2009b). His lament, "I have been in the habit for twenty years of watching Orchids, and have never seen an insect visit a flower, excepting butterflies twice sucking *O.* [*Anacamptis*] *pyramidalis* and *Gymnadenia conopsea*" (Darwin 1862; p. 34), elicits sympathy among similarly cursed orchid researchers (see Chapter 1).

Limited reproductive success is observed in deceptive species (Gill 1989). Fruit set in *Caladenia* orchids is relatively low, and often more so for sexually deceptive than for food-deceptive species (Phillips et al. 2009b; Phillips et al. 2011). This raises questions as to the evolution of the *Caladenia* s.l. system and its future. This chapter in no way attempts to rival the wonderful recent reviews that explore the evolutionary diversification of orchids (e.g. Cozzolino and Widmer 2005; Tremblay et al. 2005; Jersáková et al. 2006a; Ayasse et al. 2011; Gaskett 2011; see Chapters 2–5, 7–10). Rather, it aims to discuss a few points about the stupefying contrivances, success, and mystery of sexual deception, as in a chat with Darwin. Along with the molecular sciences (see Chapter 6), the development of ecology has been a major tool in the study of orchids since Darwin. My approach is an ecological one.

WHAT MAKES A SEXUALLY DECEPTIVE ORCHID SUCCESSFUL IN POLLINATION?

> Unless the flowers were by some means rendered attractive, they would be cursed with perpetual sterility.
> —Darwin 1862; p. 341

> What then attracts insects to these flowers?
> —Darwin 1862; p. 283

At the suggestion of food deception, Darwin initially sided with the insects:

> When we think of the special contrivances clearly showing that, after an insect has visited one flower and has been cheated, it must almost immediately go to a second flower, in order that impregnation may be effected . . . , we cannot believe in so gigantic an imposture. He who believes in this doctrine must rank very low the instinctive knowledge of many kinds of moths. (Darwin 1862; p. 46)

Darwin then tested "the intellect of moths" and found that "moths do not go to work in a quite senseless manner" (Darwin 1862; p. 47). However, even when subjected to "irritating vapours," some orchids remained devoid of nectar. Pollination by deception would have fascinated Darwin.

As with food-deceptive orchids, understanding the attractiveness of sexually deceptive orchids is fundamental to explaining evolutionary success. I suppose that it is just as fundamental to understand their lack of attractiveness to the same insects that found them attractive previously (the "refractory period"; Peakall 1990). Thanks to a fascinating body of work (Ayasse et al. 2011), we know that a single chemical (or sometimes 2—potentially a cocktail) mimicking the sex pheromone of a wingless female thynnine wasp (signaling a winged male on a piece of vegetation) is responsible for a great part of an orchid's attractiveness. The wingless females burrow in the ground to parasitize beetle larvae, climb stems of low vegetation to signal to the males, and get picked up for copulation and fed at a source of nectar/sugar (directly or via regurgitation by the male, or storage in head cavities) before being released (Given 1954; Ridsdill Smith 1970b; Alcock 1981; Bower 2001). The spider orchids' calli, of the labellum and sepal tips, produce the compounds that mimic the female wasp's pheromone (Peakall and Beattie 1996). Male wasps initially seem to find this elixir of love irresistible, although not all seem to be fooled (Peakall and Beattie 1996; Gaskett 2011). Their level of experience with treacherous orchids affects the likelihood that they will be aroused by an orchid and tricked (Peakall 1990; Peakall and Beattie 1996; Wong and Schiestl 2002; Schiestl 2004; Wong et al. 2004). Flowers of another sexually deceptive orchid, *Ophrys heldreichii*, placed in transparent containers, were of no interest to pollinators, and their visual attractiveness only came into play when volatiles released by the orchids were detected by the insects (Spaethe et al. 2007; see Chapter 3).

Although the objective here is not to review the knowledge on such volatiles (see review by Vereecken and McNeil 2010, and see Chapter 3), scent (as perceived by insects) should not be dissociated from other physical attributes of spider orchids in efforts to understand what motivates the behaviors of pollinators. It is the complete assortment of chemical, visual, and tactile stimuli that is likely to create attractiveness, although most agree that volatiles are the most important elements (Ayasse et al. 2011; Gaskett 2012).

Pollinator-baiting methods involving cut orchids paraded in various locations (Stoutamire 1983) or synthetic pheromones (Schiestl et al. 2003) are helpful in elucidating orchid taxonomy. Choice experiments and reciprocal baiting have revealed the reproductive isolation of cryptic species (e.g. Bower 1992, 1996; Bower and Brown 1997; Bower 2001, 2006, 2008; Bower and Brown 2009). Such experiments are also useful in determining the pollinators' behavior on approach (Bower 2001) and potentially the abundance and habitat preference of pollinators. Not much is known about pollinator abundance at orchid sites, yet it is essential that pollinators be there, since fruit set may depend on this factor (Ackerman et al. 1997). Although pollinators occur where orchids do not (Schiestl 2004), a basis for orchid attractiveness is to be in the right spot. The conditions must be adequate for both pollinators and orchids. Much work remains to be done on the perception of pheromones by wasps and the factors that may affect it. The study of unusual female wasps such as the scented *Tachynomyia* sp., which, from a human perspective, smells exactly like a flower of *Caladenia* syn. *Arachnorchis behrii* (Dickson and Petit 2006), would also help us to understand the timing of pheromone production.

Since, for some of us, pollinators remain ghosts in the study of many orchid species, and pheromone "visibility" to pollinators is not yet well understood, ecological observations of pollination in the field provide important means to examine the choices that pollinators make. For example, the taller of a pair of *Arachnorchis behrii* tends to be favored by pollinators, but at the population level, orchids of all heights are being pollinated (Dickson and Petit 2006). Peakall and Handel (1993) also found that the pollinators of *Chiloglottis trilabra* prefer relatively tall flowers. It is reasonable to assume that higher flowers in a clump (possibly larger and producing greater pheromone output) will be more apparent (in visual or olfactory terms), provided they occur within the range of pollinators' flights (Stoutamire 1983; Vandewoestijne et al. 2009). Pollinator-mediated selection for floral height is possible (Schiestl and Peakall 2005), although selection on morphological characters may be reduced by environmental variation (O'Connell and Johnston 1998).

Color does not affect the pollination success of *A. behrii*, for which 4 color groups were tested (Dickson and Petit 2006); color variability may be a result of relaxed selection, as in some food-deceptive orchids (Juillet and Scopece 2010; see Chapter 2). In fact, the visual mimicry of many *Arachnorchis* spp., meant to look like thynnine wasps, is analogous to "stick figures" compared with the delightful "Mona Lisa sculptures" presented by some *Drakaea* (see Jones 2006) and *Myrmechila* (see Bower 2007) species. Red, green, and white, prevalent in sexually deceptive orchids including *Arachnorchis*, are colors not easily seen by winged members of the Hymenoptera (Gaskett and Herberstein 2010). Color may serve to increase contrast between the labellum and the background or mimic the shape of a female insect on a plant for close-range attraction (Spaethe et al. 2007; Streinzer et al. 2009). It is surprising that the red-lipped orchid individuals (with a dark patch that can look like a wasp to a somewhat visually impaired human) are not more successful than the white-lipped ones (Dickson and Petit 2006). However, red does not attract many insect species (Briscoe and Chittka 2001), and it is an advantage for orchids not to attract food-searching insects that affect pollination negatively (but see "Combining Sexual and Food Deception" below). Accurate visual mimicry in spider orchids does not seem very important, at least on approach, and color seems less important than other traits in floral isolation (Schiestl and Schlüter 2009), although what humans see is different from what wasps see. We need to recognize that many aspects of coloration and ultraviolet signals could have great significance in influencing pollinator visits (Kropf and Renner 2005; Gaskett and Herberstein 2010; Gaskett 2011) and even herbivory (Gronquist et al. 2001). In light of the work by Gaskett (see Chapter 11) and others, these mysteries will no doubt be elucidated by the 200th anniversary of Darwin's first book on orchids.

ECOLOGICAL INTERACTIONS

Another aspect of attractiveness is how one shows oneself off to advantage. *Caladenia* syn. *Arachnorchis behrii* growing within the ground-hugging canopies of grass trees (*Xanthorrhoea semiplana*) are often less successful at being pollinated than those growing in the open; however, in situations when grazing is intense, they also have a better chance of remaining intact, because grass trees protect the orchids against grazers (Petit and Dickson 2005). I have observed similar protection for orchids growing among fallen branches and trees; Faast and Facelli (2009) also noted that different protective cages had different impacts on pollination; and female wasps, hidden under opaque structures, are rarely found (Peakall 1990). The masking effect of grass trees and other structures on pollinators remains mysterious. Petit and Dickson (2005) discussed the following possibilities: (1) pollinators may simply not see the orchid as easily as they may one located in the open, (2) they cannot approach the flower as easily because of the interference of grass-tree leaves, (3) the shade of grass trees reduces pheromone production, and/or (4) the plume of molecules is affected by the structure of the grass tree, lessening its effectiveness (see Streinzer et al. 2009). These hypotheses should be tested elegantly with an orchid species that may benefit from more pollinator visits than does *A. behrii*.

"Beheaded" orchid flowers, a common occurrence in the Mount Lofty Ranges of South Australia, have definitely lost all attractiveness for that year at least. In Australia, kangaroos are often presumed the guilty party when grazing comes under the spotlight, but Faast and Facelli (2009) have shown that white-winged choughs (*Corcorax melanorhamphos*) gobble up *Caladenia* syn. *Arachnorchis* flowers like some of us do chocolate pastilles. This florivory is likely to apply to the endangered *A. behrii* as well, since flowers are lost in kangaroo exclosures (Petit and Dickson 2005).

Attractiveness also depends on the availability of female and male wasps in relation to that of orchids, and how desperate male wasps will be to mate at a point in time. To be successful in attracting pollinators, orchids need of course to occur at the right time and the right place. Reproductive success in 3 sexually deceptive *Ophrys* species increases with nearest neighbor distance, decreasing population density, and increasing patch elongation (Vandewoestijne et al. 2009). To minimize the chance of stigmas being clogged with heterospecific pollen, sympatric orchid species generally use different pollinaria (pollinia + associated sterile tissues) attachment sites on their pollinators (Waterman et al. 2011). *Neozeleboria* wasp species and sex pheromone diversity may have led to *Chiloglottis* orchid diversification via pollinator switching, but phylogenetic constraints on wasp phenology and sex pheromone compounds are likely to have limited diversification (Mant et al. 2005a).

The effect of weather events on the pollination success of spider and other orchids is starting to receive increasing attention; weather factors likely have great importance in pollination (see Chapter 12). For example, warm temperature encourages adult emergence from cocoons as well as male flight (Given 1957; Ridsdill Smith 1970a,b); rain promotes the growth of fungal mycelium in *Dendrobium* pollinia and decreases its germination (Luangsuwalai et al. 2008). The right environmental conditions probably optimize reproductive output, because large-leaved plants produce more seeds than do small-leaved plants (Petit et al. 2009). Another important consideration is the composition of the plant community in which the orchids are located. Food resources need to be close to the breeding area for many members of the Thynninae (Ridsdill Smith 1970a). The plant community may also affect microclimate and herbivory, as mentioned above, as well as fungal density and species.

COMBINING SEXUAL AND FOOD DECEPTION

Orchids exhibit an almost endless diversity of beautiful adaptations.
—Darwin 1862; p. 364

In view of the complexity of the ecosystem interactions surrounding spider orchids (see "Future in a Rapidly Changing World" below) and the specificity of the pollination system, should orchids "consider" some sort of safety or emergency pollination procedure (after all, they seem to have "sensibility"; Darwin 1862; p. 285) in case the pollinator population hits a rough spot? Dixon and Tremblay (2009) reported that only 3% of *Caladenia* taxa propagate vegetatively. The attraction of undedicated pollinators, using reward or faking reward (again), is possible in spider orchids. The discovery of tiny amounts of nectar in some *Caladenia* syn. *Arachnorchis rigida* (Faast et al. 2009), for example, indicates that orchids may "double dip" into the potential pollinator fauna. Many sexually deceptive spider orchid species could be pushing the limits of their treachery by combining food deception and sexual deception. Food deception could explain the positive inbreeding coefficient of *Caladenia* syn. *Arachnorchis huegelii* (Swarts et al. 2009). The presence of 2 yellow spots ("glands"; Bates 1983, 1984a) or yellow gland-like structures (Hopper and Brown 2004) in the "throat" (base of the column) of spider orchids could lead pollen-eating insects away from the pollinia (Bates 1983) or lure foraging insects to the flowers (Petit et al. 2009), like pseudopollen or false anthers (Bates 1984a; Dafni 1984).

In view of the cost of pollen to plants, the evolution of structures mimicking pollen to attract insects is not unusual (Lunau 2000). Yellow is an attractive color to insects in range of South Australian *Arachnorchis* species (Faast et al. 2009). Such swindle is fraught with danger both for the orchid and its visitors, because nonpollinating insects of inappropriate body shape or size can get trapped in the flowers and die there (Stoutamire 1983; Bates 1984a,b; Bower 2001; Faast et al. 2009). Small or otherwise unsuitable insects can get stuck to the stigma, dry it, and eliminate at least the female function of the flower. Darwin observed an incident in *Epipactis latifolia*: "a fly, too small to remove the pollinia, had become glued to it and to the stigma, and had there miserably perished" (Darwin 1862; p. 104). On a cool and rainy day, I once observed a hoverfly (Syrphidae) stuck in an *A. behrii* flower, and its fight to disengage led to the self-pollination of the flower (Dickson and Petit 2006). Imitation stamens occur mainly in flowers pollinated by bees and syrphid flies (Lunau 2000). Considering most floral visitors of *Caladenia* syn. *Arachnorchis* occur on warm and sunny days (Stoutamire 1983; Faast et al. 2009), attracting hoverflies could make up for lost cool days with no thynnine wasp and would render the orchids less sensitive to wasp emergence time. In spite of the risks, combining food and sexual mimicry could increase an orchid's chance of being pollinated. Food deception in *Caladenia* tends to be assumed, and the boundary between food- and sexually deceptive orchids is not necessarily clear.

UNATTRACTIVE ATTENTION

A sexually deceptive spider orchid needs a relatively absentminded pollinator, although not silly enough that it will forfeit its own fitness and consequently that of future orchids. Is there an advantage for the orchid in not being perfectly attractive? Increased accuracy of wasp mimicry may repel other insects that could have been incidental pollinators or minimize opportunities for pollinator shifting. Some pollinators avoid already visited flowers (Ayasse et al. 2000; Schiestl and Ayasse 2001) or flowers at sites where they have been deceived (Peakall 1990). It is also possible that the variety of colors in some species (*A. behrii*) for example could facilitate the visit of a tricked wasp to a flower of a different color from the one that first tricked him (negative frequency dependence cited by many; e.g. Schiestl 2005; Gaskett 2012 [for floral shape]; but see Dormont et al. 2010). That the pollinator may learn to get rid of the pollinia (by visiting another flower) is another extravagant possibility, but it is more likely to do that if enough visitors have removed pollinia from flowers, leaving them female and pollen repositories rather than dispensers. Thynnine wasps are solitary, so grooming off pollinia, which happens in social insects (Roubik 2000), may not be easily achieved.

Whether or not these extravagant ideas carry some truth, the wasps have no known benefit in encouraging orchids to go on. Their goal should be to avoid being enticed by mischievous orchids in any shape or form, but male-male competition may not allow such leisurely reflection.

WHAT MAKES AN EFFECTIVE POLLINATOR?

Is it completely daft? "Insects, or at least bees, are by no means destitute of intelligence" (Darwin 1877; p. 42).

The pollinators of *Caladenia* syn. *Arachnorchis behrii* are most effective when they deposit a pollinium from another orchid (Petit et al. 2009). As demonstrated in many species and identified by Darwin (1862; p. 359), among many others:

> It is an astonishing fact that self-fertilisation should not have been an habitual occurrence. It apparently demonstrates to us that there must be something injurious in the process. Nature thus tells us, in the most emphatic manner, that she abhors perpetual self-fertilisation.

Inbreeding has detrimental impacts on most orchid species studied (Peakall and Beattie 1996; Ferdy et al. 2001; Tremblay et al. 2005; Jersáková and Johnson 2006; Jersáková et al. 2006a,b). Self-pollination reduces the probability of successful germination in *A. behrii*, in part (but not exclusively) because the seeds resulting from self-pollination are smaller than those resulting from cross-pollination (Petit et al. 2009). Some species are capable of forming seed banks (Whigham et al. 2006), although orchid seed banks are relatively short-lived (Batty et al. 2000). Large seeds resulting from cross-pollination in *A. behrii* are more likely to survive for a longer time than are small seeds (Petit et al. 2009).

In orchids that have entire pollinia deposited on stigmas, as with the spider orchids, the pollen-to-ovule ratio is expected to be 1 (Johnson and Edwards 2000). In *A. behrii*, increased pollen load (2 pollinia as opposed to 1 pollinium) does not significantly increase the number of seeds and tends to result in smaller seeds (Petit et al. 2009). Depositing a single pollinium is sufficient to be an effective pollinator.

Reproduction does not tend to be primarily resource-limited in deceptive orchids, although resource limitation can appear over time or at different life stages (Nilsson 1992; Ferdy et al. 2001; Tremblay et al. 2005; Jersáková et al. 2006b; see also references on costs of reproduction to orchids below). So what limits pollen delivery? Pollinator limitation is not equal to pollination limitation, because many pollinators could be present at the wrong time or not interested in flowers; pollinator abundance and behavior need to be understood (see Chapter 10). Vandewoestijne et al. (2009) suggested that pollinator availability is more likely to be limiting the reproductive success of sexually deceptive orchids than is pollinator learning.

Evidence of learning exists for many insects, including winged species in the Hymenoptera (Dukas 2008). To some extent, the learning ability of pollinators can determine their "effective" (a conservation biology term indicating that not all members of the population are involved in reproduction) population size from the orchids' perspective. To be operational, pollinators have to be tricked not once but at least twice: once to pick up pollinia, and another time to deliver them. The more often they are tricked, the more effective they will be as individual pollinators. The downside is that spending one's time pollinating orchids does not make one a very successful wasp, at least from a male wasp's perspective. Cost is believed to be low for pollinator species, because not all males are tricked, and their range generally extends beyond that of the orchids. However, ejaculation, demonstrated in wasps in other genera by Coleman (1929a,b), Erickson (1965a,b), Blanco and Barboza (2005), and Gaskett et al. (2008), and the loss of or negative impact on mating opportunities may have a cost to individuals (Gaskett 2011). In addition, wasp visibility to predators could be increased by carrying pollinia.

The interference of pollinia with flight (from their mass and impact on aerodynamics and wing motion) or vision in cases when pollen is deposited on the head of pollinators has unknown costs. Most insects use vision for flight regulation (Pringle 1983), which could be hindered by pollinaria. Pollinarium attachment to the head of insects is common, including eyes and mouthparts (Darwin 1862; Johnson and Liltved 1997; Johnson and Edwards 2000). Although some sexually deceptive orchid species place pollinia on the head of their pollinators (e.g. Bower 1993), which could affect their vision and the likelihood that they may be deceived again, it seems that many spider orchids tend to place pollinia on the thorax of their primary pollinators (Stoutamire 1983; Faast et al. 2009; Phillips et al. 2009b).

Attachment to the thorax is not necessarily cost free. Pollinia can also become burdens during foraging and decrease efficiency (Morse 1981).

C. Bower (personal communication) also observed wasps "struggling" under the load of pollinia from food-deceptive orchids, and Bernhardt (1995c) agreed with van der Pijl and Dodson (1966) that the staged release of *Pterostylis* anther sacs prevents the overburdening of the small gnat pollinators. Peakall and Beattie (1996) captured pollinators of *Caladenia* syn. *Arachnorchis tentaculata* carrying up to 5 pollinia.

Published information on the mass of pollinia is limited, but Vallius (2000) mentioned a mass of approximately 70 μg for a pair of *Dactylorhiza maculata* pollinia, which are generally carried by *Bombus* spp. This mass is probably negligible relative to the large bumblebees. Milkweed (*Asclepias syriaca*: Apocynaceae) pollinia represents only 0.07% of a bumblebee's mass, but an increasing pollinium load still decreased the foraging speed of the 2 *Bombus* species studied (Morse 1981). The mass of sexually deceptive spider orchid pollinia may not represent a large proportion of the pollinators' mass if large numbers of pollinia do not have a chance to accumulate on the flower visitors. However, no work of this kind has been conducted; only minimal information is available on wasp masses (e.g. Ridsdill Smith 1970b; Alcock and Gwynne 1987; Schiestl 2004), and measurements of the pollinia they carry do not seem to have been published yet.

Even if pollinia represent a small proportion of a wasp's mass, energetic cost of carrying pollinia may still be high, because the cargo reduces an insect's ability to process flowers (Morse 1981). To fly, an insect's lift must equal its weight, and the thrust must match the drag of the air on the insect's body. Aerodynamic forces generated are proportional to the wing area, the air density, and the square of the velocity of the relative wind (Pringle 1983). Carrying pollinia could have some impact on an insect's lift and thrust, because pollinia seem to have a relatively large surface area compared with a carrier's wings. Flight maneuverability, the details of which are still largely unknown for insects (Dudley 2002), could also be affected. Small insects have high wing-beat frequencies and proportionately greater flight constraints than do larger insects (Pringle 1983; Dickinson 2006), and could suffer proportionally more from carrying pollinia (note that we may expect a larger mass difference between sexes in smaller species).

The detrimental impact of carrying pollinia may be compounded for male wasps attempting to carry females; females tend to cling to vegetation, possibly to test the strength of males in mate choice; loaded males may be forced to select smaller females, which lay smaller eggs with potentially lower fitness (Alcock and Gwynne 1987). The metabolism of an active insect flight muscle is extremely high, and "in terms of the power-to-weight ratio, is comparable to that of a small aero-engine" (Pringle 1983; p. 14).

We should not discard as unimportant any hindrance to the flight process of orchid pollinators. Pollinia may also be attached to legs (Johnson and Edwards 2000), and could have a strong impact on aerodynamics and balance. The impact of environmental turbulences, demonstrated to be high for some insects (Combes and Dudley 2009), could increase the cost of carrying pollinia. Another cost of attempting to mate with an orchid is the potential increase in wing wear that may result; wing wear in bumblebees, which has many costs, including increased mortality, is associated with frequency of wing collisions, not just age (Foster and Cartar 2011). Prospective research on the aerodynamic and energetic costs of carrying pollinia is exciting. Preserved wasp specimens could be used for measurements, because body size and wet weights are strongly correlated in 2 species of thynnine wasps (Alcock and Gwynne 1987), as they are in other taxa (Shreeves and Field 2008).

Similarly, orchids may have a cost to females. A female wasp will be much less successful at attracting a male when surrounded by mimicking orchids than when signaling in an orchid-free environment (Wong and Schiestl 2002). Full attractiveness resumed by the wasp pollinator of *Chiloglottis trapeziformis* when a female dummy was located 8 m away from the center of an orchid colony (Wong et al. 2004). Wong et al. (2004) identified energetic cost as a possible negative impact of orchids on female wasps that should walk away from an orchid patch, but the increased risk of predation is another possible cost, about which nothing is known. The cost for females and males of moving from good-quality signaling and foraging areas, respectively, could be high. Although sex ratio is neutral in some species examined or biased toward females in *Hemithynnus hyalinatus* (Lloyd 1952; Ridsdill Smith 1970b), it appears that more males are available than females at one time, which represent a limited resource for males (Alcock and Gwynne 1987; Peakall 1990).

It is thus important for the male to be able to recognize an orchid as soon as possible, and for a female to call away from an orchid patch and look even more . . . wasp-like. Males seem to learn relatively quickly that orchids are not female wasps, but not as a result of chemical identification (Wong and Schiestl 2002; Schiestl 2004). Pollinators appear never to learn fully to avoid food-deceptive orchids, because flowers in general vary in rewards, and pollinators should not avoid them (Juillet et al. 2011; but see Chapters 7 and 10); conditions for sexual deception are different. It is in the interest of the orchid that a pollinator does not persist in trying to mate with it, since "vigorous" attempts may result in self-pollination

(Peakall and Beattie 1996; Bower 2001); self-pollination affects 3%–10% of flowers of *A. tentaculata* (Peakall and Beattie 1996). Visual, tactile, and location (site avoidance; Wong and Schiestl 2002; Schiestl 2004) cues are the main learning tools of the male wasps. The conditions for an evolutionary arms race are there if costs on pollinator fitness can be demonstrated (Gaskett 2011); evidence is already strong (Wong and Schiestl 2002; Vereecken 2009).

Long-range pollen dispersal is generally beneficial by minimizing the chances of inbreeding. Pollen flow and genetic studies of orchids indicate that pollen transfer can take place from a flower to itself or to another flower that may be tens of meters away (Peakall 1990; Peakall and Beattie 1996). The majority of seed dispersal seems to be limited (Chung et al. 2004; Chung and Chung 2005; Jacquemyn et al. 2007b), including in *Arachnorchis* (Peakall and Beattie 1996). However, occasional long-range dispersal, which may be facilitated by animals such as echidnas (*Tachyglossus aculeatus*; Feuerherdt 2002), will minimize genetic differentiation among populations, as will long-distance pollen dispersal (Phillips et al. 2009b; Swarts et al. 2009). So pollinators with large home ranges are preferable, and pollinators with restricted-use areas may limit pollen flow (Peakall and Schiestl 2004). Long-range pollen dispersal seems limited in many orchid species (Johnson et al. 2005; Kropf and Renner 2008), although the capacity for long-distance dispersal is high in view of pollen longevity (Johnson and Edwards 2000; Luyt and Johnson 2001; Murren 2002) and the ability of some pollinators to fly long distances (Wikelski et al. 2010), such as *Tachynomyia* spp. (Given 1957). Little is known of the behavior of many thynnine wasps; the underground, parasitoid habits of females (Given 1954; Ridsdill Smith 1970b; Bower 2001) and their relative rarity do not make them species easy to study. Lloyd (1952) suggested low reproductive capacity in the thynnine wasps of Patagonia (South America). Thynnine wasps are also victims of secondary parasitism (Burrell 1935), which makes specializing on a single species of thynnine a relatively risky business.

The best pollinators are likely to be the primary pollinators rather than minor responders. Although pollinator sharing occurs in some sexually deceptive orchid groups (Phillips et al. 2009b), pollinators by sexual deception are generally specific (Bower 1992, 1996; Bower and Brown 1997; Bower 2001; Bower and Brown 2009; Phillips et al. 2009b; Schiestl and Schlüter 2009), resulting in efficient pollination (Scopece et al. 2010). A difficulty for researchers is to identify the taxonomy of *Caladenia* and their pollinators. Genetic methods (e.g. Mant et al. 2005a,b; Ebert et al. 2009; Farrington et al. 2009; Griffiths et al. 2011) have revealed the presence of cryptic thynnine species not distinguishable morphologically. The use of gas chromatography with or without electroantennography (Schiestl and Ayasse 2002; Schiestl and Marion-Poll 2002; Gögler et al. 2011), pollinium-tracking methods (Peakall 1989a; Kropf and Renner 2008), pollinium identification from morphology and DNA (Singer et al. 2008), photographic databases that can be made available online (Singer et al. 2008), and video cameras (Micheneau et al. 2009) are improving our understanding of orchid pollination and diversification, but much remains to be done.

FUTURE IN A RAPIDLY CHANGING WORLD

The myriad of seeds found in an orchid capsule can lead to a false impression of safety.

> That a plant, not an annual, should escape destruction at some period of its life simply by the production of a vast number of seeds or seedlings, shows a poverty of contrivance, or a want of some fitting protection against some danger. (Darwin 1862; pp. 343–344)

However, the world is not carpeted by orchids;

> the great grandchildren of a single plant would nearly (in the proportion of 47 to 50) clothe with one uniform green carpet the entire surface of the land throughout the globe. What checks this unlimited multiplication cannot be told. (Darwin 1862; p. 345)

In fact, the greater *Caladenia* s.l. genus represents a third of critically endangered orchids in Australia (Dixon and Hopper 2009), and the incidence of rarity is greater for sexually deceptive than for food-deceptive species (Phillips et al. 2011).

Since Darwin, the vast majority of scientific publications on orchids has focused on the resolution of evolutionary mysteries, but many recent articles are addressing conservation concerns, including those for spider orchids (e.g. Coates and Duncan 2009; Faast et al. 2009; Petit et al. 2009; Phillips et al. 2009b; Swarts et al. 2009; Swarts and Dixon 2009; Tremblay et al. 2009a). Have spider orchids gone too far in their parasitic specialization on pollinators, so that any safety measure (e.g. the possible "double-dipping" into the pollinator community, with food *and* sexual deception) is not enough? Why is the status of some populations alarming? The ecosystem surrounding the orchid has to be just right. The pollinator must have his nectar and other food resources. The female wasp who is mimicked by the orchid must have her prey, the mycorrhizae, and a degree of soil disturbance. The orchid consumers have to be busy elsewhere or the dead trees and/or living grass trees must protect orchid patches from predators. Then there are the climatic conditions (see Chapter 12). Once protected areas are fragmented, people become both terrified and fascinated by fire. In Australia, deliberately burning vegetation to avoid economically devastating fires may be performed without full understanding of the ecological consequences. Furthermore, chemical sprays may be used in deadly combinations with man-engendered fires. Most people are city folk; the most concerned of them find comfort in the knowledge that nature is safe in the British Broadcasting Corporation (BBC) documentaries.

Characteristics of the ecosystem affect the reproductive success of orchids (Dixon and Tremblay 2009). In Western Australia, rainfall and the edaphic environment are important factors in the biogeography of *Caladenia* s.l., but probably act in concert with specific mycorrhizal fungi and pollinators (Phillips et al. 2009a; Phillips et al. 2011). Climatic parameters are known to affect orchid populations (Coates et al. 2006; Pfeifer et al. 2006; Coates and Duncan 2009; Jacquemyn et al. 2009; Hutchings 2010; Janes et al. 2010; see Chapter 12). The complex effects of climate change on flowering phenology at small scales in meadow ecosystems (Aldridge et al. 2011) are likely to apply to spider orchids and their pollinators. Liu et al. (2010) identified several mechanisms of climate change that will affect orchid species' survival. In the context of climate change and declining soil quality and pollinator abundance, it is essential to understand poorly known organisms such as pollinators and mycorrhizae, to be able to conserve orchids effectively (Waterman et al. 2011).

Common orchids may have a wider range of exploitable mycorrhizal fungi than do rare ones, as fungi are most often saprophytes (Waterman and Bidartondo 2008; Wright et al. 2009). Mycorrhizal fungus availability does not seem limiting (Brundrett et al. 2003; Feuerherdt et al. 2005), but orchid recruitment is low (Batty et al. 2001) and germination sites are variable (Whigham et al. 2006; Jacquemyn et al. 2007b). Much remains to be learned about the mycorrhizal associations of *Arachnorchis*, seedling recruitment in the genus, and environmental conditions that favor orchids and pollinators. Any benefit received by fungi from orchids is not clear, and orchids may be parasitizing them (Waterman and Bidartondo 2008).

Although some orchids seem to respond somewhat positively to fire (Coates et al. 2006; Coates and Duncan 2009), the impact of fire on pollination is unclear (see Chapter 7). Nectar availability is of crucial importance to thynnine wasps (Ridsdill Smith 1970b), and when fires stop nectar production, the area may become depleted of pollinators. Each female wasp on average will only lay very few eggs (Ridsdill Smith 1970b), so the future of orchids depends on optimal conditions for wasps. The best fungal activity for native orchids in Western Australia was recorded in coarse soil containing organic matter, which is most likely to be negatively affected by fire (Brundrett et al. 2003). Reported responses to fire of most spider orchid species are anecdotal.

Low pollination rates can lead to the temptation of hand-pollinating orchids, but pollinator limitation should be determined only after examining the reproductive costs of a population over many consecutive flowering seasons (Janzen et al. 1980), particularly since the cost of reproduction in orchids is high (Snow and Whigham 1989; Ackerman and Montalvo 1990; Calvo 1993; Willems and Dorland 2000). Hand pollination should be practiced carefully, because it is likely to have a future cost on spider orchids, and can we bet on the conditions to be right the next year (Coates and Duncan 2009; Petit et al. 2009)? Low fruit set is not necessarily an indication that orchids are unsuccessful. It may be crucial, when balanced against reproductive cost, to allow staggered recruitment over time with variable environmental conditions and resources. The probability of flowering in consecutive years for individuals of several *Caladenia* syn. *Arachnorchis* species is rare, and dormancy periods are likely to be vital to population persistence (Tremblay et al. 2009b).

Darwin would not recognize the country beyond a 3-mile radius from his protected Downe House retreat: "*extreme rurality*" has been eaten away by the city. Fragmentation due to agricultural, urban, or residential development is taking its toll. The "refractory" response (Peakall 1990) to deception shown by wasp pollinators of sexually deceptive orchids is likely to facilitate cross-pollination (Jersáková et al. 2006a), but if orchids occur in a small population in a small area (mate search ≤ 40–60 m for *A. tentaculata* pollinator; Peakall and Beattie 1996), then the pollen may be lost. Small populations may also be victims of their reproductive variability and weather sensitivity (Jacquemyn et al. 2009). Populations of *Caladenia* syn. *Arachnorchis rigida*, which are small in the southern part of their South Australian range, seem at risk there from reduced seed output (Faast et al. 2011). No matter what aggravating factor orchids are subjected to, they will survive only if enough land is available to give them resilience. This land has to be respected. In the northern Adelaide region in 2011, 7 protected areas were run by just 2 rangers. The "weed and vermin" control budget of $9,500 had to be shared among 26 reserves. Such lack of resources is not likely to favor conservation.

The complexity of orchid ecosystems makes medium- to long-term studies necessary to understand their reproduction and population dynamics (Nilsson 1992; Murren 2002; Pfeifer et al. 2006; Jacquemyn et al. 2007a; Jacquemyn and Brys 2010). Time is running short. The planet is not clothed with "one uniform green carpet" of orchids; instead, it is covered with an ever-increasing human population. Without education, without an understanding of the intricate and exquisite work of evolution, and without engagement from scientists, many orchid species will be lost this century, and the saddest thing of all is that they may not even be missed. Science can no longer afford to provide information just when it is needed; scientists need to influence governments and private entities to make decisions that may be difficult, but right for the future of humanity and biodiversity.

ACKNOWLEDGMENTS

Many thanks to Colin Bower, Robert Pemberton, Ana Francisco, and Joan Gibbs for reviewing the manuscript; to the Australian Technology Network, Native Vegetation Council of South Australia, Australian Orchid Foundation, Kersbrook Landcare Group, and University of South Australia for having sponsored my research over the years; to Erik Dahl of the Department for Environment and Natural Resources, Theresa Kaukas of Forestry SA for facilitating the research, colleagues Doug Bickerton and Manfred Jusaitis, and the many volunteers who assisted with the fieldwork and data entry, especially Catherine Dickson, Damian Morrant, Jason Tyndall, and Katelyn Ryan.

SEVEN

The Sun Orchids (*Thelymitra*) Then and Now: Large Flowers versus Small Flowers and Their Evolutionary Implications

Retha Edens-Meier and Peter Bernhardt

INTRODUCTION

Taxonomy, Geography, and Diversity of the Genus *Thelymitra*

The naturalists Johann Georg Forster (1754–1794) and his father, Johann Reinhold Forster (1729–1798), are credited with describing and naming the genus *Thelymitra* in 1776. The genus name refers to the hooded column (Figure 7.1) in the flower. A lobed and ornamented hood (see below) is a distinguishing characteristic found in most but not all species within this genus (Greek *thelys* = female + *mitra* = a hat; Jones 1988). However, it was Sir Joseph Banks (1743–1820) who first collected specimens of what is now known as *T. longifolia*, on New Zealand's North Island during Captain James Cook's first voyage (1768–1771). The Forsters collected and described *Thelymitra* using plants collected from New Zealand's South Island on Cook's second voyage (1772–1775; Cady and Rotherham 1970). We now tend to recognize more than 100 named species in this genus, with additional specimens awaiting description (see Brown et al. 2008; Jeanes 2013).

Taxonomic concepts changed rapidly in the early 21st century. For example, Moore and Edgar (1970) recognized only a dozen *Thelymitra* spp. in the flora of New Zealand, while Scanlen and St. George (2010) recognized 34 taxa (including species, hybrids, and unnamed novelties). In Australia, former species (e.g. *T. pauciflora*, *T. variegata*, *T. fuscolutea*, etc.) were reinterpreted as species complexes and subdivided into addi-

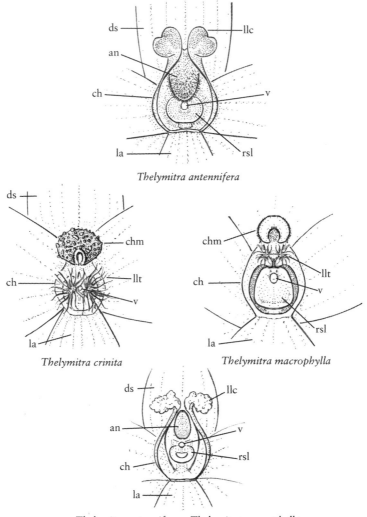

Fig. 7.1 Drawings of the interiors of the flowers of *Thelymitra antennifera*, *T. macrophylla*, their first-generation hybrid, and *T. crinita*; *an* = anther; *ch* = column hood; *chm* = single inflated column hood; *ds* = dorsal sepal; *la* = labellum; *llc* = lateral lobe of column; *llt* = lateral lobe with trichome brush; *rsl* = receptive stigma lobes; *v* = viscidium or naked rostellum. Illustration by John Myers. Used by permission.

tional species (Jeanes 2004, 2006, 2009, 2011, 2013). A modern, molecular phylogeny places *Thelymitra* spp. in the subfamily Orchidoideae, tribe Diurideae, and subtribe Thelymitrinae along with 2 other genera, *Calochilus* and *Epiblema* (Chase et al. 2003; see Chapter 5). Certain *Thelymitra* spp. are endemic to New Zealand (Molloy 1990; Scanlen and St. George

2010), Tasmania (Jones et al. 1999), and mainland Australia (Jones 2006). However, the real center of diversity of this genus occurs through moist temperate–Mediterranean Australia, although a few *Thelymitra* spp. are found as far north as tropical Queensland, Papua New Guinea, and the southern Philippines (Jeanes 2013).

Fitzgerald and Darwin Correspond on *Thelymitra* Pollination

Based on his collected botanical specimens, books, and correspondence, there is no evidence that Charles Darwin saw *Thelymitra* spp. in bloom after the HMS *Beagle* stopped at ports in Australia. In fact, the *Beagle* arrived in Western Australia so late in the season that the only species Darwin could have seen were the dark-colored, highly scented flowers in the *T. fuscolutea* complex (K. Dixon, personal communication, and see Jeanes 2006). If so, he never mentioned the genus *Thelymitra* in any of his early writings (Darwin 1862) or correspondence (see Chapter 1).

In fact, Darwin would have never addressed pollination in this genus (Darwin 1877) without the cooperation of the Australian botanist Robert David Fitzgerald (1830–1892). Fitzgerald was one of the many botanists who was so impressed by the first edition of Darwin's book (1862; see Chapter 1) that he began corresponding with him on Australian orchid species. Some letters, from Charles Darwin and his son, William Erasmus Darwin (1839–1914), survived (1875–1899) and were donated to the Sydney Library by Fitzgerald's family early in the 20th century. The letters indicate that Fitzgerald sent Charles Darwin the first parts of what would be bound together and entitled *Australian Orchids* (Fitzgerald 1875–1895).

It is clear that Darwin valued Fitzgerald's work, as he included a copy of one of Fitzgerald's illustrations in his second edition (Darwin 1877; pp. 87–88) and wrote, "I have copied the figure of *Pterostylis longifolia* from Mr. Fitzgerald's great work on the Australian Orchids, as it shows plainly the relation of all the parts." It is also apparent that he accepted Fitzgerald's observations and experiments with bell jars to see if flowers of Australian species self-pollinated when isolated from insects. Moreover, he cites Fitzgerald extensively in his second edition (Darwin 1877), using certain species as model examples of obligate self-pollination. In the earliest surviving letter of Darwin to Fitzgerald, he draws a parallel between *Thelymitra* and British specimens of self-pollinating *Ophrys apifera*: "I have been more interested about the closed flowers of the *Thelymitra*, as the case is closely analogous to the equally detestable [?] one of the Bee *Ophrys* in Europe" (C. Darwin to R. D. Fitzgerald, 16 July 1875).

For almost a century following Darwin (1877), the majority of Australian and New Zealand botanists (see Molloy 1990) and naturalists associated *Thelymitra* spp. with facultative-obligate self-pollination, even though Darwin cited only 2 species, *T. carnea* and *T. longifolia*, as self-pollinating in the bud. This was unfortunate, as Fitzgerald (1875–1895) repeatedly discriminated between *Thelymitra* species with small flowers and friable (crumbling) pollinia that fall onto the stigma as the anther opens versus large-flowered species in which the pollinarium can be removed with a pin in its entirety, as these pollinia pairs connect to a true rostellum. Fitzgerald did not record obligate self-pollination in these large-flowered species.

Cross- and Self-Pollinating "Sister" Species in *Thelymitra*

We should also note that Fitzgerald (1875–1895) was the first to document morphological similarities between some large-flowered and small-flowered *Thelymitra* spp. A species that produces large flowers and whole pollinia often resembles, most closely, a second species with tiny flowers and friable pollinia. For example, the yellow-flowered *T. antennifera* has large flowers and whole pollinia, while yellow-flowered *T. flexuosa* has tiny flowers and friable pollinia (Fitzgerald 1875–1895; Table 7.1). Fitzgerald noted that these 2 species have overlapping similarities in floral and vegetative characters. He found other paired species with large and small flowers that "mirrored" each other. Much later, Burns-Balogh and Bernhardt (1988) looked at some of these species pairs based on rehydrated flowers and spirit collections, but concluded that development of the lobes and sculptures on the column hood of small-flowered species is incomplete and arrested compared with their large-flowered sister species.

Furthermore, Sydes and Calder (1993) compared the pollination biology of small-flowered *T. circumsepta* with large-flowered *T. ixioides*. They found that *T. circumsepta* is self-pollinated, while *T. ixioides* is insect pollinated. Jeanes (2009) recently segregated the large-flowered *T. spiralis* from the small-flowered *T. uliginosa*. Peter Bernhardt (unpublished) teased open fresh flowers of *T. uliginosa* prior to pressing them for herbarium vouchers in 2009. As he was wearing a magnification visor at the time, he saw that pollen grains had germinated in the anther, and skeins of pollen tubes grew down from the anther, contacting the moist stigmas as observed previously in flowers of *T. circumsepta* by Sydes and Calder (1993).

We now understand that flower size and pollinia structure can also vary within the same *Thelymitra* species. Jones (2006) discriminated between self-pollinating (small-flowered) and cross-pollinating (large-flowered) "forms" of *T. ixioides*. He also noticed large- and small-flowered forms within the same populations of *T. malvina*.

Table 7.1 Insect vs. self-pollination (mechanical autogamy) in *Thelymitra* spp.

Thelymitra spp.	Perianth color patterns	Discernible Scent	Length of sepals (or length of longest perianth segment) (mm)	Known pollinia vectors	Breeding systems*			References
					Insect pollination	Facultative self-pollination	Subcleistogamous self-pollination	
T. adorata	Dark blue	Not reported	8–13	Unknown	N	Y	N	Jeanes 2011
T. antennifera	Yellow with brownish center	Lemon	20	*Lasioglossum* (bee); *Eurys* (wasp); *Syrphus* (fly)	Y	N	N	Brown et al. 2008; Dafni & Calder 1987; Jones 2006; Rogers 1913
T. apiculata	Purple, pinkish purple	Not reported	9–20	Unknown	Y?	N	N	Jeanes 2009
T. arenaria	Blue to purple	Not reported	7–9	Unknown	N	Y	Y	Jones et al. 1999; Jeanes 2004
T. aristata	Blue to violet	Present	10–25	Unknown	Y?	N	N	Jeanes 2011
T. benthamiana	Yellow or greenish, with reddish-brown spots and blotches	None noted	10–20	Unknown	N	Y?	N	Jeanes 2006
T. carnea	Pink to reddish	Not reported	7	Unknown	N	N	Y	Jones 2006
T. circumsepta	Greenish white to pale blue to lilac	Not reported	12	Unknown	N	N	Y	Sydes & Calder 1993
T. crinita	Pale blue to cobalt blue	Scentless	22	*Leioproctus* (bee) and *Amegilla* (bee)	Y	N	N	Edens-Meier et al. 2013
T. dedmaniarum	Mostly golden yellow, often reddish brown toward the base	Cinnamon	12–28	Unknown	Y?	N	N	Jeanes 2006

continued

Table 7.1 *continued*

Thelymitra spp.	Perianth color patterns	Discernible Scent	Length of sepals (or length of longest perianth segment) (mm)	Known pollinia vectors	Breeding systems*			References
					Insect pollination	Facultative self-pollination	Subcleistogamous self-pollination	
T. fuscolutea	Brown or purplish, with yellow or greenish spots or striations	None noted	8–20	Unknown	Y?	N	N	Jeanes 2006
T. grandiflora	Blue, lilac, mauve, or pale purplish, rarely pink	Present	8–20	Unknown	Y?	N	N	Jeanes 2011
T. ixioides	Dark blue, pink to mauve, with dark spots	Not reported	22	*Lasioglossum* (bee) and *Exoneura* (bee)	Y	N	N	Jones 2006; Sydes & Calder 1993
T. jacksonii	Dark golden brown, with large yellow blotches and striations	Sweet	10–27	Unknown	Y?	N	N	Jeanes 2006
T. kangaloonica	Dark blue, with darker longitudinal veins	Strongly scented	8–18	Unknown	N	Y	N	Jeanes 2011; Jones 2006
T. longifolia	Ranges in color from white, pink, blue	Not reported	NA	*Leioproctus*	Y	Y	Y	Lehnebach et al. 2005; Cheeseman 1880
T. luteocilium	Pale pink	Not reported	18	Unknown	N	N	Y	Jones 2006; Rogers 1913
T. macrophylla	Pale blue, pinkish blue, to dark blue	Rosewater	20	*Homalictus, Leioproctus, Lasioglossum* (all bees)	Y	N	N	Edens-Meier et al. 2013

continued

Species	Flower color	Fragrance	Chromosome number	Pollinator				Reference
T. maculata	Pale to deep pink to purple	Not reported	8–18(–25)	Unknown	Y?	N	Y	Jeanes 2009; Jones 2006
T. magnifica	Dark golden brown, with large yellow blotches and striations	Cinnamon	16–28	Unknown	Y?	N	N	Jeanes 2006
T. matthewsii	Purple with stripes	Not reported	6–10(–13)	Unknown	N	Y	Y	Jeanes 2009; Jones 2006
T. mucida	Pale to dark blue; dark veins	Not reported	7–10	Unknown	?	?	Y	Jones et al. 1999
T. nuda	Blue-mauve and bluish purple	Sweet, like lavenders and lilacs	20	*Lasioglossum* (bee)	Y	Y	N	Bernhardt & Burns-Balogh 1986a; Jones 2006
T. pauciflora-complex (n = 23)	Ranges from white to pink to blue	Large flowers of *T. malvina* fragrant	10	Unknown	Y	Y	Y	Jeanes 2004; Jones 2006
T. planicola	Medium blue, with darker blue longitudinal veins	Not reported	7–15	Unknown	N	Y	N	Jeanes 2011
T. pulcherrima	Sepals reddish spotted; petals purple spotted	Not reported	10–18	Unknown	Y?	N	N	Jeanes 2009
T. rubra	Pink to pale yellow	Fragrant	8–12	Unknown	Y?	?	Y	Jones et al. 1999; Jones 2006
T. silena	Pale blue	Not reported	10–19	Unknown	N	Y	N	Jeanes 2011
T. speciosa	Purple, with reddish-gold edges	Not reported	10–25	Unknown	Y?	N	N	Jeanes 2009
T. spiralis	Pink, reddish, purple, or blue	Not reported	8–20(–26)	Unknown	Y?	N	N	Jeanes 2009

Table 7.1 *continued*

Thelymitra spp.	Perianth color patterns	Discernible Scent	Length of sepals (or length of longest perianth segment) (mm)	Known pollinia vectors	Insect pollination	Facultative self-pollination	Subcleistogamous self-pollination	References
					Breeding systems*			
T. uliginosa	Pink, blue, mauve to purple	Not reported	6–15	Unknown	N	Y	Y	Bernhardt (pers. obs.) Jeanes 2009
T. variegata	Reddish to purple; blotched with yellow margins	Not reported	(10–)14–27	Unknown	Y?	N	N	Jeanes 2009
T. venosa	Blue with darker veins	Not reported	25	Bee?	Y?	N	N	Jones 2006; Rogers 1913
T. yorkensis	Orange, with reddish-brown edges	Cinnamon	13–26	Unknown	Y?	N	N	Jeanes 2006

Note: Y = yes; N = no; ? = questionable.

As flower size may occur along a gradient in this genus, we also suspect that self-pollination occurs along a gradient in some species (see Table 7.1). The act of mechanical self-pollination (autogamy) in *Thelymitra* spp. should therefore be subdivided into 2 overlapping modes of development. First, flowers showing facultative self-pollination open normally, but self-pollination may occur as the flower ages when pollinators fail to appear. Some populations of *T. longifolia* in northern New Zealand may fit into this category (Molloy 1990). In most cases, their flowers self-pollinate, but under certain conditions, open flowers are visited and pollinated by the small colletid bee *Leioproctus fulvescens* (Lehnebach et al. 2005). Second, the *Thelymitra* spp. with the smallest flowers are most often described as subcleistogamous. In subcleistogamy, either the anther opens and crumbling pollinia fall directly onto the receptive stigma before the bud opens, or the pollen grains germinate in the anther locules, sending penetrating pollen tubes down to the stigmas before the bud opens (e.g. *T. uliginosa* and *T. circumsepta*; see above). If the flower segments ever open fully (usually only on the hottest, sunniest days), one can already see (using a magnification device) pollinia fragments glued to the receptive stigma lobes.

There are reports suggesting that facultative and subcleistogamous self-pollination may occur in the same species (see Table 7.1). DeLange et al. (2007) found 2 breeding forms of *T. longifolia* in New Zealand. The rarer subcleistogamous plants are confined to *Nothofagus* forests. These plants produce only one white flower per scape. In contrast, facultative autogamous populations grow in a wider variety of habitats. These plants produce a scape bearing more than one flower with bluish perianth segments.

NEW ZEALAND AND AUSTRALIAN STUDIES ON *THELYMITRA* POLLINATION

Thomas F. Cheeseman (1846–1923) was the first to describe pollination in unopened buds (subcleistogamy) of *T. longifolia* in New Zealand (Cheeseman 1880). When these flowers opened on bright, warm, sunny days, he found thrips feeding on their pollen. Cheeseman (1880) believed that thrips did not pollinate orchids, unlike Darwin (1877), who believed that some flowers of *Listera ovata* were pollinated, in part, by thrips. R. S. Rogers (1913) examined the flowers of 4 Australian *Thelymitra* spp. (*T. antennifera*, *T. luteocilium*, *T. fuscolutea*, and *T. venosa*) and what we now know is the recurrent hybrid *T.* × *macmillanii* (sensu Jones 2006). He concluded that only *T. antennifera* do not self-pollinate.

Following Fitzgerald's death in 1892, an interest in insect pollination in Australian orchid pollination did not resume until the 1920s. The primary student of field research was Edith Coleman (1874–1951). She described sexual mimesis in *Cryptostylis* spp. and the pollination of some nectar-secreting orchids (Dafni and Bernhardt 1989; see Chapter 1). Although she used Darwin (1877) as a reference, she did not study *Thelymitra* spp. Rica Erickson (1908–2009) followed Coleman's studies but had nothing to add to the literature on *Thelymitra* pollination (Erickson 1965a,b).

In all fairness, the flowers of *Thelymitra* spp. are often inaccessible at the peak of their flowering seasons. Their perianth segments open in the morning and close by late afternoon, with important exceptions (e.g. *T. venosa* var. *magnifica*; Cady and Rotherham 1970). Furthermore, the perianth segments in most species usually fail to open on cool, wet, cloudy days, with important exceptions (e.g. *T. venosa*; sensu Bishop 1996). No wonder Australian naturalists continue to call them sun orchids.

Therefore, from Rogers (1913) until the present day we have learned gradually that reproduction in the genus *Thelymitra* varies between and within species. These exclusively terrestrial herbs may employ a combination of self-pollination, cross-pollination, and/or vegetative reproduction via tuberoid division (Cady and Rotherham 1970). Small native bees are also capable of transferring pollen between flowers of a self-pollinating species (Table 7.1), thus increasing genetic variability. In addition, hybrids can occur between different *Thelymitra* spp. when the same insect visits more than one species. Consequently, several "rare" *Thelymitra* spp. have been reclassified as recurrent, long-lived hybrids (Jones 1988; Molloy and Dawson 1998; Brown et al. 2008).

Floral Phenology and Flowering Patterns in *Thelymitra* spp.

With the exception of one epiphytic species (*T. fragrans*; Jones 2006), all *Thelymitra* spp. described to date are terrestrial herbs blooming from late winter (August) through spring, with a few species flowering on through November (Brown et al. 2008). As some species in New Zealand (e.g. *T. longifolia*; Johns and Molloy 1983) are distributed from sea level to tree line, flowering periods may vary between populations.

The response of flowering populations of *Thelymitra* spp. to the Australian cycle of summer bush fires in sclerophyll woodlands remains unclear. According to Jones et al. (1999), *T. cyanea*, *T. flexuosa*, and *T. polychroma* are more likely to flower profusely following bush fires and "slashing" in some Tasmanian locations. In contrast, Brown et al. (2008) noted that while some terrestrial orchid species in Western Australia bloom prolifically after a fire, none of those 32 species listed belong to the genus *Thelymitra*. Brown et al. (2008) concluded that some *Thelymitra* spp. may even be damaged or destroyed by fire, and the authors warned that sound fire management policies must be employed to conserve biodiversity (see Chapter 6).

Edens-Meier et al. (2013) found that *Thelymitra macrophylla* bloom first in September, with the flowering of *T. crinita* overlapping by early October. Both species produce flowers that live in excess of 13 days (when unpollinated) under controlled greenhouse conditions. On warm, sunny days, the flowers open after 9:00 a.m. and close between 2:00 to >4:00 p.m. in the greenhouse and at field sites. Statistical analyses showed that the removal of the pollinarium the first day the flowers opened did not shorten their floral life spans. However, hand-pollination experiments showed that crossed, selfed, and interspecific hybrid pollination of the stigma caused the perianths of both *Thelymitra* spp. to close within 48 hours and never reopen (Edens-Meier et al. 2013). While populations of *T. macrophylla* and *T. crinita* had an overlapping flowering period at the Lesmurdie site in Western Australia, they responded differently to the same climatic regime in 2009 (see Edens-Meier et al. 2013). The cycle of opening and closing of flowers of *T. crinita* is less affected by low daily temperatures compared with sympatric populations of *T. macrophylla*. This may partially explain why the rate of pollinated stigmas in *T. crinita* is significantly higher than in *T. macrophylla*, as some pollinia vectors (see below) may also remain active at lower daily temperatures.

Floral Presentation

The flowers of *Thelymitra* species are usually arranged along a terminal raceme and open acropetally. Perianth color within this genus includes various shades of whites, blues, reds, yellows, and combinations thereof. Some color patterns appear metallic to the human eye (e.g. *T. epipactoides*; see Nicholls 1969). Color variation within a particular species is also a common occurrence. In a few cases, the population may show 3 color morphs (e.g. *T. epipactoides*; Cropper and Calder 1990). More often there is a subtle and continuous gradation of pigmentation from pastel to vivid (e.g. *T. aristata*; see Nicholls 1969). Nicholls (1969) also showed variation in mitra morphology and sculpturing within the same species (e.g. *T. aristata*, *T. ixioides*, and *T. media*). In addition, the perianth segments of some *Thelymitra* spp. may offer contrasting patterns in the form of stripes, spots, or dark blotches (see Nicholls 1969). As the species name *variegata* suggests, petals and sepals of flowers within the *T. variegata* complex (sensu Jeanes 2009) are often beautifully variegated.

However, prospective observers often note that in most *Thelymitra* species, the labellum is almost identical in shape and color to the 2 lateral petals (Figure 7.1). This may have contributed to a belief that the genus was among the "most primitive" within the monandrous orchids and perpetuated itself almost exclusively via self-pollination. Yet the one morphological character that makes this genus unique, and a prospective candidate for cross-pollination, is the decoration and exaggeration of the clinandrium (sensu Dressler 1981; Figure 7.1) or mitra (sensu Burns-Balogh and Bernhardt 1988). The clinandrium consists of the filament of the solitary fertile anther fused to the 2 connate staminodia (see Darwin 1877). This simple to elaborate structure has been used continuously for identifying species (see color plates in Jones et al. 1999) and/or segregating new species (Rupp 1942; Erickson 1965b; Woolcock and Woolcock 1984; Bishop 1996; Brown et. al. 2008). In fact, novel variation in the shape, inflation, number of lobes, and sculpture of the clinandrium/mitra appears to be the primary reason why some of the rare but recurrent F_1 hybrids were once identified as separate species (Jones and Bower 2001; Figure 7.1D). Rogers (1913) is given credit for being the first to provide an adequate description of how the clinandrium and stigma work together to form a functional column in *Thelymitra* spp. (Cady and Rotherham 1970). In some *Thelymitra* spp., the clinandrium/mitra resembles a reduced, almost invisible collar (Figure 7.1B). In other species it is a much-inflated hood (see above) consisting of 3 distinct lobes (Burns-Balogh and Bernhardt 1988; Figures 7.1A, C).

When the hood has only a thin and sheath-like midlobe, this sheath may be smooth or well decorated with papillae that contrast in color with the petals and/or sepals (Figure 7.1A). Sometimes additional papillae are found on the exposed dorsum of the naked, fertile anther, as in *T. antennifera* (Figure 7.1A). The 2 lateral arms (staminodia) of these "collared" species are often a contrasting color to the perianth, and highly sculptured, with smooth to rugulose and/or fringed margins resembling spikes, antlers, or feathered wings (Figure 7.1B). In contrast, species with distinct to prominent 3-lobed hoods usually have 2 lateral lobes forming slender arms with fringed margins, or they are smooth and terminate in trichome brushes resembling white or pastel pom-poms (Burns-Balogh and Bernhardt 1988; Figures 7.1A, C). Some species are strongly scented to the human nose, and the fragrance has been compared to vanilla extract, rosewater, spices (Edens-Meier et al. in preparation; Rupp 1942; Bates and Weber 1990; Jones 2006), or the perfumes of domesticated flowers

(Bernhardt and Burns-Balogh, 1986a). Pigmentation patterns and scent descriptions in the large-flowered *Thelymitra* spp. appear indicative of cross-pollinated systems.

Breeding Systems

Obviously, small-flowered, self-pollinating *Thelymitra* spp. must be self-compatible. Growing evidence in some large-flowered *Thelymitra* spp. shows that they, too, are self-compatible, but do not self-pollinate mechanically when pollinators fail to appear as the flower ages (Cropper and Calder 1990; Sydes and Calder 1993). Laboratory experiments and field observations showed that the breeding system is identical in *T. crinita* and *T. macrophylla*, both large-flowered species (Edens-Meier et al. 2013). Hand pollinations showed that their flowers are self-compatible, with pollen tubes reaching the ovules within 6 days following the deposition of pollinia fragments onto wet stigmas. Late-acting self-incompatibility is suspected in *T. antennifera*, based on the growth of pollen tubes within the styles following controlled hand-pollination experiments (Dafni and Calder 1987). However, the result of this experiment may actually be the result of delayed fertilization. Edens-Meier et al. (2013) found that pollen tubes of *T. macrophylla* and *T. crinita* failed to reach ovules 3 days after hand pollination. In other orchid species, pollen tubes may become arrested at the bases of the styles, presumably until ovule development (the embryo sac) is complete (see review in Edens-Meier 2010 and see Chapter 10).

Insect Pollination in *Thelymitra* Revealed at Last

Cady and Rotherham (1970) published photos of hoverflies (identified as *Eristalis tenax*; Syrphidae) foraging on the open flowers of *T. media*, a blue-flowered species with a trichome brush mitra. While insects in this family eat pollen and drink the nectar of many flowering plants, there is no evidence that Cady and Rotherham caught the flies and examined them for the presence of the pollinia of *T. media*.

Jones (1981) observed fast-flying bees landing on the column hoods of *T. media* and *T. aristata* (also blue with a trichome brush mitra), and noted that the bees flew off with pollinaria attached to the last 2 abdominal segments. He saw the bees curl up on the terminus of the mitra (midlobe) and make thrusting movements. Jones therefore interpreted the midlobe and lateral trichome brush sculptures as dummy females that attracted male bees. Unfortunately, there is no reference in this description to the collection of the bees or their gender. Bernhardt and Burns-Balogh (1986a) observed and collected a few bees foraging on *T. nuda* (syn. *T. megcalyptra*; see Chapter 5). The petals and sepals of this orchid are bluish purple, while the mitra has a smooth yellowish-brown midlobe and 2 white trichome brushes. As in the case of *T. media* and *T. aristata* (above), pollinia fragments and viscidia were deposited toward the tip of the dorsum of the abdomen, but all bees collected were female. Pollen load analyses showed they were generalist (polylectic) foragers. Dafni and Calder (1987) investigated the pollination of T. *antennifera* following a summer bush fire. This species has a reduced, sheath-like midlobe (Figure 7.1A). They collected female bees, hoverflies (*Syrphus damaster*), and a small solitary wasp (*Eurys* sp.) carrying pollinaria (Table 7.1). Most insect specimens wore the viscidia on their abdomens, while a few carried them on their mouthparts.

Dafni and Calder (1987) and Burns-Balogh and Bernhardt (1988) concluded that most *Thelymitra* spp. are mimics of nectarless flowers that produce pollen collected by female bees provisioning their offspring. In some cases (Dafni and Calder 1987), the flowers of some *Thelymitra* spp. may mimic a guild of co-blooming species with yellow and cream flowers. In other cases (Bernhardt and Burns-Balogh 1986a), orchid pigmentation mimics blue-purple models (Table 7.1). It is believed that these nectarless orchids mimic both native, petaloid monocots and some eudicotyledons (Jones 1988), based on pollen load analyses of bees caught on the orchid flowers (Bernhardt and Burns-Balogh 1986a; Edens-Meier et al. 2013). Female bees deceived by flowers in this genus tend to belong to genera (*Amegilla, Leioproctus, Lasioglossum, Nomia*) known for their polylectic foraging habits, taking nectar and/or pollen from some co-blooming flowers and pollen from co-blooming species that never secrete nectar (Bernhardt and Burns-Balogh 1986a; Dafni and Calder 1987; Cropper and Calder 1990; Edens-Meier et al. 2013; Table 7.1). In all these studies, viscidia and pollinia fragments, when found, are almost invariably deposited toward the tip of the bees' abdomens, as described originally by Jones (1981).

How frequently do these bees visit the large-flowered *Thelymitra* species? Specifically, how often do these visits result in the removal of entire pollinaria and/or pollinia deposition on receptive stigmatic lobes? Thus far, all published fieldwork studies agree that bee visits are infrequent (Bernhardt and Burns-Balogh 1986a; Dafni and Calder 1987; Cropper and Calder 1990; Edens-Meier et al. submitted). In fact, Cropper and Calder (1990) concluded that low visitation rates by *Nomia* spp. are responsible, in part, for the decline in populations of *T. epipactoides*. Edens-Meier et al. (2013) found that natural rates of insect-mediated pollinaria removal for

T. crinita and *T. macrophylla* (Lesmurdie, WA) are almost identical and at a low rate of 16%–17%. However, rates of pollinarium removal by bees are far greater than rates of pollinarium deposition on stigmas. In *T. macrophylla*, the rate of pollinarium removal was 10 times higher than the rate of pollinia deposition in one season (Lesmurdie, WA). While the deposition of pollinia on stigmas of *T. crinita* was significantly higher than *T. macrophylla* at the same site, natural rates of insect-mediated pollination were less than 12% in both species. The "gullibility" of bees visiting *Thelymitra* spp. may depend on a broad number of variables, including bee taxonomy. For example, Edens-Meier et al. (2013) observed one *Amegilla chlorocyanea* (Apidae) visiting the flowers on several inflorescences of *T. crinita* and leaving pollen on several stigmas before it left the site.

No *Thelymitra* spp. to date have been found to secrete nectar, although flies in the family Syrphidae have been observed repeatedly drinking stigmatic fluids (Dafni and Calder 1987; Bernhardt, personal observation; and see above). These fluids may account for the presence of some pollinaria of *T. antennifera* attached to the heads of some insects if the insects drink from the stigmatic lobe and their head contacts the rostellum (Dafni and Calder 1987). Edens-Meier et al. (2013) noted that syrphid flies observed and collected on flowers of *T. macrophylla* were also observed and collected on the large, blue but nectarless flowers of *Orthrosanthus laxa*. These flies appeared to consume the pollen of *O. laxa* (Edens-Meier et al. in preparation). The viscidium of many *Thelymitra* spp. often glistens due to the copious amounts of fluids secreted by the receptive lobes of the stigmas (Edens-Meier et al. submitted). However, the nutritional content of the stigmatic fluid remains unknown, and we were unable to find references to bee pollinators drinking this fluid. Therefore, most *Thelymitra* spp. with large flowers appear to be food mimics (Tremblay et al. 2005). The mitra/clinandrium unit appears to have evolved as an adaptation to position the abdomen of the naïve pollinator so it will contact the viscidium and the stigma lobes.

Most of these large-flowered species appear to be mimicking flowers with radial symmetry, so the orchid's labellum is usually identical in shape and pigmentation pattern to the lateral petals. This looks like a retrograde trend in floral evolution when we consider labellum-column evolution in the vast majority of monandrous orchid lineages (see Chapters 2–6, 8–10). In the case of *Thelymitra*, though, the peloric perianth appears to be selectively advantageous. The relationship of the mitra/clinandrium to the largely rotate perianth is as adaptive as the relationship between the staminodium and the saccate labellum in cross-pollinated species of slipper orchids (Cypripedioideae, see Chapter 10). Why? In the large-flowered *Thelymitra* spp., it seems unlikely that the viscidium would become attached to a bee's abdomen if the same insect landed first on the labellum and then crawled toward the mitra/clinandrium. There appears to be a correlation between ornamentation and exaggeration of the mitra/clinandrium and the shape of the labellum in this genus. Two alpine to subalpine species (*T. cyanea* and *T. venosa*) have labella that fail to match the shape of the lateral petals, and they also have the least-ornamented and least-exaggerated mitras (Burns-Balogh and Bernhardt 1988).

A novel mode of insect pollination in *Thelymitra* spp. was described recently in small-flowered species in New Zealand. It is novel because the interpretation disagrees with Cheeseman (1880). Scanlen (2008) observed thrips visiting the anthers of *T. cyanea*, *T. nervosa*, and *T. pulchella*. His photographic evidence strongly suggests that these insects could have carried appreciable quantities of pollen to the stigma, provided that these *Thelymitra* spp. produced crumbling (friable) pollinia. The thrip may die following the act of pollination, as its body becomes glued to the stigma. More research is needed before the evolutionary significance of this thrip-flower interaction can be interpreted. It may represent an incipient pollinator shift, but it more likely represents a repetitive and predictable consequence of small pollen-consuming insects crawling too close to an adjacent gluey stigma.

Pollination and Hybridization

Edens-Meier et al. (2013) found that reciprocal hand pollinations of greenhouse collections between *T. crinita* and *T. macrophylla* showed that interspecific crosses are compatible. Pollen tubes were found to enter ovules within 6 days. In the Western Australian bush, putative F_1 hybrids between these 2 species have now been located and identified at 2 sites (see Edens-Meier et al. 2013).

Compared with other genera in the family Orchidaceae, with the exception of some Euro-Mediterranean lineages (Dafni and Bernhardt 1989) and some genera in the Cypripediodeae (see Chapter 10), collections and reports of interspecific hybrids seem unusually high in the genus *Thelymitra*. In Australia, a number of "rare species" have been reclassified as recurrent hybrids. These include but are not limited to *T.* × *chasmogama*, *T.* × *irregularis*, *T.* × *macmillanii*, *T.* × *merranae*, and *T.* × *truncata* (see Bates and Weber 1990; Bernhardt 1993; Jones 2006). Some of these hybrids are difficult to recognize because they are found infrequently and, when they are, often grow intermixed with the parent species they most closely resemble.

Furthermore, these hybrids show such a suave intergradation of floral and vegetative characters that they may be mistaken for one or the other parent species. For example, the infrequent hybrid between *T. crinita* and *T. macrophylla* at the Lesmurdie, WA, site was produced by 2 large-flowered and blue parent species. The hybrid offspring showed intermediate characters (see Figure 7.1; Edens-Meier et al. 2013), and the mixed parentage became clear only after consultation with Dr. Andrew Brown and microscopy work on pollen grain configuration comparing parent species with the putative F_1 hybrid (Edens-Meier et al. submitted). In contrast, hybrids between *T. antennifera* and blue-flowered *Thelymitra* spp. (*T. luteocilium* and *T. macrophylla*) are so dissimilar in perianth color and mitra sculptures that they are easy to discriminate from either parent species and from each other (Bates and Weber 1990). Edens-Meier et al. (submitted) found extensive numbers of *T. antennifera* × *T. macrophylla* hybrids at the Tenterden, WA, site intermingled with *T. antennifera*. Some flowers bearing the typical mitra of *T. antennifera* (Figure 7.1) had red blotches on their perianth segments, suggesting possible backcrosses between hybrids and *T. antennifera* (Edens-Meier et al. submitted).

The persistence and demography of hybrid populations differ broadly, based on geography and habitat. In South Australia, *T.* × *chasmogama* (*T. luteocilium* × *T. nuda*) is associated with sites disturbed by grazing, quarrying, and/or roadwork. This hybrid often forms clumps of closely packed flowering stems by vegetative reproduction (Bates and Weber 1990). South Australian records indicate that *T.* × *merranae* (*T. ixiodes* × *T. holmesii*) may form populations in which they outnumber the parent species, *T. ixiodes*.

Recurrent hybrids are not restricted to recombinations of large-flowered *Thelymitra* spp., as even subcleistogamous species open on some days, and foraging bees are polylectic (see above). The evidence suggests that a number of recurrent hybrids are the result of crosses between large- and small-flowered species or even crosses between 2 small-flowered species. This appears to be especially common in New Zealand (Molloy and Dawson 1998; Scanlen and St. George 2010). McAlpine (1978) argued that when *T.* × *truncata* backcrossed with its large-flowered parent, *T. ixiodes*, in the Blue Mountains of New South Wales (Australia), it produced the taxon *T. ixiodes forma merranae*. Unfortunately, this publication offers no morphometric or isoenzyme evidence.

Evolution of Large-Flowered Species

This combination of classical and recent information allows us to expand on the role of adaptations in cross-pollination of large-flowered

Thelymitra spp. It also allows us to compare pollination systems in this lineage with other genera in the family Orchidaceae, and suggests new hypotheses to test. For example, while it is interesting to note that different large-flowered *Thelymitra* spp. may open and close their flowers in response to cloud cover and/or temperature regimes (Edens-Meier et al. 2013), the adaptive significance of this mechanism has never been tested. Large-flowered *Thelymitra* spp. are among the few members of the huge orchid family to do this. We grant that sun orchids stay open during climatological periods, and during spring–early summer hours, when Australian bees are most likely to be active (Dafni and Calder 1987; Bernhardt 1989a, 1995b, 1996).

Flowers that close during rain might protect the exposed stigmatic fluids from dilution, rinsing grains from the stigma before they can germinate, and protect exposed pollinia from premature hydration. However, why should these flowers almost always close from mid- to late afternoon until the following morning? We observed this rather inflexible cycle in *T. antennifera*, *T. crinita*, and *T. macrophylla* in Western Australia, even when a warm, sunny day ended in a dry, warm afternoon and evening. This flowering behavior can't be dismissed as just another nyctinastic movement, as in the leaves of so many leguminous plants. Could closing perianth segments during late hours offer the selective advantage of protecting pollen and stigmatic lobes from various florivores? We suggest testing the hypothesis of whether pollen-robbing flies (Syrphidae), thrips, and other unidentified "pollinia pests" remain active under cloudy, cool, and crepuscular conditions. Another hypothesis is that closed flowers are "invisible" flowers. If a food-mimic flower opens and closes every day, wouldn't it be more difficult for a foraging bee to discriminate this "new" flower from existing flowers that offer edible rewards (Edens-Meier et al. in preparation)?

In contrast, there is precedence for orchid flowers closing and not reopening after the act of successful pollination. The literature on pollen-pistil interactions in some epidendroid orchids shows that removal of the pollinarium and/or deposition of pollen on the stigma of a flower interrupts and shortens the floral life span (Huda and Wilcock 2012). This is generally interpreted as selectively advantageous, as the plant saves resources used to produce nectar, pigments, and odors, and water used to keep fleshy floral parts turgid. Male fitness should increase when a flower shuts following pollinarium removal, while female fitness should increase after the flower shuts following pollinia deposition on its stigma. In the cases of *T. crinita* and *T. macrophylla*, experiments by Edens-Meier et al. (2013) showed that while removing the pollinaria of 2 *Thelymitra* spp. did not significantly shorten individual floral life spans, the act of depositing pollen on their respective stigmas caused the perianth segments to close and wither within less than 2 days. This response was identical whether a flower received its own pollen or that of another member of the same species (see above). Much the same response was obtained by controlled experiments performed on flowers of *Mystacidium venosum* (Luyt and Johnson 2001) and *Chloraea alpina* (Clayton and Aizen 1996), so a similar mechanism occurs in other orchid species whether they belong to the same or different subfamilies. We argue that in cross-pollinated *Thelymitra* spp., shutting down the flower after pollinarium removal would be most maladaptive. As Edens-Meier et al. (2013) found, and in previous studies by other authors, the conversion rate of flowers into fruit in large-flowered *Thelymitra* spp. is extremely low (Dafni and Calder 1987; Cropper and Calder 1990; Sydes and Calder 1993). Shutting these bisexual flowers after pollinaria removal would continue to lower their female fitness.

This naturally takes us to the paradox of the act of insect pollination in large-flowered *Thelymitra* spp. While the field studies of Edens-Meier et al. (2013) in Western Australia expanded the range of pollinaria vectors to the genera *Homalictus* and *Amegilla*, the employment of such bees as pollinators doesn't increase fruit set in either sun orchid species compared with previous Australian studies (Table 7.1). What do these bee genera in 3 different families (Apidae, Colletidae, and Halictidae) represent aside from infrequent visitations to large-flowered *Thelymitra* species? All appear to be genera dominated by species that have polylectic females known to collect pollen from a wide variety of unrelated angiosperm species. In fact, some collect pollen from nectarless flowers such as *Acacia* (Bernhardt 1989a), *Dianella* (Bernhardt 1995b), and *Hibbertia* (Dafni and Calder 1987; Bernhardt 1996). When the pollen loads of these bees are analyzed and identified, it is common to find the grains of 3 to 5 plant species derived from 3 to 5 plant families in the same load (Dafni and Calder 1987; Bernhardt 1989a, 1995b, 1996; Edens-Meier et al. submitted).

We do not argue that different bee-pollinated *Thelymitra* spp. may mimic guilds of flowers with similar colors and/or odors. We do not argue that some *Thelymitra* spp. may one day prove to be specific (Batesian) mimics of certain flowering species that offer pollen and/or nectar. Some orchid species in the related genus *Diuris* appear to be specific mimics of the flowers of some papilionoid legumes (Dafni and Bernhardt 1989; see Chapter 5). However, pollen load analyses of bees in this chapter and in the work of other authors (see Table 7.1) indicate that the major pollinators of *Thelymitra* spp. also visit flowers with color patterns that do not match the *Thelymitra* sp. in bloom. For example, Edens-Meier et al. (2013) found that bees collected on blue-flowered *T. macrophylla* at Lesmurdie, WA, were found to carry mixed loads of grains derived from members of the family Myrtaceae (white to pink) and some papilionoid legumes (yellow, orange, deep purple). At that site, none of the Myrtaceae in bloom and only one of the species of papilionoid legumes produced blue to purple flowers.

We suspect that while large-flowered *Thelymitra* spp. mimic guilds of co-blooming plants, their modes of floral presentation may also compete for the most limited resource of all, naïve pollinators. Remember, pollinators appear to receive no benefits from visiting a large-flowered *Thelymitra* sp., but flowering periods and distributions of these orchids overlap in Western Australia (see Brown et al. 2008). Therefore, we propose that variation in the odor, pigmentation patterns, and mitra sculptures of different *Thelymitra* spp. stimulates interspecific competition for the limited pollinator resource.

Field observations (Edens-Meier et al. 2013) indicate that it is not merely a matter of competing for the small and fluctuating reserve of naïve female polylectic bees. Darwin (1862, 1877) was the first to document insect-mediated removal of pollinaria in orchid flowers around his home, and did so over several seasons (see Chapters 1 and 2). Edens-Meier et al. (2013) used this technique in co-blooming, sympatric populations of *T. crinita* and *T. macrophylla* (Lesmurdie, WA). In both species, rates of pollinaria removal were low, but rates of pollinaria removal by insects were always many times higher than rates of insect-mediated pollinia deposition on receptive stigmas. Consequently, natural selection favors populations of mimetic flowers that repeatedly attract insects visiting more than one flower of the same species. During the single season of observation, pollen deposition on stigmatic lobes of *T. crinita* significantly surpassed pollen deposition on stigmas of *T. macrophylla*. It appears, though, that most bees visiting either species visited once and never returned, or pollination rates would have been far higher.

Morse (1981) studied the foraging activity of 2 bumblebee species (*Bombus*) that visited nectar- and pollinaria-producing flowers of *Asclepias syriaca*. In both bee species, foraging speeds declined as pollinaria numbers increased on their limbs and mouthparts. If the deposition of pollinaria produced by a nectar-rich flower slows down large bumblebees, what does the deposition of pollinaria, from a flower offering no rewards, onto the abdomens of far smaller bees do to both foraging speeds and floral constancy? Consider the potential drag produced by one or more pollinaria on the bodies of bees less than 9 mm in length. Most large-flowered *Thelymitra* spp. studied to date depend on small- to medium-size bees. The current exception was *T. crinita*, visited infrequently by the large (bumblebee size) *Amegilla chlorocyanea*. This insect did not appear to notice extra weight or drag and visited several inflorescences of *T. crinita*. Small- to medium-size halictids appear to be the most common pollinaria vectors of large-flowered *Thelymitra* spp. (Table 7.1), but large bees in the Apidae may be the preferred "bonus pollinators" (Edens-Meier et al. 2013).

We also wonder how much secondary pollination occurs when syrphid flies visit open, large flowers of *Thelymitra* spp., as some of these insects do carry some grains of the host flower even though they do not always appear to remove the entire pollinarium (Dafni and Calder 1987; Edens-Meier et al. submitted). This might occur if the fly probes the pollinia when it is fixed to the viscidium, and/or when a bee leaves fragments of pollinia on a wet, receptive stigma. After all, the flower does not close immediately after the stigma is pollinated.

Evolution of Small-Flowered Species

If we accept the work of modern orchid taxonomists, then 45 to >50% of all *Thelymitra* spp. are facultative-obligate self-pollinators (Jones and Bower 2001; Jeanes 2004, 2006, 2009; Jones 2006; Brown et al. 2008) in which the reduced, poorly developed mitra may be little more than a vestigial organ. Would Darwin have been amazed by the overwhelming numbers of self-pollinating *Thelymitra* spp. if he had lived late into the 20th and early 21st centuries (Jeanes 2004, 2006, 2009; Bates 2010)? Probably not, as self-pollinating species are no longer considered rare, not even in the orchid flora of Europe. While Darwin (1877) explained the biomechanics of insect pollination in flowers of *Epipactis*, this Eurasian genus also shows a broad gradation of cross- and self-pollinating taxa (Squirrell et al. 2002). Mechanical self-pollination is probably common in other orchid lineages in which one can see an obvious difference in flower size and floral ornamentation/presentation.

If we accept Stebbins's rule that most self-pollinating species derive ultimately from cross-pollinated ancestors (Stebbins 1974; and see Chapter 5), then the major trend in *Thelymitra* is toward the evolution of species with small, usually self-pollinating flowers. In fact, the taxonomic evidence in the species-rich *Thelymitra pauciflora* complex suggests strongly that while all self-pollinating species must have a cross-pollinated ancestor, (ultimately) many self-pollinating species must also evolve from self-pollinating ancestors (Jeanes 2004; Bates 2010). Why is this so? Should we blame it, as usual, on a hot and drying Australian continent in which natural selection favors the survival of weedy species with high levels of homozygosity that best adapts them for stable but harsh environments? Probably not, as this fails to explain the secondary center of diversity of self-pollinating species in cool, moist, montane New Zealand (Molloy and Dawson 1998). We wonder instead whether small, self-pollinating species dominate this genus because food mimesis and/or guild mimesis produces so few ripe fruit in the remaining large-flowered species. As natural selection favors the reproductively fit, self-pollinating sun orchids make fruit annually, while their cross-pollinated sibling species may fail to set fruit every year (see Darwin 1877).

There is no obvious synchronous coevolution between the large-flowered *Thelymitra* spp. and their bees (see above, and see Chapters 9 and 10). In fact, natural selection should favor female bees that do not fall for the large, pretty, perfumed, but empty flowers more than once (if at all). As cross-pollination rates drop over time, self-pollination rates should rise or extinction rates must increase. Is it possible that some small-flowered, self-pollinating species lost their large-flowered sibling species long before the arrival of Darwin or Fitzgerald? The small-flowered, tardily opening *T. circumsepta* (sensu Bernhardt 1993) may produce a stalked, brushy appendage on the basal rim of its mitra (opposite the stigma). This appendage varies in size from plant to plant. Some taxonomists preferred to classify as *T. retecta* plants with flowers that fail to produce this appendage (Bernhardt 1993). In either case, though, there is no large-flowered "mirror species" described to date that has a mighty, visible brush attached to the basal rim of the mitra.

In fact, we should also look for better ways to determine the extent of self-pollination in small-flowered species. Historically, flower size, mitra development, perianth opening versus closing cycles, and pollinia composition (entire vs. friable) were preferred characters used to determine whether a *Thelymitra* species was facultatively self-pollinating or subclesitogamous (see above). Lehnebach et al. (2005) offered another character that is time-consuming but probably very useful when considering reproductive trends within a lineage. They compared breeding systems in 4 species (representing 3 genera) native to New Zealand. One of the characters they used was the number of pollen grains in the whole pollinarium (P) versus the number of ovules in the ovary (O). The P/O ratio for *T. longifolia* was 24:1. This is regarded as a relatively low ratio of pollen grains to ovules in angiosperms and is usually indicative of a self-pollinating system (Cruden 1977). Hand pollinations of the same flowers showed that >70% set fruit by mechanical autogamy, while >83% set fruit when cross-pollinated by hand. As we know that this species is visited by colletid bees on hot, sunny days (Lehnebach et al. 2005), it is reasonable to classify some New Zealand populations as facultative self-pollinators in the absence of insect pollinators.

And About Those Hybrids . . .

Why are interspecific hybrids so common in this genus, especially in Western Australia (Jones 2006; Brown et al. 2008)? If these flowers (both large and small) don't recognize and reject their own pollen, it's unlikely they can recognize and reject the grains of co-blooming species following a hot, sunny day, when bees and/or syrphid flies attempt to forage on more than one *Thelymitra* species. That could easily explain recurrent hybrids between large- and small-flowered species (see above). The role of introgression in speciation within this genus was first suggested by Burns-Balogh and Bernhardt (1988). These authors suggested that a history of introgressive speciation produced the broad and variable mitra forms and sculptures we still use to segregate and identify species to this day. This interpretation has not enjoyed universal acceptance. After all, the short paper by McAlpine (1978) lacked convincing data. Far more convincing evidence comes from New Zealand. A history of hybridization between both large- and small-flowered *Thelymitra* spp. appears to have produced 4 species currently segregated as *T. decora, T. hatchii, T. pulchella,* and *T. tholiformis.*

The species Australians describe as *T. carnea* may not be the same species described in the flora of New Zealand. In New Zealand, *T. carnea* may be a stabilized hybrid between *T. flexuosa* and *T. pauciflora.* How did these "species" evolve? Molloy and Dawson (1998) provided evidence that these endemic species in New Zealand were the result of interspecific hybridization followed by chromosome evolution known as amphidiploidy. In each case, the genealogy of the new species was derived ultimately from hybridization between a parent species with less than 30 chromosomes ($2n$) and a parent with either a similar number of chromosomes (e.g. *T. longifolia;* $2n = 26$) or a greater number of chromosomes. The "new" species always had more chromosomes than either putative parent ($2n = 54$ to 66). This new species self-perpetuates by facultative self-pollination. Chromosome numbers should be counted and compared in temperate to Mediterranean Australia, where the majority of *Thelymitra* spp. proliferates.

Some would argue that this form of reticulate evolution cannot contribute to the evolution of new species in *Thelymitra*, because the first generation of hybrids is sterile or shows a high degree of sterility. Yes, but sterility in these hybrids has been measured in terms of male sterility (McCrae and Molloy 1998; Edens-Meier et al. submitted). We simply do not know yet whether the stigmas of these hybrids receive and process pollinia and produce fertilized ovaries. In particular, some of the putative hybrids between *T. antennifera* and *T. macrophylla* (Tenterden, WA) produced normal pollinaria, and a few appeared to have been removed by insects (Edens-Meier et al. 2013). Ultimately, only modern genetic analyses will tell us whether the specimens of *T. antennifera* with bright-red splotches on their perianth segments represent backcrosses between *T. antennifera* and its putative hybrid with *T. macrophylla* (Edens-Meier et al. 2013).

CONCLUSION

Let's consider the lineage we know as the genus *Thelymitra* and Darwin's argument regarding the selective advantage of reproduction in bisexual flowers. Even though Darwin lacked genetic evidence, he was quite correct in stating that "she [Nature] abhors perpetual self fertilisation" (Darwin 1877; p. 293). He used descriptions, as provided by Fitzgerald, of self-fertilization in 2 *Thelymitra* spp. (*T. carnea* and *T. longifolia*; Darwin 1877; p. 127) as atypical examples of self-fertilizing taxa in a largely cross-pollinated family. Darwin (1877; pp. 291–292) states, "*Thelymitra* offers indeed the only instance known to me of two species within the same genus which regularly fertilise themselves." He thought that 2 self-pollinating species in the same orchid genus were unique in his day. Moreover, he believed that since self-fertilizing flowers had all the necessary parts for cross-fertilization, the former were "descended from species or varieties which were formerly fertilized by insect-aid" (Darwin 1877; p. 291). He ponders further:

> Whether any species which is now never cross-fertilised will be able to resist the evil effects of long-continued self-fertilisation, so as to survive for as long an average period as the other species of the same genera which are habitually cross-fertilised, cannot of course be told. (Darwin 1877; p. 292)

This suggests future studies of paired species within the same genus in which self-pollinating orchid populations are compared with insect-pollinated ones, examining both for evidence of in-breeding depression. Darwin (1877; p. 292) also states:

> It is indeed possible that these self-fertile species may revert in the course of time to what was undoubtedly their pristine condition, and in this case their various adaptations for cross-fertilisation would be again brought into action.

It's apparent that Darwin did not believe that the process of self-pollination was healthy for populations over time. Sure, self-pollination was better than extinction, but was self-pollination a short-term "fix" by Nature? Was this "fix" supposed to be corrected in the future through gradual modifications leading back to a healthy cross-pollination system? According to results in Chapter 5, this has not happened within the tribe Diurideae. Darwin used the genus *Thelymitra*, as well as other orchid genera, in his argument for the process of natural selection.

Had Darwin been able to communicate with botanists working in the late 20th and early 21st centuries, his anticipated initial surprise might have changed to delight. In fact, cross-pollination in large-flowered *Thelymitra* spp. remains a slow but ongoing field of study illuminating interactions between nonrewarding flowers and their insect pollen vectors. These insects show varying capacities to discern, learn, and reject. Surely, this is an excellent and elegant model system for Darwin's own theories on natural selection and the struggle for existence (Darwin 1859). Based on evidence from New Zealand (Molloy and Dawson 1998; Lehnebach et al. 2005), even self-pollinating species avoid perpetual self-fertilization. This permits some degree of cross-pollination and the evolution of new species via hybridization.

ACKNOWLEDGMENTS

We wish to acknowledge that funding for this research was provided by the National Geographic Society (#8530–08). With deepest gratitude and appreciation, we sincerely thank Dr. Kingsley Dixon for being such a gracious host and welcoming us into the Kings Park and Botanic Garden research facility. We extend special thanks to Andrew Brown for helping us locate field sites and for making all necessary botanical identifications. We are also grateful to Keith Smith for guiding us to field sites in Albany, WA. We express our gratitude to Dr. Terry Houston at the Western Australian Museum for identifying our flies. We acknowledge and sincerely thank Dr. Eric Bunn and Dr. Siegy Krauss for providing us with necessary supplies. We acknowledge Dr. Miles and Dr. Ryan for their assistance in helping us with GPS locations and the process of preparing government reports. Bob Dixon is thanked for providing timely assistance in locating and delivering supplies and equipment. We thank Ms. Keran Keys for all that she did to provide what we needed for working efficiently within the lab. Finally, we are grateful to Larry Meier for providing assistance and encouragement.

PART IV

Darwin and His Colleagues:
Orchid Evolution in the Tropics

Pollination Biology and Evolutionary History of Angraecoid Orchids: From Darwin to the Present Day

Claire Micheneau, Jacques Fournel, and Thierry Pailler

INTRODUCTION

Pollination systems provide unique opportunities to establish clear connections between the ecological bases of phenotypic differentiation and the establishment of prezygotic barriers, reproductive isolation, and speciation (e.g. Bradshaw and Schemske 2003; Rundle and Nosil 2005; Rieseberg and Willis 2007; Whittall and Hodges 2007; Hodges and Derieg 2009; Ramírez et al. 2011b). No other plant family shows as wide an assortment of mutualistic interactions with insect pollinators as the Orchidaceae. With the 1862 publication of *On the Various Contrivances by which British and Foreign Orchids Are Fertilised by Insects, and on the Good Effects of Intercrossing*, Charles Darwin first demonstrated that orchids provide strong evidence for adaptations that promote cross-pollination. It is now acknowledged that within the orchid family, pollinators have probably played a crucial role in the unprecedented diversification of the extant floral forms (e.g. van der Pijl and Dodson 1966; Tremblay 1992; Johnson et al. 1998; Johnson and Steiner 2000; Schiestl et al. 2003; Tremblay et al. 2005; Ramírez et al. 2011a,b), resulting in more than 25 000 species (Chase et al. 2003; Govaerts et al. 2012).

Darwin understood the major role of interspecific interactions in the diversification of species, and used orchid pollination systems to further support his theory of evolution by natural selection, which had been much criticized. Of particular importance were his observations of the

Madagascar star orchid, *Angraecum sesquipedale*. Darwin introduced the concept of coevolution in which adaptive traits by reciprocal selection act simultaneously on 2 species or populations. This predicted the existence of specific long-tongued moths in Madagascar, and explained the development of long corollas in flowering plants (Darwin 1862; see also Wallace 1867 and paragraph 4 in the section "Pollination Biology: The Famous Prediction of Darwin 1862" below; see also Chapter 1). Darwin theorized that the evolution of both long nectar tubes and long moth tongues was the result of reciprocal and gradual increases, "a race in gaining length" (Darwin 1862; p. 202). It was a response to mutual benefit driven by natural selection, defined as the gain in reproduction efficiency for the orchid, and gain in food foraging for the sphinx moth (Darwin 1862, 1877). There is little evidence of coevolution driving macroevolutionary transitions in orchid-pollinator interactions, especially because orchid pollination systems are mostly asymmetric. Most are specialized flowers dependent on generalist pollinators, which means unidirectional selection or only diffuse coevolution (Wasserthal 1997; see also Micheneau et al. 2009; Ramírez et al. 2011a,b). Nonetheless, Darwin's innovative idea of reciprocal evolution driven by natural selection is one of his great contributions to evolutionary biology (e.g. Nilsson 1988; Anderson and Johnson 2008; Pauw et al. 2009).

ANGRAECOID ORCHID SYSTEMATICS: MORPHOLOGY VERSUS CYTOLOGY AND MOLECULAR DATA

Since Darwin's observations on the genus *Angraecum* over 150 years ago, the angraecoid orchids (Epidendroideae, Vandeae) have never ceased to fascinate naturalists and evolutionary biologists. These orchids constitute a large group of monopodial epiphytic species, including more than 760 taxa (Govaerts et al. 2012). The angraecoid orchids were formerly placed in 2 subtribes (see Schlechter 1926; Dressler 1981, 1933b; Chase et al. 2003; Carlsward et al. 2006). Species placed in Angraecinae (approximately 15 genera, 410 species, mostly from Madagascar; Table 8.1) and Aerangidinae (approximately 32 genera, 350 species, mostly from Africa; Table 8.1) are very similar in morphology. Therefore, it's not surprising that recent insights from molecular data provided evidence to support recognition of a single subtribe, Angraecinae. This now includes all angraecoid orchids (i.e. Angraecinae s.l.; Carlsward et al. 2006). The majority of these orchids are found in Africa, Madagascar, and nearby islands, with 2 genera in tropical America (*Campylocentrum* and *Dendrophylax*) and

Table. 8.1 Angraecoid genera according to their distribution

Subtribes	Genus[a] [# of species]	Distribution[b]		x[c]	DNA[d]
Primary Africa					
Aerangidinae	Tridactyle [47]	Af [47]		4	I[e]
Aerangidinae	Rhipidoglossum [35]	Af [35]			I[e]
Aerangidinae	Aerangis [53]	Af [32]	Ma [21]	4	I[e]
Aerangidinae	Diaphananthe [22]	Af [22]		4	I[e]
Aerangidinae	Angraecopsis [22]	Af [21*]	Ma [2*]	4	I[e,f]
Aerangidinae	Microcoelia [30]	Af [21*]	Ma [10*]	4	I[e]
Aerangidinae	Ancistrorhynchus [18]	Af [18]		4	I[e]
Aerangidinae	Cyrtorchis [18]	Af [18]		4	I[e]
Aerangidinae	Chamaeangis [11]	Af [11]		4	I[e,f]
Aerangidinae	Mystacidium [10]	Af [10]		4	I[e,f]
Aerangidinae	Solenangis [8]	Af [8*]	Ma [1*]	4	I[e,f]
Aerangidinae	Rangaeris [7]	Af [7]		4	I[e]
Aerangidinae	Bolusiella [6]	Af [6*]	Ma [1*]	4	I[e]
Aerangidinae	Margelliantha [6]	Af [6]			
Aerangidinae	Ypsilopus [5]	Af [5]			I[e]
Aerangidinae	Cribbia [4]	Af [4]			I[e]
Aerangidinae	Dinklageella [4]	Af [4]			
Aerangidinae	Eggelingia [3]	Af [3]			
Aerangidinae	Eurychone [2]	Af [2]		4	I[e]
Aerangidinae	Nephrangis [2]	Af [2]			
Aerangidinae	Summerhayesia [2]	Af [2]		4	
Aerangidinae	Calyptrochilum [2]	Af [2]		3	I[e]
Aerangidinae	Cardiochilos [1]	Af [1]			
Aerangidinae	Chauliodon [1]	Af [1]			
Aerangidinae	Distylodon [1]	Af [1]			
Aerangidinae	Listrostachys [1]	Af [1]		4	I[e]
Aerangidinae	Ossiculum [1]	Af [1]		3	
Aerangidinae	Plectrelminthus [1]	Af [1]		4	I[e]
Aerangidinae	Podangis [1]	Af [1]		4	I[e]
Aerangidinae	Rhaesteria [1]	Af [1]			
Aerangidinae	Sphyrarhynchus [1]	Af [1]		4	I[e]
Aerangidinae	Taeniorrhiza [1]	Af [1]			
Aerangidinae	Triceratorhynchus [1]	Af [1]			
Total		Af [297]	Ma [35]		
Primary Madagascar					
Angraecinae	Angraecum [221]	Af [52**]	Ma [171**]	2, 1	I, II[e,f]

continued

Table. 8.1 *continued*

Subtribes	Genus[a] [# of species]	Distribution[b]		x^c	DNA[d]
Angraecinae	*Jumellea* [61]	Af [2]	Ma [59]	1	II[e,f]
Angraecinae	*Aeranthes* [43]	Af [2]	Ma [41]	1	II[e,f]
Aerangidinae	*Microterangis* [7]		Ma [7]		I[e]
Angraecinae	*Neobathiea* [5]		Ma [5]		I[e,f]
Angraecinae	*Oeonia* [5]		Ma [5]		I[e,f]
Angraecinae	*Cryptopus* [4]		Ma [4]	1	I[e,f]
Angraecinae	*Lemurella* [4]		Ma [4]		I[e]
Angraecinae	*Sobennikoffia* [4]		Ma [4]		II[e]
Aerangidinae	*Beclardia* [2]		Ma [2]		I[e,f]
Angraecinae	*Oeoniella* [2]		Ma [2]	1	II[e]
Angraecinae	*Ambrella* [1]		Ma [1]		
Angraecinae	*Lemurorchis* [1]		Ma [1]		II[e]
Total		Af [56]	Ma [306]		
America					
Angraecinae	*Campylocentrum* [64]		Am [64]		I[e,f]
Angraecinae	*Dendrophylax* [14]		Am [14]		I[e,f]
Total [766]		Af [353]	Ma [341]	Am [78]	

Notes: Cytotaxonomic group is based on the shape of the rostellum and the monoploid number of chromosomes (i.e. the number of unique chromosomes in a single complete set), and molecular phylogeny (DNA).

The number of species is given in brackets (not including subspecies and varieties).

Asterisks show the number of species that have a widespread distribution across mainland Africa and western Indian Ocean, with * corresponding to 1 species, and ** to 2 species,
[a] Chase et al. 2003 (subtribes Aerangidinae and Angraecinae).
[b] Govaerts et al. 2012; Af: refers to species occurring in Africa and the Gulf of Guinea Islands (São Tomé and Príncipe); Ma: refers to species occurring in Madagascar and surrounding islands (Comoros and Mascarenes).
[c] x refers to the groups defined by Arends and van der Laan (1983) on the basis of the shape of the rostellum and the basic number of chromosomes: Group 1: rostellum nonelongated, x = 19; Group 2: rostellum nonelongated, x = 21, 23, 24, 25 (i.e. *Angraecum* from Africa); Group 3: rostellum elongated, x = 17, 18; Group 4: rostellum elongated, x = 21 to 27.
[d] I refers to clade I; and II refers to clade II (see also Figure 8.1).
[e] Carlsward et al. 2006.
[f] Micheneau et al. 2008a.

1 additional taxa extending up to Sri Lanka. Flowers vary in shape, color, and size, but most of them are white, and characterized by the presence of an elongated nectar spur. Pollinaria are characterized by 2 "hard," pale-yellow pollinia attached to a sticky viscidium by distinct caudicles (see Dressler 1993b and Figure 8.1A).

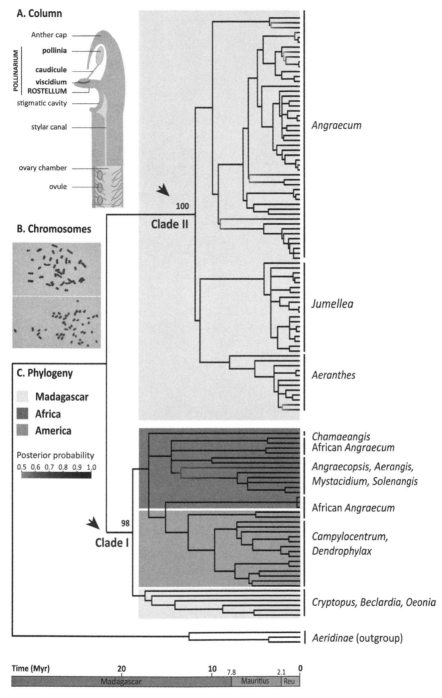

A. Column

POLLINARIUM
- Anther cap
- pollinia
- caudicule
- viscidium
- ROSTELLUM
- stigmatic cavity
- stylar canal
- ovary chamber
- ovule

B. Chromosomes

C. Phylogeny

Madagascar
Africa
America

Posterior probability
0.5 0.6 0.7 0.8 0.9 1.0

100
Clade II

98
Clade I

Angraecum

Jumellea

Aeranthes

Chamaeangis
African *Angraecum*

*Angraecopsis, Aerangis,
Mystacidium, Solenangis*

African *Angraecum*

*Campylocentrum,
Dendrophylax*

Cryptopus, Beclardia, Oeonia

Aeridinae (outgroup)

Time (Myr)   20   10   7.8   2.1   0
Madagascar   Mauritius   Reu

Fig. 8.1 Phylogeny of the angraecoid orchids, with illustrations of characters used in historical taxonomy. (*A*) Illustration of the angraecoid orchid column (modified from Roguenant et al. 2005). (*B*) Illustration of chromosome shape (modified from Arends and van der Laan 1983). (*C*) Molecular phylogeny of the angraecoid orchids, showing species repartition (modified from Micheneau et al. 2010). Used by permission of Springer Science + Business Media.

Our inventory of the angraecoid orchids is far from exhaustive, especially regarding Central Africa, where key biodiversity hotspots such as the Congo Basin are still largely unexplored. Nevertheless, the angraecoid orchids from mainland Africa are characterized by a rich diversity of genera but with a relative few number of species per genus (36 genera; less than 50 species per genus; 11 monospecific genera; Table 8.1). By contrast, the Madagascar group, also including the surrounding archipelagos (the Comoros, Seychelles, and Mascarene Islands) comprises a lower number of genera, but with a higher number of species per genus (18 genera; approximately 170 species for the genus *Angraecum*; 3 monospecific genera; Table 8.1).

Despite the importance of the group, species relationships within the angraecoids are not clearly understood. The main difficulties arise because the majority of species have been defined using relatively few floral characters. Therefore, species diagnoses are rudimentary. Most of the floral features used to distinguish taxa in the past (e.g. column shape, spur length, lip shape) are thought to be highly homoplasic due to convergent evolution in response to moth pollination.

Similarly, the distinction between the old subtribes Angraecinae and Aerangidinae relied on only 2 characters: rostellum shape (Schlechter 1926) and chromosome number (hereafter x; Summerhayes 1966). The Aerangidinae were described as having an elongated rostellum (entire or bifid) and $x = 25$. The Angraecinae, in contrast, were identified by a rostellum that was variable in shape and $x = 19$. However, further studies investigating cytotaxonomy of the group have shown that neither the shape of the rostellum nor the number of chromosomes could constitute relevant synapomorphies for morphological distinction between the 2 subtribes (Arends 1986; Arends and van der Laan 1983). Moreover, *Angraecum* encompasses species with $x = 19, 21, 23, 24,$ or 25 (Table 8.1). Furthermore, chromosome number is not an easy character to assess or work with, because these tropical orchids have tiny chromosomes (Figure 8.1B). Despite the relevant aspects of cytological data for the systematics of the group, and the long history of plant chromosome investigation, chromosome number is still unknown for the majority of angraecoid species.

Corroborating cytological investigations and what is now apparent in taxonomy and biogeography, molecular studies have highlighted the artificial nature of current taxonomic concepts (Carlsward et al. 2003, 2006; Micheneau et al. 2008a; Table 8.1, Figure 8.1C; see Chapter 5). Based on molecular data, the angraecoid orchids consist of 2 well-supported clades (i.e. Clade I and Clade II in Figure 8.1C), but these clades do not follow the originally proposed subtribes. The first clade includes (1) the American lineage (subtribe Angraecinae); (2) genera that occur predominately in Africa, such as *Aerangis, Angraecopsis, Mystacidium, Chamaeangis*, or *Solenangis* (subtribe Aerangidinae); (3) the *Angraecum* species (subtribe Angraecinae) that are found exclusively in Africa (e.g. *A. moendense, A. infundibulare, A. birrimense, A. eichlerianum, A. doratophyllum, A. subulatum, A. distichum, A. aporoides*); and (4) a second small group of species restricted to Madagascar and characterized by flowers with a lobed lip, such as *Cryptopus* and *Oeonia* (subtribe Angraecinae) and *Beclardia* (subtribe Aerangidinae), as well as species for which the taxonomy has to be revised (i.e. new species or new genera; Figure 8.1C). The second clade comprises taxa predominately occurring in Madagascar, such as the genera *Jumellea* and *Aeranthes* (subtribe Angraecinae), as well as the bulk of *Angraecum* species, which are mostly exclusive to Madagascar (Figure 8.1C).

With a clear-cut delineation of species occurring in Africa, and those occurring in Madagascar and nearby islands (Table 8.1, Figure 8.1C), the genus *Angraecum* does not form a monophyletic entity. Except for evidence from molecular data, no morphological characters have been recognized clearly to support the split of the genus *Angraecum*. Some species, which belong to the same section (being morphologically very similar), could be found in either the first or the second clade (e.g. sects. *Arachnangraecum, Angraecoides, Pectinaria*). This situation makes the interpretation of evolutionary patterns within the angraecoids particularly difficult, and makes the taxonomic framework challenging, especially for those sections that have species occurring in both mainland Africa and on islands in the western Indian Ocean.

Fortunately, the angraecoid orchids are currently the subject of intensive research focusing on taxonomy and molecular phylogeny. The anticipated phylogenetic insights should help clarify taxonomic concepts, as well as our understanding of species relationships within this charismatic group of orchids.

MASCARENE ANGRAECOIDS: ANGRAECUM SECTION *HADRANGIS*

The angraecoid flora of such Mascarene islands as Reunion, Mauritius, and Rodrigues are all situated in the western Indian Ocean at 800, 1000, and 1500 km from the eastern coast of Madagascar, respectively. These orchids show a strong affinity with species from Madagascar, including morphological and molecular evidence. This suggests that the young volcanic islands of the Mascarene Archipelago (approximately 7.8 mya

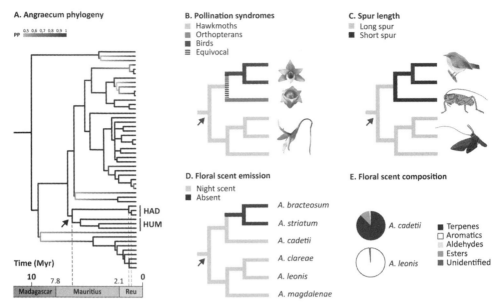

A. Angraecum phylogeny

PP 0.5 0.6 0.7 0.8 0.9 1

HAD
HUM

Time (Myr)
10   7.8   2.1   0
Madagascar | Mauritius | Reu

B. Pollination syndromes
- Hawkmoths
- Orthopterans
- Birds
≡ Equivocal

C. Spur length
- Long spur
- Short spur

D. Floral scent emission
- Night scent
- Absent

A. bracteosum
A. striatum
A. cadetii
A. clareae
A. leonis
A. magdalenae

E. Floral scent composition

A. cadetii
A. leonis

- Terpenes
- Aromatics
- Aldehydes
- Esters
- Unidentified

Fig. 8.2 Phylogenetic position of *Angraecum* section *Hadrangis* and ancestral reconstruction of pollination syndromes and pollinator-linked floral traits. (*A*) Age estimates of the Madagascan *Angraecum* clade (modified from Micheneau et al. 2010). (*B–D*) Ancestral reconstructions using parsimony of pollination syndromes (*B*), spur length (*C*), and floral scent (*D*) in *Angraecum* sect. *Humblotiangraecum*, HUM (Madagascar, Comoros), and sect. *Hadrangis*, HAD (Mascarenes, *arrow*). (*E*) Nocturnal scent composition of *A. cadetii* (orthopteran-pollination) and *A. leonis* (sphingophily) according to chemical classes. No compounds were detected for *A. bracteosum* and *A. striatum* (ornithophily), as modified from Micheneau et al. 2010.

for Mauritius and 2.1 mya for Reunion) were colonized primarily from Madagascar (88 mya since its separation from India). The highest species richness is found on Reunion, the youngest and largest island of the Mascarenes (2500 km²), and the closest to Madagascar. Reunion is mountainous (3070 m), and harbors a rich diversity of habitats (Strasberg et al. 2005) that are more likely to offer favorable conditions for species diversification.

Among the rare cases of intra-archipelago adaptive radiation, sect. *Hadrangis* of genus *Angraecum* is of particular interest. Although the closest relatives in Madagascar have long-spurred flowers emitting a strong scent at nightfall (moth-pollination syndrome), the 3 species of sect. *Hadrangis* (*A. striatum* and *A. bracteosum*, strictly from Reunion, and *A. cadetii* from both Reunion and Mauritius) have medium-size white flowers with short, wide spurs (Figure 8.2). These latter features are uncommon within *Angraecum*. The section *Hadrangis* was revised to include only these 3 species (Bosser 1987), making it endemic to the Mascarenes. Today, the 3 species are often found within the same habitat and in sympatry, sometimes growing on the same host tree (Claire Micheneau, personal observations). However, floral forms are clearly separated and species boundaries are well defined (Figure 8.2), suggesting hybridization events are uncommon in the wild. This raises the interesting questions of how these species diversified on the Mascarenes and how their genetic integrity is maintained, especially in sympatric populations.

POLLINATION BIOLOGY: THE FAMOUS PREDICTION OF DARWIN (1862)

In Africa and Madagascar, the long-spurred angraecoid orchids have developed high levels of specialization with their pollinators, and these orchids have received much attention from evolutionary biologists (e.g. Darwin 1862, 1877; Wallace 1867, 1871; Nilsson et al. 1985, 1987; Nilsson 1988, 1992; Nilsson and Rabakonandrianina 1988; Wasserthal 1997, 1998; Martins and Johnson 2007). However, the first person who fully understood the evolutionary interest of long-spurred *Angraecum* species was Darwin, and he did so with great enthusiasm: "I must stop. I have just received such a box full from Mr Bateman with the astounding *Angræcum sesquipedalia* with a nectary a foot long—Good Heavens what insect can suck it" (Darwin, letter to Hooker, 25 January 1862), and "Bateman has just sent me a lot of orchids with the *Angræcum sesquipedale*: do you know its marvellous nectary 11½ inches long, with nectar only at the extremity. What a proboscis the moth that sucks it, must have! It is a very pretty case" (Darwin, letter to Hooker, 30 January 1862).

Upon observing the flower morphology of the long spur of *A. sesquipedale*, and after having tried himself to "fertilize" the flower, Darwin predicted that the only pollinator of this orchid would be a moth with a "wonderfully long proboscis" (Darwin 1862; p. 201). More than 20 years after his death, the predicted sphingid, *Xanthopan morganii* var. *praedicta* with a proboscis as long as Darwin had imagined, was finally described from Madagascar (Rothschild and Jordan 1903). Darwin never had the opportunity to personally observe the moth. It is a shame, as in his correspondence to Alfred Russel Wallace he wrote about the famous illustration of his predicted pollination: "I like the Figure but I wish the artist had drawn a better sphinx" (Darwin, letter to Wallace, 12–13 October 1867; see Figure 8.3).

The famous prediction of the pollinator of *A. sesquipedale* is widely known, and has been reported many times (see review in Arditti et al. 2012). Nevertheless, Darwin's careful observations of flower shapes, as

Fig. 8.3 *Angraecum sesquipedale* and its predicted long-tongued hawkmoth pollinator, as illustrated in Wallace 1867.

well as his innovative reasoning, led him to speculate on the existence of coevolution, a new concept that drastically overturned natural sciences. At that time, however, Darwin's explanations on the development of the long nectar spur in *Angraecum* invoked much criticism: "I am glad you had the courage to take up *Angraecum* after the Duke's attack" (Darwin, letter to Wallace, 12–13 October 1867). "I shall be glad to know whether I have done it satisfactorily to you, and hope you will not be so very sparing of criticism as you usually are" (Wallace, letter to Darwin, 1 October 1867).

"Mr Darwin's great work" (Wallace 1867; p. 471) also found strong support, especially from Alfred Russel Wallace, who dedicated an essay, "Creation by Law," in favor of Darwin's views (Wallace 1867). The note at the end of the text demonstrates Wallace's fervent support for Darwin's prediction:

> I have carefully measured the proboscis of a specimen of *Macrosila cluentius* from South America in the collection of the British Museum, and find it to be nine inches and a quarter long! One from tropical Africa (*Macrosila morganii*) is seven inches and a half. A species having a proboscis two or three inches longer could reach the nectar in the largest flowers of *Angræcum sesquipedale*, whose nectaries vary in length from ten to fourteen inches. That such a moth exists in Madagascar may be safely predicted; and naturalists who visit that island should search for it with as much confidence as astronomers searched for the planet Neptune,—and they will be equally successful! (Wallace 1867)

HAWKMOTH-POLLINATED SPECIES

Flower Morphology

Most long-spurred angraecoid orchids display floral adaptations that characterize the hawkmoth pollination syndrome. This includes floral traits linked to pollinator attraction (i.e. white flower and strong scent). It also includes pollinator fidelity as determined by large amounts of sugar-rich reward situated in deep spurs accessible only to long-tongued pollinators. Typically, hawkmoth-pollinated orchids have a long spur (from 4–5 cm to >20 cm; Table 8.2; Figure 8.4A) and offer a large amount of sucrose-rich nectar to their pollinators (Table 8.2). These flowers live a long time if not pollinated. For example, *Mystacidium venosum* flowers last 24 days (Luyt and Johnson 2001). They produce strong and sweet scents at dusk that are characterized by the presence of aromatic volatiles with benzene rings. This represents a class of chemicals known to attract nocturnal members of the

Table 8.2 Correlation of spur length in angraecoid orchids with pollinator records, occurrence of auto(self)-pollination and floral characters

Spur length	Spur > 20 cm	Spur ~ 10 cm		Spur ~ 5cm	Spur < 2 cm	
Pollinator groups	Hawkmoths	Hawkmoths	[Auto-pollination]	Hawkmoths	Birds	Orthoterans
Genera	*Angraecum*[a]	*Angraecum, Aerangis, Jumellea, Neobathiea, Rangaeris*[b]	*Jumellea*[c]	*Aerangis, Mystacidium*[d]	*Angraecum*[e]	*Angraecum*[f]
Distribution	Madagascar	Africa – Madagascar	Reunion	Kenya – South Africa	Reunion	Reunion – Mauritius
Pollinator species	*Xanthopan morganii Coelonia solani*	*Xanthopan morganii, Panogena lingens, Agrius convolvuli, Coelonia fulvinotata*	na	*Hippotion celerio, Daphnis nerii, Nephele accentifera accentifera*	*Zosterops olivaceus, Z. borbonicus*	*Glomeremus orchidophilus*
Pollinaria attachment	Proboscis	Head/proboscis	na	Proboscis	Base/Tip of the beak	Top of the head
Flower Color	White	White	Cream-white	White	White	Cream-white
Spur length (cm)[g]	29.8	13.7	13.8	4.5	1.0	0.6
Spur shape	Filiform	Filiform[h]	Filiform	Filiform	Conical	Conical

Spur opening	Narrow	Narrow	Narrow	Narrow	Large[i]	Very large
Nectar volume (µL)[g]	136.0	13.9	6.1	4.3	6.3	14.5
% sugar[g]	13.4	15.3	10.7	16.5	9.7	12.3
Floral scent (human nose)	Strong in the evening	Strong in the evening	Not detectable	Strong in the evening	Not detectable	Weak at night
Fruit set (%)[g]	20.0	18.1	62.8	13.1	12.6	26.0

[a] Compiled data from Wasserthal 1997 (study species are *Angraecum sesquipedale* and *A. sororium*).

[b] Compiled data from Nilsson et al. 1985, 1987; Nilsson and Rabakonandrianina 1988; Wasserthal 1997; Martins and Johnson 2007 (study species are *Angraecum arachnites, A. compactum, Aerangis articulata, A. brachycarpa, A. ellisii, A. fuscata, A. kotchyana, A. thomsonii, Jumellea teretifolia, Neobathiea grandidierana,* and *Rangaeris amaniensis*).

[c] Data from Micheneau et al. 2008b (study species is *Jumellea stenophylla*).

[d] Compiled data from Luyt and Johnson 2001 and Martins and Johnson 2007 (study species are *Aerangis confusa* and *Mystacidium venosum*).

[e] Compiled data from Micheneau et al. 2006, 2008c (study species are *Angraecum striatum* and *A. braectosum*).

[f] Compiled data from Micheneau et al. 2010 (study species is *Angraecum cadetii*).

[g] Mean is given, compiled from the literature in combining sites and years if applicable.

[h] Could be spirally twisted.

[i] *Angraecum bracteosum* display a callus on the lip, which restricted flower's entrance.

Fig. 8.4 The different pollinator groups recorded in the angraecoid orchids. (A) *Coelonia fulvinotata* visiting flowers of *Aerangis brachycarpa* (modified from Martins and Johnson 2007; used by permission). (B, C) *Angraecum striatum* and its bird pollinator *Zosterops borbonicus* (gray white-eyes) with pollinaria attached on its beak (photographs taken from video captures, Reunion Island, by Claire Micheneau and Jacques Fournel). (D) *Angraecum cadetii* and its raspy cricket pollinator, *Glomeremus orchidophilus*, with pollinaria attached on its head (photographs taken from video captures with night-shot option, Reunion Island, by Claire Micheneau and Jacques Fournel).

Lepidoptera (e.g. Huber et al. 2005; see Chapter 9). In addition, structural modifications of the column, the lip, and/or the spur are often observed and are thought likely to favor deep contact between the moth's body and the orchid's column. This contributes to increased pollination efficiency and specificity in the following ways: (1) Extensions of the rostellum may flank the entrance of the spur, presumably to force the head of the pollinator into close contact with the orchid column (Wasserthal 1997); (2) Similarly, particular appendices of the lip may function as mechanical guides to orient the base of the proboscis into the spur (e.g. *Aerangis ellisii*; Nilsson and Rabakonandrianina 1988); (3) The location of the nectar in the spur (i.e. the nectar column height) is also of crucial importance for favoring efficient pollination by hawkmoths (e.g. Darwin 1862; Nilsson et al. 1985). In most long-spurred flowers, the nectar is located only at the very base of the spur, forcing the pollinator to deeply probe the spur to reach the nectar and "to drain the last drop" (Darwin 1862; p. 203). This should ensure the successful removal and deposition of pollinaria (Nilsson et al. 1985; Luyt and Johnson 2001); (4) Finally, a noticeable key feature has been highlighted in *Aerangis thomsonii*, *A. kotschyana*, and *Angraecum arachnites*. These species display spirally twisted spurs. Interestingly, the spur's torsion is thought to encourage close contact between the spur and the hawkmoth's tongue during nectar consumption. As a result of this floral adaptation, moths should be encouraged to probe deeper into the spur, stay longer in the flower, and experience closer contact with the orchid's reproductive organs (see Martins and Johnson 2007).

Breeding System

Hawkmoth-pollinated angraecoid orchids are typically self-compatible (e.g. Darwin 1862; Nilsson et al. 1985, 1987; Nilsson and Rabakonandrianina 1988; Wasserthal 1997; Martins and Johnson 2007; see also Chapter 10), with similar fruit set by self- and cross-pollinated flowers. A noticeable exception was found in *Mystacidium venosum*, in which self-pollination significantly decreased fruit weight, fruit length, and the percentage of seeds bearing embryos in fruits. Self-pollination in this species has a strong negative effect on seed production (Luyt and Johnson 2001).

Pollinator Identity

As Darwin predicted, the orchid species with the longest tubes in Madagascar (often longer than 30 cm in length) are the star orchids, *Angraecum sesquipedale* and *A. sororium*, and they are pollinated exclusively by hawkmoths with the longest tongues (often exceeding 20 cm). *Xanthopan morganii* var. *praedicta* and *Coelonia solani*, respectively, are the key pollinators (Wasserthal 1997). The lepidopteran fauna of Madagascar (approximately 60 species) is comparatively rich in long-tongued representatives. However, only one species, *Panogena lingens*, is the most commonly recorded pollinator. It has been observed carrying pollinia of no less than 6 orchid species, including *Angraecum arachnites*, *A. compactum*, *Neobathiea grandidierana*, *Jumellea teretifolia*, *Aerangis fuscata*, and *A. ellissii* (Nilsson et al. 1985, 1987; Nilsson and Rabakonandrianina 1988; Wasserthal 1997). *Agrius convolvuli* and *Coelonia fulvinotata* also have been observed pollinating long-spurred *Aerangis* and *Rangaeris* species in Madagascar and Kenya in Africa (Nilsson and Rabakonandrianina 1988; Martins and Johnson 2007). In South Africa, *Mystacidium venosum* (spur approximately 5 cm) has been reported to be pollinated by the hawkmoth *Nephele accentifera accentifera* (Luyt and Johnson 2001). In Kenya, *Aerangis confusa* (spur 5 cm) is pollinated by *Hippotion celerio* and *Daphnis nerii* (Martins and Johnson 2007).

Pollinator Specificity

Studies conducted so far have shown a high degree of specialization between long-spurred angraecoid orchids and their hawkmoth pollinators, as only a few pollinators have been recorded per orchid species. However, in these specialized pollination systems, mutualistic relationships are very asymmetric, leading to the orchid being much more dependent on the moth (for sexual reproduction) than the moth on the orchid (for food). Moths are opportunistic foragers and can exploit nectar from a wide variety of flowers (e.g. Nilsson et al. 1987).

Hawkmoth Behavior

In most cases, hawkmoths have been observed foraging on long-spurred flowers at dusk through evening between 5:15 p.m. and 7:30 p.m. (e.g. Nilsson et al. 1987; Luyt and Johnson 2001; Martins and Johnson 2007). Flower visits are rapid, often lasting less than 2 seconds. Hawkmoths have been described approaching the orchids in a "zigzag flight upwind" (Nilsson et al. 1987; p. 313), and are attracted over long distances by the sweet and strong scent emitted by the flowers. At shorter distances, as flowers become easily distinguishable with their white color, hawkmoths hover to face the orchid lip (see Figure 8.4A) before inserting their extended proboscis into the spur. Their head presses against the orchid's column while they collect nectar.

Pollinaria are attached to the dorsal or ventral surface of the proboscis or directly to the hawkmoth's head, depending on the orchid species (Table 8.2). A pollinarium can remain firmly attached and undamaged for a long time, allowing hawkmoths to disperse pollen over long distances (e.g. Nilsson et al. 1985, 1987; Nilsson and Rabakonandrianina 1988; Nilsson et al. 1992; Wasserthal 1997). Some moths are caught carrying pollinaria of different orchid species (Nilsson et al. 1987), and up to 20 pollinaria have been counted on a single moth's tongue (Luyt and Johnson 2001). Pollinaria also undergo a bending movement of 90 degrees a few minutes after removal from the anthers. The drying process makes them orient correctly, facilitating the receipt of the pollinia on a stigma with conspecific flowers. Depending on the time spent by pollinators on an orchid inflorescence, and the time taken by caudicles to dry and move (see Chapters 1 and 2), pollinarium bending is thought to avoid selfing between flowers found on the same plant, a process known as geitonogamy (e.g. Darwin 1862, 1877; Wasserthal 1997; Johnson and Edwards 2000; Luyt and Johnson 2001; Peter and Johnson 2006).

Pollination Efficiency and Reproductive Success

Data on pollination rates (rates of pollen removal and deposition) were available for only *Aerangis ellisii* with a spur length of approximately 16 cm in Madagascar (Nilsson and Rabakonandrianina 1988), and *Mystacidium venosum* with a spur length of approximately 5 cm in South Africa (Luyt and Johnson 2001). Their rates of pollinaria removal were very similar (51% and 53.5%, respectively), while pollen deposition was 24% and 27.5%, respectively. Fruiting success is quantified in more species (Table 8.2 and Figure 8.5). Combined records show that fruit set in hawkmoth-pollinated species is about 17% on average per flowering season, and ranges from only 6% in *Aerangis brachycarpa* (Martins and Johnson 2007) to 41% in *Angraecum arachnites* (Nilsson et al. 1985).

From Madagascar to the Mascarenes: Surprising Pollinator Shifts

Although angraecoid orchids from Africa or Madagascar are known for their white, spectacularly long-spurred flowers specialized for hawkmoth pollination, insular species from the volcanic islands of the Mascarenes show divergent floral morphologies. Presumably as a result of dispersal from Madagascar to the Mascarenes, where the specific long-tongued pollinators were likely absent, their reproductive strategies have changed (Micheneau et al. 2006, 2008c, 2010). Although new adaptive features for

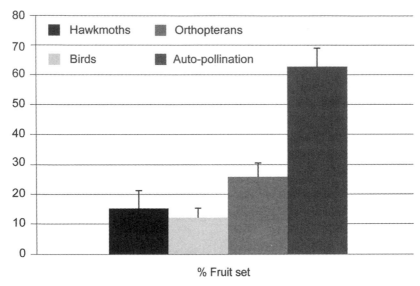

Fig. 8.5 Comparison of reproductive success between hawkmoth, bird, orthopteran, and auto-pollination in the angraecoid orchids. Data were compiled from Nilsson et al. 1985, Nilsson and Rabakonandrianina 1988, Wasserthal 1997, Luyt and Johnson 2001, and Martins and Johnson 2007 for hawkmoth pollination; from Micheneau et al. 2006, 2008c for bird pollination; from Micheneau et al. 2010 for orthopteran pollination; and from Micheneau 2005; Micheneau et al. 2008a for auto-pollination.

a specific group of pollinators can be viewed as responsible for the success of a lineage, extreme specialization can lead to a loss of evolutionary potential that could result in extinction (see Ackerman and Roubik 2012), or decreasing a population's ability to adapt to future environmental change. To successfully establish on oceanic islands like the Mascarenes without their native pollinators, highly specialized plant lineages required the development of novel reproductive strategies. This included pollinator and/or mating system shifts (e.g. Carlquist 1974; Barrett 1996; Olesen and Valido 2003; Sakai et al. 2006). Consistent with a probable historical absence of long-tongued pollinators in the Mascarenes (e.g. Pailler et al. 1998), evolutionary shifts in the angraecoids included from outcrossing to selfing, as well as pollinator shifts from hawkmoth to bird and to orthopteran pollination. This last pollination syndrome represents the first supported case of orthopteran pollination in extant angiosperms (Micheneau et al. 2010).

Shifts from outcrossing to selfing are thought to be widespread in many insular species in both the genera *Angraecum* and *Jumellea* (Micheneau 2005; see also Jacquemyn et al. 2005 and Micheneau et al. 2008ab). Intra-island pollinator shifts, however, are less common within the group, and are exemplified in the *Angraecum* sect. *Hadrangis*, where unusual cases of bird and orthoperan pollination are reported.

BIRD-POLLINATED SPECIES

Flower Morphology

Within sect. *Hadrangis*, the 2 *Angraecum* species pollinated by birds, namely *A. striatum* and *A. bracteosum*, are endemic to the island of Reunion and show convergent adaptations observed in other bird-pollinated orchids. However, the flowers of these orchids remain pure white and not the vivid pink, red, yellow, or orange associated with bird pollination (Micheneau et al. 2006; Figure 8.4B,C). *Angraecum striatum* and *A. bracteosum* display multiflowered inflorescences of unscented and short-spurred flowers, which are rather fleshy in texture. In addition, *A. bracteosum* has a strong fold on the lip (callus), restricting the opening of the flower. The presence of such a callus is common in bird-pollinated orchids, and is thought to force the birds to push their beak against the orchid column to reach the nectar in the spur. This modification favors adhesion of the pollinarium on their beak (e.g. van der Pijl and Dodson 1966; van der Cingel 2001). Flowers last from 10 to 20 days if not pollinated (Micheneau et al. 2006). The spur is conical, short (approximately 8 to 10 mm), and rather wide at the entrance in *A. striatum* (3.9 mm), but is narrower in *A. bracteosum* (1.9 mm, restricted by the presence of the callus on the lip). Spurs contain 5 to 8 µL of dilute nectar (approximately 10% sucrose in both species; Table 8.2).

Breeding System

Angraecum striatum and *A. bracteosum* are fully self-compatible, but require pollinating agents to set fruit (Table 8.3). No significant differences were found in fruit production between self- and cross-pollinated flowers (Micheneau et al. 2006, 2008c).

Pollinator Identity

Birds observed to pollinate short-spurred *Angraecum* species in Reunion are white-eye species (Zosteropidae) endemic to the island, namely the gray white-eye (*Zosterops borbonicus borbonicus*) and the olive white-eye (*Z. olivaceus olivaceus*). Gray white-eyes are common pollinators of *A. striatum* (Micheneau et al. 2006; Figure 8.4B,C). These birds may also occasionally remove pollinaria of *A. braectosum*, but effective pollination is thought to

Table 8.3 Percentages ± standard deviation (*n* plants/*n* flowers) of pollination and reproductive success in the *Angraecum* sect. *Hadrangis*

Species	*A. striatum*[a]	*A. bracteosum*[b]	*A. cadetii*[c]
Pollinator group (Genus)	Birds (Zosterops)	Birds (Zosterops)	Orthopterans (*Glomeremus*)
Pollinator species	Z. borbonicus	Z. olivaceus, Z. borbonicus (rare)	G. orchidophilus
Pollination success			
Pollinia removal			
2003	43.4 ± 33.1 (28/303)	39.4 ± 26.9 (20/458)	51.1 ± 36.1 (21/99)
2004	72.8 ± 32.8 (21/100)	23.6 ± 22.0 (19/471)	46.1 ± 42.9 (17/74)
2005	57.4 ± 27.2 (16/164)	26.1 ± 34.2 (17/417)	42.3 ± 35.4 (17/83)
Pollinia deposition			
2003	17.6 ± 19.4 (28/303)	21.6 ± 14.3 (20/458)	36.7 ± 34.3 (21/99)
2004	19.2 ± 32.6 (21/100)	6.6 ± 8.7 (19/471)	23.6 ± 19.1 (17/74)
2005	12.4 ± 16.0 (16/164)	10.4 ± 13.6 (17/417)	25.2 ± 33.3 (17/83)
Fruit set			
Natural conditions			
2003	10.7 ± 17.5 (28/303)	6.0 ± 6.5 (20/458)	24.7 ± 33.6 (21/99)
2004	16.8 ± 32.1 (21/100)	3.5 ± 7.5 (19/471)	22.5 ± 18.0 (17/74)
2005	12.1 ± 16.0 (16/164)	5.7 ± 10.3 (17/417)	18.7 ± 33.6 (17/83)
Pollinator excluded	0 (23/63)	0 (8/88)	0 (8/45)

Note: Data from 3 consecutive years at the study site of La Plaine des Palmistes (Reunion Island), where the 3 species are growing in sympatry.
[a] Micheneau 2005; Micheneau et al. 2006.
[b] Micheneau et al. 2008c.
[c] Micheneau et al. 2010.

be rare due to morphological incompatibility (see below). Olive white-eyes, in contrast, are the main pollinators of *A. bracteosum* and have never been observed foraging on *A. striatum* (Micheneau et al. 2008c). Most white-eyes have a generalist diet that includes fruit, insects, and nectar (Gill 1971). Interestingly, orchid pollination by birds in the family Zosteropidae was not described previously. On Reunion, olive white-eyes have unusually long and slender beaks, averaging 15.6 mm in length (Gill 1971). Their beaks match perfectly with the narrow opening of *A. bracteosum* flowers

(Micheneau et al. 2008c). Gray white-eyes display shorter and wider beaks, averaging 11.6 mm in length (Gill 1971). Their beaks match perfectly with the size of the flower entrance of *A. striatum* (Micheneau et al. 2006).

In the early 1970s, Gill (1971) hypothesized that olive white-eyes, *Z. olivaceus*, became established well before gray white-eyes on the Mascarenes. The olive white-eyes are believed to have developed a specialized nectarivorous diet mainly because the ecological niche was available. There were no nectar-feeding birds present on these islands when ancestors of these birds colonized the archipelago (Gill 1971; Warren et al. 2006). The establishment of the second white-eyes species on the Mascarenes, *Z. borbonicus*, is also thought to have increased competitive behavior for food foraging, leading *Z. olivaceus* to "become an even more specialized flower-feeder" (Gill 1971; p. 54). Interestingly, *A. bracteosum*, the only species in which these specialized nectar-feeding birds have been observed as effective pollinators, is also the only species of sect. *Hadrangis* to display a strong fold on its lip, a distinctive floral adaptation to bird pollination.

Pollinator Specificity

According to our observations, but also based on morphological evidence, it seems that bird-orchid interactions on Reunion involve specialized pollination systems, i.e. gray white-eyes are pollinators of *A. striatum* and olive white-eyes are pollinators of *A. bracteosum* (Micheneau et al. 2006, 2008c). This is rather unusual considering the fact that orchid species pollinated by birds are typically pollinated by more than one bird species (e.g. van der Pijl and Dodson 1966; Rodríguez-Robles et al. 1992; Johnson 1996; Singer and Sazima 2000; van der Cingel 2001; Johnson and Brown 2004). However, it seems likely that our knowledge of orchid-bird interactions is underestimated, as pollinations are difficult to observe in the wild due to birds' furtive behavior (van der Pijl and Dodson 1966).

Bird Behavior

White-eyes have been observed to be active on *Angraecum* flowers all day long, from 6:00 a.m. to 6:00 p.m., whatever the weather conditions. Both bird species display the same behavior while feeding on orchid nectar. They typically land on a leaf or inflorescence and probe the majority of opened flowers within their reach (Figure 8.4B,C) for a duration ranging from 9 to 27 seconds (Micheneau et al. 2006, 2008c). As in most pollination events involving angraecoid orchids, birds press their beak firmly against orchids' columns while drinking nectar and pick up pollinaria as they retreat from the flowers. Due to differences in floral morphology, flower opening size, and spur length, pollinarium attachment on a bird's beak contrasts strongly between the 2 orchid species. This seems to preclude hybridization events. Foraging on *A. striatum* involves pollinarium attachment on the base of the beak. In contrast, pollinaria are always fixed to the extreme tip of the beak in *A. bracteosum* (Table 8.2). Birds are able to remove a large number of pollinaria during a single visit, but it seems likely that some pollinaria are lost as these birds clean their beak by wiping it on trees or ferns (Claire Micheneau, personal observation).

Pollination Efficiency and Reproductive Success

Pollinarium removal and deposition have been recorded for the 2 species during 3 consecutive years in different sites on Reunion (Micheneau et al. 2006, 2008c). When sites and years are combined, the rates of pollinarium removal average 39% and pollinarium deposition is 19% in *A. bracteosum*, but ranges between 27 and 54% in *A. striatum*. Fruit set averaged 8% in *A. bracteosum* and 17% in *A. striatum* (see also Figure 8.5). In each study site, the 2 bird species were highly active within orchid populations. When sampled, 88% of individual plants had at least one flower with its pollinarium removed in *A. bracteosum* and 84% in *A. striatum* (Micheneau et al. 2006, 2008c).

The divergent occurrence of the 2 white-eye species on Reunion may explain why pollination rates and fruiting success were better in *A. striatum* (pollinated by gray white-eyes, *Z. borbonicus*) than in *A. braceosum* (pollinated by olive white-eyes, *Z. olivaceus*). Gray white-eyes are among the most common passerine birds on the island. An estimate of 465 000 individuals was made in 1983 (Barau et al. 2005). They occur in a range of habitats, including inhabited disturbed zones from sea level to 2300 m. Olive white-eyes are much less common, with an estimation of 150 000 individuals (Barau et al. 2005). These birds are more likely to be found in preserved habitats from 500 (rarely 200) to 2400 m. In addition, *A. bracteosum* has the lowest fruit set (8%) ever reported in a bird-pollinated orchid species (e.g. Rodríguez-Robles et al. 1992; Johnson and Brown 2004).

ORTHOPTERAN-POLLINATED SPECIES

Flower Morphology

Flowers of the orthopteran-pollinated species of sect. *Hadrangis*, namely *A. cadetii*, are very peculiar with a rather primitive shape, making the orchid species easily distinguishable from all other congenerics (see Figures 8.2 and 8.4D). Pollinia are quite large, flat, and triangular. This pollinia morphology is also atypical in the genus *Angraecum* compared with the small, ball-like pollinia of the bird-pollinated sister species. Sepals and petals are cream-white, small, and fleshy. Flowers last approximately 25 days if not pollinated (Micheneau 2005). The spur is conical, short (6.3 mm), and wide at the entrance (5.2 mm), and may contain a large amount of visible nectar (15 μL per flower, 12% sucrose; Table 8.2). Flowers only emit their scents at night (almost undetectable by human olfaction). These floral emissions are characterized biochemically by the presence of monoterpene compounds (see Micheneau et al. 2010).

Breeding System

Angraecum cadetii is fully self-compatible, but requires a pollinating agent to set fruit (Table 8.3). There is no significant difference in fruit production between self- (96%) and cross-pollinated flowers (100%) (Micheneau et al. 2010).

Pollinator Identity

Surprisingly, observations on Reunion using camcorders revealed that *A. cadetii* is pollinated at night by medium-size raspy crickets, *Glomeremus orchidophilus*, from the subtribe Gryllacridinae in the order Orthoptera (sensu Flook et al. 1999; Figure 8.4D). These orthopterans regularly visit the orchid's flowers throughout its blooming season, typically from mid-end of January to March (austral summer). Our observations highlighted the first recorded case of regular pollination by an orthopteran among extant angiosperms (Micheneau et al. 2010). The orthopteran species was yet unnamed when discovered pollinating the orchids. We decided to name it *Glomeremus orchidophilus* (Hugel et al. 2010). Nicely, *G. orchidophilus* has been included as 1 of the Top 10 New Species described in 2010 by the International Institute for Species Exploration at Arizona State University (http://species.asu.edu/Top10).

Members of the order Orthoptera are well-known herbivores, and do not visit flowers as regular pollinators (e.g. Darwin 1862; Knuth 1909; van der Pijl and Dodson 1966; Kevan and Baker 1983; Proctor et al. 1996; van der Cingel 2001; see also supplemental information in Micheneau et al. 2010). Some families within the Orthoptera are known for their specialized diet of nectar, pollen, or floral tissues (i.e. Zaprochilinae and Phasmodinae; see Hugel et al. 2010). Records of these insects visiting flowers and serving as pollen vectors are rare, and no actual pollination mutualism was previously known.

Pollinator Specificity

On Reunion, *G. orchidophilus* appears to be the only pollinator of *A. cadetii*, and the role of these orthopterans in the reproduction of the orchid seems well established. In addition, we did not observe these insects damaging flower organs, suggesting they are not floral predators (Micheneau et al. 2010). The flowers of *Angraecum cadetii* were visited frequently both night and day by a broad number of animals, including birds, geckos, and nocturnal arthropods such as spiders, centipedes, cockroaches, moths, mosquitoes, and other crickets. None were able to remove pollinaria (see Micheneau et al. 2010). Only raspy crickets show a close match between the size of the insect's head and the orchid's spur opening. We observed raspy crickets carrying pollinaria and contacting the flower's stigma (Micheneau et al. 2010). The peculiar floral bouquet of *A. cadetii*, dominated by monoterpene volatiles (Micheneau et al. 2010), probably plays a major role in attracting raspy crickets. These insects typically use their tremendously long antennae, as do most gryllacridids, to explore their surroundings at night (Claire Micheneau, personal observation; Sylvain Hugel, personal communication).

Orthopteran Behavior

Raspy crickets visit flowers of *A. cadetii* only at night between 7:50 p.m. and 4:40 a.m., and return to the same patch of orchids night after night. In contrast to the furtive behavior of both sphingids and birds, visits by orthopterans can last up to 41 seconds in a single flower. During this period, raspy crickets probe deeply into the orchid's spur. Their heads become completely hidden in the center of the flower. Pollinaria of the orchid is removed as raspy crickets retreat from flowers. The pollinarium is firmly attached by its viscidium to the top of the insect's head (Micheneau et al. 2010). Members of the Gryllacridinae have a diet that includes arthropods and plant material such as seeds, fruit, or flower parts (Hale and Rentz 2001). They are known to be active during the night and to return to their nest in the morning where they stay during the day (Hale 2000; Hale and Rentz 2001). It is expected that memory capacities developed by gryllacridids to relocate their nests each morning (Hale and Bailey 2004) might have equally facilitated their abilities to repeatedly visit the same floral food sources, as observed *in situ*. Our raspy crickets visited the same group of orchids night after night. The flowers provided a rich and localized source of nectar for several days. (see also Hugel et al. 2010; Micheneau et al. 2010).

Pollination Efficiency and Reproductive Success

Even more surprising are the high levels of pollination and fruiting success recorded in natural populations of *A. cadetti*. When sites and years are combined, fruit set is at 26% (Micheneau et al. 2010; see also Table 8.2 and Figure 8.5). At the study site of La Plaine des Palmistes, where the 3 species of sect. *Hadrangis* are sympatric, reproductive success of the orthopteran-pollinated orchid (22%) was higher than those recorded in the bird-pollinated sister species (5 and 13% for *A. bracteosum* and *A. striatum*, respectively; Table 8.3).

SELF-POLLINATED SPECIES

Flower Morphology

Self-pollination (or auto-pollination sensu Catling 1990) provides reproductive assurance when pollinators are absent, and is particularly common on oceanic islands where insect faunas are often under-represented (see Ackerman 1985; Barrett 1985; Catling 1990; Barrett 1996; Eckert et al. 2006; see Chapters 5 and 7). On Reunion, the proportion of self-pollinating orchids is particularly high, with almost half the species self-pollinating (Jacquemyn et al. 2005). For comparison, 5%–20% of orchid species (Catling 1990) and <1% of Vandeae are thought to self-pollinate (Dressler 1993b). This high proportion of self-pollinating species was not expected for the angraecoids (Dressler 1993b), which have long been viewed as a group being predominantly pollinated by lepidopterans. Several self-pollinating species found in Reunion, mainly in the genera *Jumellea* and *Angraecum*, look similar to their hawkmoth-pollinated sister species occurring in Madagascar. These self-pollinating flowers have spurs greater than 10 cm, with a visible amount of nectar (Table 8.2 and Figure 8.6). However, none of these self-pollinating but long-spurred species emit a strong and sweet scent at dusk. Furthermore, their floral longevity is comparatively short, lasting only 4–5 days (Micheneau et al. 2008b). The lack of floral scent associated with a short floral life span makes these flowers unattractive to pollinators. It seems unlikely that any of these orchids are cross-pollinated, even occasionally.

Fig. 8.6 The auto-pollinated *Jumellea stenophylla*. Photographer: Claire Micheneau, Plaine des Palmistes, Reunion Island.

Breeding System and Reproductive Success

Studies conducted on the long-spurred *Jumellea stenophylla* (Figure 8.6) endemic to Reunion show that this species is clearly able to reproduce in the absence of pollinators (Micheneau 2005; Micheneau et al. 2008b). Fruiting success averaged 64% when sites and years are combined (Figure 8.6). Typically, fruit production of self-pollinated species is much higher than those recorded in pollinator-dependent species, averaging 77% in the orchid family (Tremblay et al. 2005).

Mechanisms of Self-Pollination

Self-pollination is not rare in orchids, and many divergent self-pollination mechanisms have been described for the family (see Catling 1990). In *J. stenophylla*, the column lacks a true rostellum, and pollinia fuse onto the stigma very early after the flower opens. Pollinarium removal by an animal seems impossible during its short floral life span (Micheneau et al. 2008b).

To what extent have high levels of self-pollination in long-spurred Mascarene species evolved *in situ* on the islands or were they results of pioneer individuals that were already able to self-pollinate before dispersal from Madagascar to the Mascarenes is still unclear. The ability to self-pollinate may have facilitated the establishment of populations on young volcanic islands (Baker 1955, 1967; Stebbins 1970), especially in species highly specialized to a given pollinator group (e.g. Carlquist 1974; Woodell 1979; McMullen 1987; Barrett 1996; Anderson et al. 2001). On the other hand, evolutionary transitions from outcrossing to selfing are still poorly understood (e.g. Schueller 2004), as the time required for such transitions in orchids is unknown.

CONCLUSION

More than 150 years have passed since Darwin received the box of orchids from J. Bateman containing the star orchid *Angraecum sesquipedale*. Following Darwin's information and inspiration, the angraecoid orchids have fascinated evolutionary biologists over the centuries, offering "very pretty cases" (Darwin, letter to Hooker, 30 January 1862). They remain the most intriguing of biological systems, especially orthopteran pollination. To realize that the evolutionary potential of these orchids and their specific pollinators may already have been lost due to the massive destruction of their habitats in Madagascar, Mauritius, and parts of Africa should contribute as a relevant factor to pursue their study.

Darwin's idea that orchids provide evidence for adaptations to insect pollination has found strong support among researchers and naturalists. Pollination systems have provided some of the best examples for testing the adaptive nature of floral traits. This justifies increasing current research on molecular patterns underlying ecological speciation, because the broader influence of pollinator behavior on flower variation deeply frames the role of natural selection in the origin and maintenance of species.

ACKNOWLEDGMENTS

The authors warmly thank Retha Meier and Peter Bernhardt for the invitation to contribute this chapter. We dedicate this chapter to our dear colleagues and good friends who have contributed to previous and ongoing research on the systematics and the pollination biology of the angraecoid orchids.

Plate 1 Unidentified soldier beetle bearing pollinaria of *Neottia ovata* (syn. *Listera ovata*), Downe Bank (Darwin's "Orchis Bank"), Kent, England (see Chapter 1). Photographer: John Palmer of the Kent Wildlife Trust. By courtesy of Irene Palmer.

Plate 2 Ophrys insectifera (syn. *O. muscifera*) with 3 males of the wasp *Argogorytes mystaceus*, Downe Bank (Darwin's "Orchis Bank"), Kent, England (see Chapters 1 and 3). Photographer: Grant Hazlehurst of the Kent Wildlife Trust. Used by permission.

Plate 3 Two color morphs of *Dactylorhiza sambucina*, southern Italy (see Chapter 2). Photographers: Antonia Scrugli and Sal Cozzolino. Used by permission.

Plate 4 Hawkmoth, *Theretra capensis*, wearing multiple pollinaria of *Bonatea speciosa*. Note the attachment of pollinaria to the moth's eyes. South Africa (see Chapter 4). Photographer: Steven Johnson.

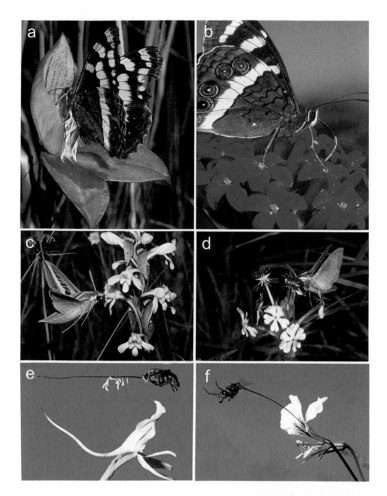

Plate 5 Orchids and other plants belonging to butterfly, hawkmoth, and long-proboscid-fly pollination guilds in South Africa. Both the orchid *Disa uniflora* (*a*) and succulent shrub *Crassula coccinea* (*b*) are adapted for pollination by the satryiine butterfly, *Aeropetes tulbaghia*. Both the orchid *Satyrium longicauda* (*c*) and *Zaluzianskya natalensis* (*d*) are adapted for pollination by medium-size hawkmoths. The long-proboscid fly *Moegistorhynchus longirostris* pollinates both flowers of the orchid *Disa draconis* (*e*) and the small, shrubby *Pelargonium longicaule* (*f*). South Africa (see Chapters 4 and 11). Photographer: Steven Johnson.

Plate 6 Close-ups of the labellum of *Drakaea glyptodon* (note connection to greenish-yellow column foot) and the wingless female of *Zaspilothynnus trilobatus*, Western Australia (see Chapters 5 and 11). Photographer: Rod Peakall. Used by permission.

Plate 7 Male *Zaspilothynnus trilobatus* collides with the column of *Drakaea glyptodon* after attempting to fly off with the dummy female, Western Australia (see Chapters 5 and 11). Photographer: Rod Peakall. Used by permission.

Plate 8 Male wasp (*Neozeleboria* sp.) regurgitates nectar and attempts to feed dummy female on labellum of *Chiloglottis reflexa*, Mount Wilson, NSW, Australia (see Chapters 5 and 11). Photographers: Ray and Elma Kearney. Used by permission.

Plate 9 Female *Leioproctus* sp. grasps the ornamented anther of *Thelymitra antennifera*, Western Australia. Note that the bee's abdomen is now positioned to contact the stigma (see Chapters 5, 7, and 10). Photographer: Andrew Brown. Used by permission.

Plate 10 Female *Leioproctus* sp. examines the trichome brushes of *Thelymitra macrophylla*, Western Australia. Note also pollen (from non-orchid species) on the scopal hairs of the hind leg (see Chapters 5, 7, and 11). Photographer: Andrew Brown. Used by permission.

Plate 11 Male bee, *Eufriesea ornata*, carrying multiple pollinaria of *Stanhopea costaricensis*, Panama (see Chapter 9). Photographer: David Roubik.

Plate 12 Male bee, *Euglossa tridentata*, emerging from the pollination orifice of the bucket orchid, *Coryanthes hunteriana*, Panama (see Chapter 9). Photographer: David Roubik.

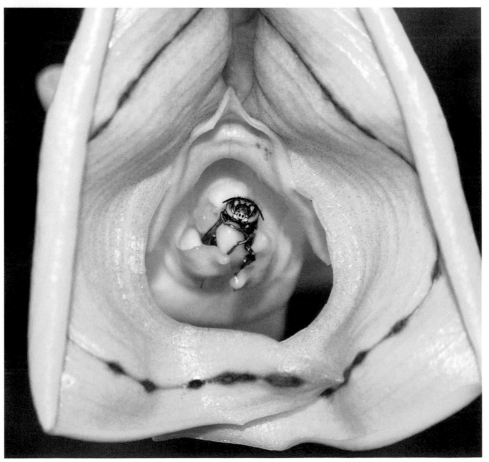

Plate 13 Male bee, *Euglossa igniventris*, collecting fragrance from the labellum of *Coryanthes mastersiana*, Panama (see Chapter 9). Photographer: David Roubik.

Plate 14 Male bees of *Euglossa despecta* collecting fragrance from flowers of *Mormodes flavida*—red form, Panama (see Chapter 9). Photographer: David Roubik.

Plate 15 Female hoverfly, *Ocyptamus antiphales*, emerges from 1 of the 2 basal openings of *Phragmipedium pearcei*, Ecuador. This hoverfly usually oviposits on aphids (see Chapter 10). Photographer: Alex Portilla. Used by permission.

Plate 16 Male wasp, *Lissopimpla excelsa*, on *Cryptostylis erecta*, wearing pollinaria of this species on the tip of the abdomen, NSW, Australia (see Chapter 5 and 11). Photographer: Rod Peakall. Used by permission.

NINE

Orchids and Neotropical Pollinators since Darwin's Time

David W. Roubik

All these appliances are subordinate to the aid of insects. Without their aid, not a plant belonging to this tribe, in the species of the twenty-nine genera examined by me, would set a seed.
—Darwin 1862; p. 210

INTRODUCTION

Charles Darwin was an orchid enthusiast, and he also was an entomologist. Many botanists, pollination biologists, and evolutionary ecologists followed his lead, and that influence is strong to this day. Lacking in Darwin's time, where intellectual curiosity and a record-breaking ecotourism voyage were combined with "pragmatic" natural history and a gentrified lifestyle (Chapter 1), were laboratories dedicated to chemical floral fragrance analysis, nectar chemistry, and molecular phylogenetics. There were no high-speed waterproof cameras or megapixel high-definition videos. Unlike now (Meisel and Woodward 2005), there was apparently no one talking about orchid conservation in the wild. Most lacking, however, were a concise natural history and population biology of tropical orchids.

It is one thing to describe floral mechanisms that may cause outcrossing pollination, but quite another to surmise field conditions over the past 20 million years. Did Darwin in fact do both? In a way, he did, and perhaps this compilation of works performed by others would not overly amaze him. However, most orchid species are tropical members of the Epidendroideae, and they live in the same area with members of the Vanilloideae (Govaerts et al. 2012) that flower in the forest (Gentry and Dodson 1987). They are primarily epiphytes, they often attract pollinators by their smell, and their individual species, when flowering, are scattered in space and time through a canopy that reaches over 60 m, often in majestic forests through tropical America, Asia, Africa, the Indo-pacific, and Australia (see Chapter 8). Further, at their roots, they are an asparagus-like kind of plant. They are monocot angiosperms, and they are the largest group within the largest plant family on earth. In this chapter, I review pollination, pollinators, and evolution of relatively large-flowered Neotropical orchids, primarily epidendroids on which Darwin's insight coalesced, and try to elaborate on some of the timely foundations that Darwin's writings provided. Recognition of the faults and triumphs of past research is required, and my attempt to codify the topic will follow that mode.

The orchid pollination book, published 3 years after the *Origin* and then revised 15 years later, as we will review below, seemingly hid fundamental aspects of pollination biology for a century. For some of Darwin's contemporaries, orchid flowers were essentially for human pleasure. Well, if orchids and their ecology are the pleasure referred to, then I agree. Darwin widened the horizons of his peers by explaining how a mechanism called natural selection could produce the intricacies of the living world. He had no idea, however, how precisely that natural selection could influence the traits seen in living organisms. But the orchid pollination experiments related with zeal by Darwin, from growing houses or "hothouses" in the temperate zone, lacked pollinators—which mostly were bees, among Neotropical genera that Darwin knew. And the plants Darwin saw in flower lacked an environment containing mutualists, competitors, natural enemies, and other key ecological players. Those key players, we know now, include an ant, a bee, a fungus, and a tree.

Tropical epidendroid ecology includes not only ants, bees, floral fragrances, tree substrates for epiphytes, and fungi—but flies, beetles, hummingbirds, butterflies, wasps, moths, sun, shade, moisture, nutrients, air currents, and the decay that feeds saprophytic plants (Dressler 1981; Williams 1982). The Neotropical genera including *Coryanthes*, *Catasetum*, *Maxillaria*, *Mormodes*, *Stanhopea*, *Cycnoches*, *Clowesia*, *Trichocentrum*, *Oncidium* s.l., *Encyclia*, and *Gongora* have a wide berth in specialized literature, aptly summarized in many aspects (see e.g. Dressler and Dodson 1960; Dodson 1962a,b; van der Pijl and Dodson 1966; Dressler 1968; Hills et al. 1972; Dressler 1981, 1993a,b; Williams 1982; Ackerman 1983a–c; Kaiser

1993, 2011; Williams and Whitten 1999; Roubik 2000; Bateman et al. 2003; Chase et al. 2003; van den Berg et al. 2005; Carlsward et al. 2006; Lubinsky et al. 2006; Singer et al. 2006; Whitten et al. 2007; Tupac et al. 2007; Rasmussen 2008; Kull et al. 2009; Ramírez et al. 2011b; Torretta et al. 2011; Ziegler 2011; Pacek et al. 2012). Those genera among the 576 Epidendroideae are still bursting with new facets and questions since Darwin explored some of their curious pollination characteristics. They largely *are* the orchids, and we may follow their lead.

Darwin (1877) relied rather heavily on observations from a tropical botanical garden on the island of Trinidad, published by Crüger (1864). On that island, Hermann Crüger (1818–1864) worked as an apothecary and plant collector, and later a botanical garden director. Perhaps Darwin failed to gain all that he could from Crüger's observations, but at face value, that Darwin-Crüger connection produced something tantamount to a report that ruminants have 4 stomachs, without mention of their cud chewing, vegetation consumption, or methane production. Orchid bees were highlighted here and there (see below), some as undescribed (not given a scientific name yet), and linked, with visual justification (see Color Plate 11), to the "humble bees" (*Bombus*, or bumblebees) known to visit and pollinate orchids in England. Darwin really said nothing about floral odors. Crüger mentioned them on 3 pages in Darwin (1877), with no further comment, but Darwin did modify the title of his book according to the input of Crüger and F. Müller, among other tropical scientists (see Chapter 1).

At the concluding chapter, Darwin (1862, 1877) asks, plainly, what attracts insects to orchids?—given that those like *Catasetum*, *Gongora*, and other tropical epidendroids (none native in England) produce no floral nectar. Their odors are actually mentioned, but not nearly as often as *gnaw* and its derivatives—in reference to what bees were apparently doing at orchid flowers. If a small adjustment had been made, changing the word *gnaw* to perhaps *claw* or *scratch*, then the observations could have seemed accurate. As it is, they are wrong, and continue through Darwin's closing comments in the second edition of his book. Crüger is quoted considering bee consumption of something on orchid flowers, as though food, and this gets repeated several times and incorporated as a cornerstone of orchid ecology. Darwin called it "bold speculation" that he had thought bees were attracted to food at tropical epidendroids such as species of *Mormodes*, *Stanhopea*, *Catasetum*, and *Gongora*, but he received "full confirmation" that bees sought nutrition in the floral labellum by chewing the flower himself (Crüger 1864). He indeed tasted flowers of some *Cycnoches*, *Mormodes*, and *Catasetum*, and verifies their "nutritious taste." Darwin (1877; p. 275) mentions that the strong scent of orchids could attract insect pollinators as well as could the floral color. And that is as far as he dared to speculate, with no support other than that given by Crüger for the nutrition hypothesis—and Crüger was wrong, although he suspected odors attracted bees to *Coryanthes* (see van der Pijl and Dodson 1966; Williams 1982; and particularly Gerlach 2011). The bees neither taste nor chew orchid flowers. In the Neotropics, they come to gather orchid floral fragrances, and occasionally to collect their non-pollen resources, or nectar.

I would like to use the telegraphic style of Darwin and the naturalist/ ecotourist reports of his day to outline—or, better stated, headline—the present discussion, compared and contrasted with Darwin's volumes of long ago.

1. What is different in orchids? Individual flowers often last several to many days, and plants are very long lived. The family of plants is the largest on earth. In the flower, the anther is hidden, pollen is massed and is sometimes within a hard shell, and it is not consumed by any animal. The seeds and pollen are extraordinarily numerous per flower, but they are tiny. Because they are so tiny, the seeds have no life support; they depend completely on fungal mutualists or certain ants for germination. Decaying substrates are needed for saprophytic orchids. Reproduction includes some gender-dimorphic flowers, and while some hermaphroditic flowers are self-pollinating, most benefit from outcrossing.

2. Orchid mutualists/pollinators include ants that consume extrafloral nectar and potentially discourage herbivores that would consume the plant, flower, or fruit. Bees with extremely long tongues are dedicated orchid pollinators. They are male euglossine bees that take no nectar, with a few exceptions among other bees that do take nectar. Their livelihood and that of orchids are supported by diverse tropical plants that provide their food. Male euglossine bees gather and store orchid fragrances to attain status at a mating perch. Other bees and insects, and hummingbirds, pollinate epidendroids, seeking floral nectar, or are deceived, or strongly attracted to, rewardless and even some rewarding flowers, through floral mimicry.

3. Evolution of orchid flowers seems either precise or imperfect in adaptations to pollination, morphology, behavior, and phenology. This variability signals rapid, vigorous evolutionary elasticity. The euglossine bee pollination syndrome evolved at least 3 times, and the bees were co-opted, not coevolved. Fragrance rewards in flowers, which attract euglossine bees, evolved from deceit flowers, which offered no food or resource. Pollinarium placement sites and examples of multiple bee pol-

lination visits show that orchids are successful in attracting and partitioning their pollinators. Orchid fragrance thievery is common; bees take a resource and do not pollinate, and some stingless bees rob flowers (chew them open to extract food) for nectar, while orchids generally have no nectar production or accessible pollen presentation. The robbers are seen as pivotal in promoting floral deceit and the evolution of pollinaria. Nonetheless, the main robber genera have produced pollinators of some orchids. Long-term studies on asynchronous pollinator and floral phenology show that pollination is sometimes secondary to Darwinian/Wallacean fitness. Nonetheless, despite morphological variation of some epidendroids, attractants maintain their chemical signature. Pollinating bee species, however, vary in their choices, in space and time.

ORCHID FLORAL BIOLOGY: VISITORS, ATTRACTANTS, REWARDS, AND DEFENSES

No Fast Food in Orchid Flowers

The 100-year hiatus in tropical orchid pollination research ended in the early 1960s when field observers noticed that male bees on flowers were collecting something and putting it on their legs (see "Bees," below). Dodson and Frymire (1961) and Vogel (1963a,b, 1966) observed and understood that the bees were not finding or consuming food or drink, but gathering chemicals. Indeed, no nectar or food of any kind is given to bees visiting most Neotropical orchid flowers discussed by Darwin, with some notable exceptions in social bee-, fly-, lepidopteran-, and hummingbird-pollinated flowers (Dressler 1981; Singer et al. 2006; Davies 2009, and references therein; Torretta et al. 2011; Pacek et al. 2012). What has happened to the flowers of orchids, to enhance their nonfood value to bees? Why can't they offer nectar and pollen, as do other flowers? The disappearance of pollen rewards, present in orchids long ago (Johnson and Edwards 2000; Davies 2009), will be discussed below (but see also Chapter 5). In fact, many orchids produce some kind of reward, usually in nectaries or with trichomes, plant cells, or hairs on the floral labellum. These rewards may include proteins, resins, or oils, rich in amino acids and aroma, or sugary nectar (Davies 2009; Steiner et al. 2011; Davies and Stpiczyńska 2012). But for many of the Neotropical large-flowered orchids, this is not the case.

The "androeuglossophilous" flowers, examined later in detail, are visited by hundreds of orchid bee species (Euglossini) and few other animals—neither bees, nor arthropods, nor vertebrates. The euglossine orchid bees are *sole* pollinators of approximately 700 orchids, and comprise around 250 species (Williams 1978, 1982; Dressler 1982; Williams and Whitten 1983; Roubik and Hanson 2004). The ratio is therefore about 3 to 1, considering orchids and the bees (Roubik and Ackerman 1987), and the numbers of orchids tied to euglossine bees thus total about 2.5% of total orchid species. Notably, orchid bee species are largely aseasonal, while orchid species usually flower for only part of the year. The "smelly" flowers were taken into the euglossine bees' pollination repertoire 3 separate times, and seem to represent plant exploitation of a preexisting euglossine bee behavior (Ramírez et al. 2011b). Orchid chemical odors, produced in scent glands (osmophores) in the floral labellum, or under the hypochile of the lip (Vogel 1966; Dressler 1981; Curry et al. 1991; Eltz et al. 2005a; Gerlach 2011), are what attract those pollinators and cause their responses, and ostensibly cause evolution (Zimmermann et al. 2009). The result might produce new orchid species (Hills et al. 1972; Williams 1982), but does not seem to lead to orchid bee speciation (Ackerman 1983c; Ramírez et al. 2011b; Ackerman and Roubik 2012). Floral odors are perhaps the fundamental attractant for all pollinators except birds, which are attracted primarily to red color but do not require the signal of floral odors (Dressler 1981; Dobson 2006; Raguso 2008; Willmer 2011; but see also Chapter 11). However, 2 groups of Neotropical bee-pollinated orchids (*Vanilla* and the Oncidiinae) produce oil or are rewardless food mimics, and their flowers have no obvious fragrance.

For the orchids, fragrant, volatile chemicals have brought bees into one of the most intricate and intimate evolutionary dances, one that they could not now easily exit. Yet orchid bees are certainly a tiny minority among orchid pollinators (Dressler 1981). They did not come into this pollination syndrome, called the euglossine bee syndrome (Dressler 1968; Williams 1982), by design, or by coevolution (see below). Had Darwin understood this point, he might have had an even more convincing argument of cause and effect for contemplation by the scientists and clergy of his time. Orchid bees were potentially "entrapped," because they were present for nearly 12 million years before they began to visit orchid flowers—they are an Eocene group, while their host/mutualist orchids are Miocene (Ramírez et al. 2011b). Furthermore, survival of orchid bees is, de facto, permitted by the non-orchid pollen and nectar flowers that support such pollinators.

Coryanthes is the well-known Neotropical epidendroid bucket orchid, with a variety of visually noteworthy flowers and pollinating *Euglossa* (Gerlach and Dressler 2003; Gerlach 2011; Ziegler 2011; see Figure 9.1A,B

Fig. 9.1 (*Left*) An Africanized honeybee and weevils drowned in the "bucket of water" of *Coryanthes panamensis*, central Panama. (*Right*) Multiple males of *Euglossa* sp. collecting fragrances at flowers of *C. panamensis*, central Panama. Photographs: David Roubik.

and Color Plates 12 and 13). Darwin pointed out facets of pollinarium placement and transfer in the orchids and the bees that were probably doing the work, and for *Coryanthes*, pollination requires that bees make a laborious passage through a pool of water, twice (once to get the pollen, and once to transfer it), where they tend to sink (Figure 9.1A and Color Plates 12 and 13). *Contra* Darwin (1877; p. 173), the 20 ml or so of liquid in the large bucket of *Coryanthes* registers near zero on a sucrose-calibrated refractometer, and the nonzero component is not sugar; it may be saponin (Gerlach 2011) or another wetting agent. Darwin was equivocal on sugars and bucket orchids. Quoting Crüger (1864), he introduces another speculation on the same page—that bees vying for position to collect something from the orchid are perhaps "intoxicated by the matter they are indulging in," and thus sometimes tumble into the bucket and crawl out the small opening at the other end, where the stigmatic cleft removes pollinia, or a pollinarium is placed on an exiting bee. Darwin refers to the *Coryanthes* "bucket of water" correctly on page 265 of his pollination book's second edition. He also expresses the opinion that the large amount of water was not to provide a diluted sugar resource but was to wet the wings of the bees, and not allow them to leave without receiving or transferring pollinia. The liquid is not a pollinator reward in any sense, but the fragrance offered on the floral lip, which attracts the bees,

Orchids and Neotropical Pollinators since Darwin's Time   *235* is. Furthermore, that bees fall into the bucket because of their interaction with water dripped from the flower (Endress 1996) is doubtful, unless they merely slip in it while touching part of the "bucket." Bees do not have the space to hover within the bucket and cannot be fouled in flight by having a drop fall on them (Figure 9.1A,B and Color Plates 12 and 13), but may, indeed, collide with each other and wind up in the soup (Crüger 1864, quoted by Darwin 1877; p. 175).

The bees on Trinidadian epidendroid species were, I suspect, the euglossine species *Eulaema cingulata* (formerly *E. pseudocingulata*), 3 endemic *Eulaema*, and *Euglossa imperialis* and *allosticta*, the last 2 with tongues neatly folded under the abdomen and extending beyond its tip. They all "scratch and sniff." They work the surface of the orchid labellum directly beneath their mouthparts. When standing on the flower, they move their splayed basitarsal foreleg brushes up and down in alternating sweeps, much like a mime pretending to encounter a transparent wall, but there is negligible movement of the mandibles and no "gnawing" of the floral lip or labellum. The bee mouthparts—mainly the sharply pointed mandibles—might open a tiny bit and display a liquid, which is the oily liquid they themselves produce to capture the volatile orchid fragrances (Whitten et al. 1989; Eltz et al. 1999, 2003, 2005a, 2007, 2008). The antennae are in constant motion as their tips touch the flower's surface. Their mandibles do not come in contact with the flower.

Extrafloral Resources for Mutualists

Many Neotropical epidendroid species do have nectar, but not in their flowers. The sugary liquid associated with flowers in *Coryanthes*, *Catasetum*, *Gongora*, and *Cycnoches*, for instance, is secreted at various places near the flower and on the growing inflorescence, and only when floral buds, bracts, or stems are produced (Figure 9.2A,B). This implies that insects feeding on extrafloral nectar may aid orchid reproduction, but not as pollinators—as Darwin foresaw by discounting assertions that pollen arriving somewhere near the column and stigma of a flower could effect pollination. Darwin did mention nectar on the peduncle, sepals, or bracts in *Oncidium* s.l., *Notylia*, and *Vanilla*, and noted extrafloral nectaries on *Gongora*, but mentioned that the nectar only appears in the last genus after fruit set or pollination. His meaning was that the nectar is not produced to attract pollinators, yet he made no speculation regarding a function in attracting the ants, which might repel herbivores.

As we have learned, ants use extrafloral nectar and may rely on it, to the point that they nest in or near, and nurture, orchids. Orchid ex-

Fig. 9.2 (*left, right*). Ants on inflorescences of *Catasetum* sp., Panama. Photographs: David Roubik.

trafloral nectar is consumed by ants (Jeffrey et al. 1970; Fisher and Zimmerman 1988; Fisher 1992; Damon and Pérez-Soriano 2005), and ants are known plant defenders (Bentley 1977). The feeding ants may sting, bite, or jointly attack other organisms that approach, and thus can feed on an extra source of protein using orchid nectar as bait. Although present in small number, ants may shift in species from day to night on an inflorescence, and are there 24 hours a day (personal observation). This bodes well for a protection argument. But there is more—ants may build satellite colonies of orchid/ant gardens, and take advantage of a reliable source of carbohydrate through the year.

Ant colonies may inhabit epiphytic orchids in their dried pseudobulbs (Damon and Pérez-Soriano 2005), their rootlet masses (Gerlach 2011), or in an independent nest, near enough to provide some benefit to orchids, either by defecations and their nitrogen or fertilizer content, or by attacking herbivores. However, there is evidently little correlation with ant presence at orchid inflorescences and proximity of ant colonies (Fisher and Zimmerman 1988), and this seems generally true from my simple observations that, whenever an epidendroid orchid is about to flower in a tropical forest, ants arrive in a short time—but they are not living there. An exception is seen, usually, with *Coryanthes* of various species, which establish huge plants with dozens of pseudobulbs each, and perhaps a network of ant colonies, and flowers almost constantly present: *Cremato-*

gaster, Azteca, and *Camponotus* species are among those to take advantage of this arrangement (Gerlach 2011; personal observation), enriching the rootlets with their fecal debris and that brought from the forest. Darwin believed extrafloral nectar was a plant waste product. The recent authors suggest that ants defend flowers, and nectar satisfies the ants. There is no field observation validating either opinion, or, if someone has witnessed and recorded an ant chasing away a beetle, a cockroach, slug, stingless bee (see below), caterpillar, or katydid, among other animals that really do gnaw on a floral bud or flower, I have failed to find the reference. Finally, the tiny seeds of *Epidendrum* and possibly *Coryanthes* are attractive to ants and are carried from orchid fruit into ant nests, where they may germinate, be incorporated into ant gardens, and enjoy protection of "pugnacious body guards."

Orchid Bees for Pollination

The orchid or euglossine (Latin for "true-tongued") bees take nectar averaging 40% sugar (Roubik et al. 1995; Ramírez et al. 2002; Roubik and Hanson 2004; Borrell 2007). Most of that nectar comes from flowers with long, tubular corollas that other, shorter-tongued bees do not have legitimate access to, such as species within the families Apocynaceae, Rubiaceae, Bignoniaceae, Costaceae, Gesneriaceae, Verbenaceae, Marantaceae, Zingiberaceae, and Fabaceae (Kennedy 1978; Ackerman 1985; Ackerman et al. 1982; Roubik et al. 1995; Kay and Schemske 2003; Roubik and Hanson 2004; Roubik and Moreno, unpublished data). Euglossine bee non-orchid resources are essential to orchids that receive pollination visits from these bees. The euglossine female bees often collect pollen from hard, poricidal anthers—with a single pore from which pollen is ejected—but only when a bee grasps the anther with its legs and holds it to the venter while vibrating intensely, thus creating the "buzz" that "buzz-pollination" entails (see e.g. Buchmann 1985). These flowers can hold their pollen without hazards from insects, sun, wind, or rain, because it is protected, and typifies species within the families Solanaceae, Melastomataceae, and some Fabaceae. When we consider the lack of nectar for the euglossine bees in all but some *Sobralia* spp. and a few other nectar-making, large-flowered orchids (Ackerman et al. 2011), we must understand that the abovementioned pollen and nectar flowers are, indirectly to be sure, responsible for the existence of epidendroid orchids. Jersáková et al. (2008) maintain that orchid flowers supplemented with nectar are more readily pollinated than those lacking nectar. Yet orchids such a *Cattleya,* which do have nectar, are not bee pollinated. Orchids are an exception to the pollination rule of reward, reproduction, and regulation. Why?

In the evolutionary history of orchids, flowers that deceived their pollinators by appearing to have nectar and/or pollen were sought by the male and/or the female bees. Since males seek no pollen (because they do not feed offspring), this evidently gave rise to fragrance flowers (Chase and Hills 1992; Whitten et al. 2000, 2007; Chase et al. 2003; Ramírez et al. 2011b). Floral deceit involving resources or rewards is fairly common among angiosperm flowers and is based on floral mimicry (Dafni 1984; Ackerman 1986; Ferdy et al. 1998; Singer 2002; see Chapters 2, 4, 7, 10). Broadly defined, such mimicry can involve flowers with a similar shape, size, color, texture, pattern, position within the forest floor to canopy, and smell (see Smith et al. 2006 and Chapter 11). A "model" flower has food or a resource, and its mimic does not. As already noted, deceit is the pollination strategy of perhaps one-third of all orchid species (e.g. Renner 2005). However, this was presumably not as dependable for the euglossine-orchid connection as the system that has replaced it. The benefits to plants came ostensibly from higher fruit set in environments where nutrients and pollinators are scarce, and orchid population density low (Neiland and Wilcock 1998; Tremblay et al. 2005; Pérez-Hernández et al. 2011). The benefits to the pollinator very likely came from the elimination of competitors, little by little, as the orchid flowers responded to Darwinian selection and modified their rewards, or their presentation, to match fewer and fewer species.

When only one species is the principal pollinator, that animal has a resource that is not preempted by another species. The evolutionary door is thereby opened toward fidelity, specialization, and dependable pollination service (see Ackerman and Roubik 2012). This viewpoint is contrary to the "evolutionary race," which Darwin believed led to the modification of a floral trait to match the morphology of an important pollinator (see Chapter 8). There is little resemblance to a race. Instead, this coevolutionary counterpoint reveals a fundamental mechanism of close mutualism: resource reliability. It may be an example of either imperfect adaptation, or perfect adaptation, depending on point of view. From an evolutionary perspective, it is a win-win situation.

Pollinators of a proto-orchid were looking for nectar in nectarless flowers, and were among many generalist or unspecialized insect visitors, both male and female (Chase and Hills 1992). They were perhaps gullible, but they were not reliable, as such. Williams (1982) has suggested that obsolete nectaries and nectar repositories are present in the *Exaerete*-pollinated *Aspasia*, and that it rewarded euglossine bees with nectar before producing odors collected by males. This argument makes sense, so long as odors are more economically produced than nectar. Furthermore, nectar has hidden costs—by attracting perennial, social bees to steal or rob the resource, discussed further below.

Complex floral fragrances are present in the species representing the tribes Stanhopeinae, Catasetinae, and some Zygopetalinae, given as a reward to male orchid bees. Fragrances are sequestered by the bees in their hind legs and gain importance in successful mating (see below). We do not know if any other bees used orchid fragrances for similar purposes (but see Norden and Batra 1985), because there is no fossil record concerning whether orchids or other flowers provided fragrances used in bee mating behavior. We do know that euglossine bees are attracted by odor to a variety of other monocot and eudicot plants, including the Araceae, Melastomataceae, Euphorbiaceae, Solanaceae, and Gesneriaceae (Whitten et al. 1986; Ramírez et al. 2002; Roubik and Hanson 2004; and personal observation). We are unable to trace the lineages of orchid bee fragrance-seeking behavior back to euglossine bee origins, which occurred roughly 40 million years ago, but there are interesting hypotheses as to how and when fragrance collection might have arisen (Eltz et al. 2007; Ramírez et al. 2011b).

Fragrance Studies

At present, individual orchid flowers do not produce more odor constituents than other kinds of flowers (Kaiser 2011). The list of fragrant floral substances with chemical names is usually between 30 and 50 per orchid flower, but new techniques imply that earlier studies failed to detect around 75% of the different odor molecules (Perraudin et al. 2006). Even tropical trees, excluding their flowers, produce an average of 37 fragrant compounds per species (Courtois et al. 2009). In addition, research on orchids and other flowers demonstrates that a single flower can produce different odors at different times, and different individuals of the same botanical species also vary in the production of attractants or rewards (Kaiser 1993, 2011; Knudsen 2002; Moya and Ackerman 2008; Ackerman et al. 2011; Steiner et al. 2011). The variation that occurs within a species may eventually force us to consider altering taxonomy and species concepts for orchids, and also, perhaps, for orchid bees that build their singular fragrance blends, analogous to attractive, species-specific pheromones (see Zimmermann et al. 2009).

Having a flowering orchid from the field is the first step in building a reference library of its fragrances, and comparing it with existing refer-

Fig. 9.3 Fragrance-capture methods for *Catasetum* sp. taken from the field on Coiba Island, Panama. (*Left*) Open flowers of *Catasetum* sp. (*Right*) Flowering stems of *Catasetum* sp. enclosed in an oven bag used for headspace methods to collect fragrances in the sample column. Photographs: David Roubik.

ence libraries of named chemicals. The procedure for collecting orchid fragrances in the field, or from any living flower, involves "headspace" techniques. A container is placed over flowers, and allows air to be drawn from the outside environment. With the aid of a small pump, a large volume of that air is passed through an exhaust tube that captures the airborne fragrances; a "control," with no orchid present but having the same environmental air, is run at the same time (Figure 9.3A,B). The floral fragrances are then identified through gas chromatograph/mass spectrometry analysis (see Hills et al. 1972; Whitten et al. 1986; Kaiser 1993, 2011).

The genus *Catasetum* is perpetually intriguing and is highlighted by Darwin and in subsequent work; it serves to establish benchmarks in floral biology and orchid fragrance analysis. This is a gender-dimorphic flowering plant (see Ashman et a1. 2005). The individual male or female flowers of the dioecious *Catasetum* species (Schromburgk 1837) wait for 1 to 3 days after opening to produce a fragrance, and may continue to produce odors for up to a month—unless visited and pollinated (van der Pijl and Dodson 1966; Hills et al. 1972; Janzen 1981a,b; Williams 1982). Some single-day orchid flowers, such as hermaphroditic *Gongora*, also cease to produce odors after pollen is received, but not after the pollinarium is removed (Martini et al. 2008). This is completely the reverse of other orchids, which soon cease to produce odors and also wilt as soon as pollinia are taken by an animal visitor (Huda and Wilcock 2012).

Why *Catasetum* flowers, upon opening, do not immediately emit their fragrances is a mystery. Both sexes of a *Catasetum* make the same odors (Hills et al. 1972), also true for species of *Cycnoches* (Gregg 1983) and conceivably for *Mormodes* (Chase and Hills 1992), the other dioecious (Darwin 1862) orchid flowers. The *Catasetum* species produce a core fragrance of caraway seed–like smells, made also in alternative floral fragrance resources—the aroids—used by the same pollinating bees (Kaiser 2011; p. 227). Kaiser lists fragrance components of 2 species, each totaling 30 or more. In Panama, Jette Knudsen and I have sampled 7 plants of *Catasetum maculatum* from the Province of Veraguas, and have found 77 different chemicals, which include fatty-acid derivatives, terpenoids, monoterpenes, sesquiterpenes, benzoids, N-containing aromatics, and many more unknowns than all the preceding. If it seems that this particular *Catasetum* is hedging its bets by attracting different pollinators in different settings, this is perhaps unlikely, because the species attracts only a few species of *Eulaema* (van der Pijl and Dodson 1966; Schmid 1969; Williams and Whitten 1983; Roubik and Hanson 2004). Sesquiterpenes, monoterpenes, and other volatile constituents in blended chemical plant products are repellents and deterrents, to both herbivores and certain flower visitors, while aromatics primarily attract bees and other potential pollinators (Dobson 2006; Raguso 2008; Courtois et al. 2009; Schiestl and Schlüter 2009; Junker and Blüthgen 2010; Schiestl 2010). The male *Euglossa cybelia* responds to an attractant, but although sensing the odor, does not respond to a potential repellent (Schiestl and Roubik 2002).

Orchids apparently do not combine their fragrances to produce one spectacular attractant for pollinators. A synchronous, communitywide flowering of an orchid is very rare (Dressler 1981). Orchids such as *Maxillaria camaridii*, at my rain forest home in Panama, open in synchrony on several plants, smell strongly in early morning, then wilt the next day. Their flowers attract a variety of beetles in the family Chrysomelidae (beetles and bees sometimes find attractive fragrances in the same flowers; Knudsen and Mori 1996), which mate on the flowers. I see halictid bees and stingless bees entering flowers and they may transport pollinia, but this flower is perhaps self-pollinated—with flowers of such short longevity (see Davies and Turner 2004; Martini et al. 2008). Genus *Maxillaria* is so large and diverse that further research into natural species groups and their phenology or fragrances will probably serve to reorganize the taxonomy and delimit genera (Whitten et al. 2007), as happened for the genus *Catasetum*, which produced *Dressleria* and others.

How much fragrance a bee gets from its activity at an orchid labellum, how often that stop results in the pollination of an orchid, and how much the flower "pays" for making an attractant chemical rather than a showy flower that stands out to the pollinator like a neon sign, or nectar or pollen that are needed by a pollinator for its personal fitness, are still quite unknown. Like Darwin, we lack observations to propose a comprehensive natural history.

Orchid Pollination Mechanics

Darwin never witnessed the drop-through (*Stanhopea*), slide-down (*Gongora, Mormodes*), scratch-and-sniff (*Coryanthes, Catasetum*), or pseudo-deceit mimicry (*Oncidium* s.l.) orchid pollination we know about now. Although he tried to put us in touch with the act of pollination, he was more formidable in explaining the traits for reproduction and self-perpetuation that orchid flowers possess. Darwin comprehended that a large pollen-to-ovule ratio means that although the pollen grains and ovules in a single flower already number in the millions, there is nonetheless strong competition between male flowers. His explanation was that the pollen grains are crucial and well protected so that they will be dispersed and deposited neatly in a stigmatic cleft, thus leading to fertilization of the many seeds required by the single-event pollination of an orchid. Conversely, he thought of orchid seeds as ideally suited for dispersal, due to their light seed coats and minuscule size, but he missed the importance of fungi or ants in making such light and tiny seeds viable (in contrast, see Dressler 1981; Tupac et al. 2002, 2007; Rasmussen 2008; Gerlach 2011). And these seeds die in spectacular number, without establishment in suitable sites. Because seed set can also be extremely low (e.g. Martini et al. 2008), the long life of an orchid seems essential. Coupled with a very precise and relatively quite narrow pollinator relationship, one might expect that long-lived orchids have the smallest chances of reproduction at a given time, and have the smallest effective population sizes (see Tremblay et al. 2005; Pérez-Hernández et al. 2011). Their success requires a generous helping of what we will term and explore as pollinator fidelity.

A Prelude to the Bees

A long, antagonistic relationship with bees not only may have turned some bee robbers (those that damage floral structures) or thieves (hopelessly bad pollinators) into pollinators, but also seems likely to have promoted the evolution of pollinaria (encased pollen sacs) and nectarless flowers among orchids. Orchids discussed by Darwin seem to have solved

Fig. 9.4 (*Left*) *Trigona ferricauda* in the process of damaging a labellum. (*Right*) Floral damage on *Sobralia* caused by nectar-robbing *Trigona ferricauda*, Colon Province, Panama. Photographs: David Roubik.

a fundamental problem by being long lived and investing in pollen protection, although the use of deceit or a lack of nectar was unexplained. The existence of pollinaria also is paradoxical because it is so rare, occurring only in the asclepiads (Apocynaceae) and most Orchidaceae (Johnson and Edwards 2000; Willmer 2011). The closest thing in floral strategy is perhaps the poricidal anther in certain eudicots, which also encloses and protects the pollen grains, at least from damaging ultraviolet radiation, pollen-predaceous thrips (Willmer 2011), and inclement weather. Poricidal anthers, with loose pollen within them, are chewed open and have their pollen removed by *Trigona* (Meliponini) (Renner 1983; Roubik 1989, and references therein). *Trigona* also chew open flowers of orchids to obtain nectar (Figure 9.4A,B), a behavior in which they excel (Roubik 1982, 1989). Interestingly, the earliest physical evidence of orchid flowers that indicates a pollination relationship with small Meliponini is a flower of *Goodyera* (subfamily Orchidoideae) entrapped in amber with a stingless bee, circa 20 million years ago (Ramírez et al. 2007), with the pollinia sectile globs, small masses connected in a pollinium (Dressler 1981), on the bee.

In a nutshell, I propose that because most orchid flowers are long lived, among the Meliponini, *Trigona* together with a few *Partamona* that damage and perforate flowers to this day, there are stingless bees that have evolved into orchid pollinators, and have influenced the trajectory of orchid evolutionary adaptations to flower visitors (Roubik 2000). Moreover, the longest-lived Neotropical orchid flowers (individual flower life, and flowering period of the local population; see Ackerman and Roubik 2012) would be unable to persist as potential pollen donors or receivers if their nectar and pollen were available to nonpollinating or robbing stingless bees. Even for a flower lasting a single day, or only for a few hours, as in members of the genus *Sobralia* (Figure 9.4B), there is no escape from the flower-perforating and destroying stingless bees within Neotropical forests. Most tropical flowers last for one day, in some contrast to temperate-zone flowers, and the outstanding exceptions are orchid flowers, the only ones besides Apocynaceae with encased pollen masses, and usually no nectar (Endress 1996; Ashman and Schoen 1994; Ashman et al. 2005; Willmer 2011; p. 489–493).

Orchids have a long and perhaps unusual relationship with the stingless bees, particularly in the Neotropics, where Darwin focused on the large flowers with showy displays. Stingless bees obviously find and may recruit to such floral displays (Roubik 1989). The fact that they are social, and have an acute foraging memory, means that there are always a lot of starving colonies to feed, and their numbers are in the tens of thousands in each hectare of forest (Kajobe and Roubik 2006). If they did not influence the evolution of floral traits in orchids, it would be surprising.

Special Orchid Features and Futures

Several botanical terms applied to orchid flowers have changed over the century and a half since Darwin's book, and need brief explanation here. As an entomologist, Darwin borrowed certain entomological terms for botanical structures, such as *antenna* for the sensitive appendage cirrhus inside the flower of a *Catasetum* sp., which triggers a catapulting pollinarium that sticks on a potential pollinator. He also used the word *pollinium* for each of 2 different structures—the pollinarium (see van der Pijl and Dodson 1966; Dressler 1981), which is the entire pollen-bearing unit of an orchid flower, with hard pollinia for the individual pollen sacs or pollinia, of which there are 2 to 8 in a flower (depending on the genus), and a viscidium and stipe; or the massed pollen in loose bunches, with no hard outer case, with single or double anthers and 1 or 2 pollen masses or pollinia as, for example, in the genus *Vanilla*, the subfamily Cypripediodeae (see Chapter 10), or *Sobralia* spp. The parts of the pollen unit also were viewed with an entomological bent as the pedicel and disc, now referred to, respectively, as the stipe and viscidium. The term *caudicle* has endured (see Chapters 2 and 3), and is the flexible portion of some pollinia, which ruptures to allow pollen grains to contact a stigma. Further nomenclature puts into perspective the fact that the hard pollinaria carried by many orchid bees and other pollinators, such as bees, flies, wasps, birds, or butterflies, are complemented by the soft masses of pollinia, including sectile bundles within a single mass, which can be called a pollinium but not a pollinarium.

As a botanist, Darwin adroitly contended that 3 genera were in fact 1, in the case of *Catasetum* spp. and their gender-dimorphic flowers, which also produce some sterile hermaphroditic flowers (Schromburgk 1837; Knoll 1858; Hoehne 1933). Some confusion occurs in his persistent use of different generic names for the flowers that he explained are the male and female from the same plant. He did emphasize the inheritance of useless traits in the sterile hermaphrodites (Darwin 1877; p. 201), also sometimes on the same plant as male and female flowers, and offered this example as a regression to a progenitor—strictly the product of individual inheritance and genes rather than a knowledgeable design. At present, even the DNA molecular "bar coding" techniques employing the chloroplast "matK" sequence cannot distinguish among different species within a genus belonging to the Epidendroideae (personal observation), nor is the nuclear ribosomal internal transcriber (nrITS sequence) much better (Gerlach 2011). We have yet to eclipse Darwinian reasoning with pure fact.

Outcrossing or Species Isolation?

Darwin emphasized outcrossing in orchids. Others since the 1960s emphasize isolating mechanisms provided by fragrances or other floral traits, and their role in orchid and orchid bee specialization and speciation (Williams 1982; Gerlach and Schill 1991; Ferdy et al. 1998; Neiland and Wilcock 1998; Eltz et al. 2006, 2008; Zimmermann et al. 2006, 2009). Darwin also observed that orchid flowers first remove pollinia from flower-visiting animals, and then place their pollinarium, or the combined mass of pollen, on a bee as it leaves the flower—a fundamental mechanism for avoiding self-pollination. He further pointed out that there is an anther cap covering a pollinarium, which falls off after 20 minutes or so (see also Williams 1982; Dressler 1981). This cap is a second mechanism to avoid self-fertilization. The individual pollinia also shrink with time, making them fit within the stigmatic cleft or slit, thus usually avoiding fertilization on the same inflorescence or plant (Gerlach 2011). Protandry by floral morphology is achieved by species of *Clowesia* and *Dressleria* spp. (Chase and Hills 1992) in which the anther covers the stigmatic cleft; thus, a hermaphrodite flower is male before the pollinarium is taken, and female afterward. Other researchers indicated the diversity of pollinarium placement sites used on an individual bee, such that as many as 5 different orchids can be serviced by the same bee at the same time (Dressler 1981, 1993b; Roubik and Hanson 2004; Ramírez et al. 2011b).

The large size, color, and evocative shape of orchid flowers, in addition to their fragrance, attract and potentially deceive pollinators (see Naug and Arathi 2007). The smell was added or accentuated among the other physical attractants (Raguso 2008), and carried the bees a step further. They are perennial seekers of the rare orchids in the Neotropical epiphyte jungles. And strong attraction is essential to induce a bee that has been slapped by a catapulting pollinarium to again seek the same fragrance and enter a similar flower. Romero and Nelson (1986) noted bees carrying up to 3 pollinaria of some *Catasetum* spp., but observed that some did not take the opportunity to visit a second male flower, after having recent experience with a catapulting pollinarium. Who would question the deep motivation of an orchid bee that visited more than one flower of a *Catasetum* sp., with a large (among the largest of all) pollinaria, and also its forceful ejection, which was immediately visible to Darwin and his contemporaries? Ejected pollinaria ended up on the glass windows of hothouses. Finally, as Darwin suggests in the final line of his book, it has been proved that orchid flowers may be either self-incompatible, thus requiring outcrossing, or self-compatible; but that the nature of selfing makes it likely that smaller or fewer seeds or fruit survive, which is similar to many eudicots (Roubik 2002; Huda and Wilcock 2012).

Darwin (1877) did not enunciate the "deceit" mechanism of orchid pollinator attraction, but he did address bees visiting nonrewarding flowers as a point of some contention between his experimental results, and express skepticism that bees knew what they were doing—they apparently visited flower after flower without receiving any reward. Darwin did point out that his experimental data led one to believe bees mostly abandon a nonrewarding *Catasetum* sp., and that bees with a pollinarium of *Coryanthes* approach yet another bucket orchid (Figure 9.6). How do we know that the bees are so highly motivated—and also pollinate—by the transfer of a pollinium between different flowers? If one closely examines *Euglossa* spp. that visit *Coryanthes* spp. (see Roubik and Hanson 2004), there are stipes and viscidia still on the bee. The viscidium (Figure 9.5) cannot readily detach. Pollinia do not fall off when carried from a flower (being firmly glued and protected by an anther cap). The anther cap also is carried from male flowers such as those of *Catasetum* spp., so that it is not necessarily functioning solely to promote outcrossing. Interestingly, Darwin noted that artificial self-pollination of an orchid could occur, with pollinia placed on stigmata, but the anther cap, flowering phenology (timing), and bee behavior would normally prevent this. Orchids do hybridize, and the orchids that seem best suited to outcrossing do self-pollinate (van der Pijl and Dodson 1966; Dressler 1981; Romero and Nelson 1986; Huda and Wilcock 2012; and see Chapter 7), but not very often.

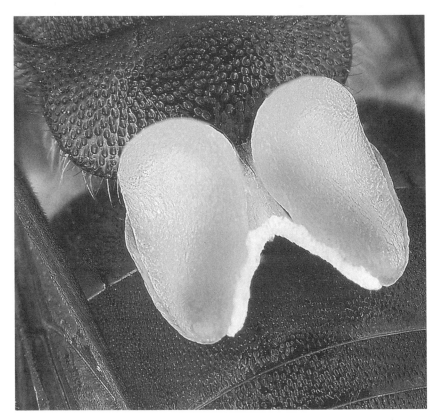

Fig. 9.5 The pollinia and white posterior rim; the caudicle of *Coryanthes* sp., riding on a male *Euglossa mixta*, USNM. Photograph: David Roubik.

Long after Darwin, it was proposed that the weight of the pollinarium was registered by bees visiting *Catasetum*, and was a deterrent to further visitation of a floral morph, although the smell of the flower was the same (Hills et al. 1972; Romero and Nelson 1986). It may indeed seem that a male orchid bee, much like a large football player with a pair of "Hoovers" (vacuum cleaners) strapped on his back, would like to be rid of pollinaria (Figure 9.6; and see related comments in Chapter 6). That pollinarium was thought to be a sizable portion of a pollinator's weight (Romero and Nelson 1986), but this seems somewhat inaccurate. The size or weight of the pollinarium is unlikely to matter to a euglossine bee, given its flight power, its body weight, or the reward that the fragrance represents. For example, the pollinator of *Catasetum maculatum* (Janzen 1981a,b) is *Eulaema marcii* (formerly called *E. cingulata*), and the pollinarium is extremely large. That bee weighs approximately 611 mg; the pol-

Fig. 9.6 Important pollinators of *Catasetum, Coryanthes,* and *Gongora* in Panama, showing multiple pollinaria from multiple floral visits. Large bees are male *Eulaema marcii,* carrying pollinaria from *C. maculatum, C. viridiflavum,* or the much smaller *Gongora* sp.; the smaller bee, *Euglossa dressleri,* carries 3 pollinaria from *Coryanthes,* 2 of which are missing 1 pollinium, taken by the stigmatic cleft of 2 flowers. Photograph: David Roubik.

linarium weighs 28 mg when it has been on the bee (and dried), and the stipe and viscidia, with no pollinia, weigh 18 mg (personal observation; Figure 9.6). Pollen and its 2 pollinia covers, on the pollinator, weigh 10 mg. Yet bees can visit and carry 2 pollinaria (Figure 9.6). After visiting the orchid, and perhaps having viable pollinia attached for a number of days, the bee must carry an added 5% of its own weight.

Orchid bees can apparently maintain normal activity carrying a load (a tracking transmitter), similarly placed on the thorax, equal to their total weight (Wikelski et al. 2010), and bees in general can bring home a foraging load equal to their weight (Roubik 1989). Smaller pollinaria of *Catasetum* spp., such as that of *C. minor*, weigh less than a milligram (personal observation). They probably do not impede or influence bee visitation to dimorphic orchids, yet their flowers are strongly dimorphic, which had been suggested (Romero and Nelson 1986) to occur mainly among the larger *Catasetum* spp. with the heaviest pollinaria.

The viscidium, stipe, and pollinia of *C. maculatum* appear aerodynamically suitable for a flying bee (Figure 9.6), and the connection between the stipe and viscidium is flexible (Darwin 1862); thus, resistance to wind and final reception by the stigmatic cleft are optimized as the bee switches its position and moves either up or down. Darwin alluded to the flexibility of the stipe in different contexts, emphatically in the requirement placed on bees to climb within the flowers of *Catasetum* upside down. Some visitors, the larger *Eulaema* spp., do not do this (Figure 9.7)—and this particular behavior seems to demonstrate "imperfect adaptation," which helps to confirm the process of natural selection that Darwin and Wallace presented.

The persistence of male orchid bee visitation to complicated flowers is remarkable. Shown in the middle of Figure 9.6 is a *Euglossa dressleri* that has visited 5 flowers of *Coryanthes*. How do I know that? There are 3 viscidia, with 1, 1, and 2 pollinia. The visible stipe in the figure has a single pollinium. The viscidium and its light-colored strap (Figure 9.5) remained on bees that lack 1 or 2 of the pollinia placed there by an orchid flower. These bees have had their pollinia removed by the stigmata of *Coryanthes*. The bees are indeed strongly motivated, and an independent and considerably more bizarre example is their behavior of seeking and removing fragrances from the legs of dead male bees (Eberhard 1997; Roubik 1998; Roubik and Hanson 2004). The scent reward seems to outweigh many obstacles, and its evolution in orchids has produced unparalleled results.

We still cannot know exactly where the pollen goes after removal by bees (see Janzen 1981a,b). Once the genetics of pollen and plants in the area are studied, there can be very good guesses as to which parents sired which seeds (Austerlitz et al. 2004), and more directly, there is a foreshadowing of radio-tracking a euglossine bee on its trips to several flowers (Wikelski et al. 2010). For now, what we do know is that pollinia were removed from a bee by a flower, and whether a bee carrying pollen has visited multiple flowers of the same or different species. Janzen (1981b) proved the case for effective pollen-dispersal and fragrance-seeking be-

Fig. 9.7 Three male *Eulaema*, 1 smaller *E. cingulata*, and 2 larger *E. meriana*, on the open flowers of female *Catasetum discolor* in Suriname, South America. No bee is carrying a pollinarium, 2 are oriented incorrectly for pollinarium placement (an upside-down approach is required), and all 3 are mopping up odors and sequestering them. Photograph: David Roubik.

havior of male bees, despite perhaps learned (and then forgotten) aversions to having pollinaria slapped onto their back. Learning and, more important, memory depend on the speed at which the stimulus is administered (Fernandez et al. 2009). In this context, the incredible rapidity of a catapulting pollinarium from flowers of *Catasetum* spp. (Nicholson et al. 2008) might minimize a bee's memory of a negative experience, as it likely diminishes the inefficiency of pollen wastage and promotes the evolution of unisexual flowers (Romero and Nelson 1986). I hasten to add that although Darwin (1877) wrote about and studied 7 *Catasetum* spp., he did not know whether they flowered at the same time in nature, or if they were visited by the same bee species. He undoubtedly sensed this weakness and thus concentrated on the outcrossing of orchid individuals, enforced by their occasional unisexual flowers (of which he recognized 2 of the 3 Neotropical catasetines), by their structure, and by comparative morphology, rather than the wonderfully complex and perfect contrivances by which they avoided wasting their floral sexual phase. After the next section, this theme will be continued.

BEES

The Neotropical Bees, Which Darwin Never Saw

Orchid bees of the tribe Euglossini in the Apidae (1 of the 7 bee families) serve as an example of what can be found in nature, to our constant delight, when one takes a good look (Figure 9.8). On very close inspection, the euglossines prove capable of tremendous flight power and endurance, exhibiting impressive flight in their daily undertakings (Janzen 1971; Kroodsma 1975; Dillon and Dudley 2004; Darveau et al. 2005; Wikelski et al. 2010). There could not be a better pollinator for widely spaced tropical plants growing in a rain forest. Due to their size ranges and differing morphology and physiology, the orchid bees can survive and forage throughout tropical forests and even certain much drier and open habitats (Wille 1958; Roubik 1993; Roubik et al. 2003), and they often fly during rain.

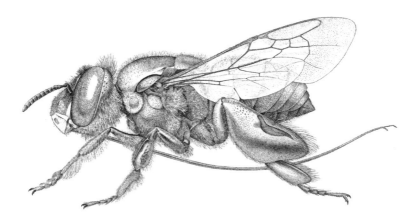

Fig. 9.8. A male *Euglossa ignita* showing the folded tongue extending beyond the tip of the body, and the large slit extending along the edge of the hind tibia, where volatile chemicals are stored or removed. Illustration: J. M. F. de Camargo; gift of the artist.

Taxonomy sometimes muddies the waters in Darwin's (and nearly everybody else's) biological investigations (Roubik 2013). The diversity and abundance of the tribe Euglossini (common taxonomic usage is stated in Michener 2007) are basically high in rainy and lowland- to middle-elevation Neotropical environments. On the other hand, there is no congruence seen between orchid species or genera pollinated by the bees, and the numbers of total orchid species in the habitat. That is, for one example, as many orchid bee species are present in lowland moist or wet forest, and cloud forest of 800-m elevation, or in the middle of the Amazon versus the core of Central America—Costa Rica and Panama, where some 50% of the national territory is still covered with native vegetation (Roubik 2004). On Barro Colorado Island, in the middle of a nature reserve in central Panama, there are 50 orchid bee species, but only 27 species of euglossine-pollinated orchids (Ackerman 1985). In the higher elevations of cloud forest (ca. 1200-m elevation and above) in both Central and South America, the numbers of orchid species in many groups tend to increase substantially (Meisel and Woodward 2005), while those used by euglossine bees do not. Indeed, we see fewer than a dozen orchid bee species above 1200 m at 9° north latitude, as in the edges of the tropics in Mexico and higher-latitude habitats of South America (Wittmann et al. 1988; Roubik and Hanson 2004; Michener 2007). Other pollinators of orchids service the bulk of their species in the richest tropical orchid assemblages (see van der Pijl and Dodson 1966; Gentry and Dodson 1987).

The orchid pollination niche regarding euglossine bees is mainly filled by the small colonies (roughly 2 to 20 individual females) that produce the needed pollinating male *Eulaema* and *Euglossa*, active all year (Dressler 1981; Roubik and Hanson 2004). Among the orchid bees are 2 cleptoparasitic genera (*Aglae* and *Exaerete*, 9 species), 1 solitary genus (*Eufriesea*, about 65 species), 1 mostly social genus (*Euglossa*, of around 130 species), and 1 occasionally social genus (of large, bumblebee-like insects, *Eulaema*, with around 20 species). The euglossine bees have no bee that lives in permanent or relatively long-lived colonies that have a queen and worker castes, but some *Euglossa* have dozens of female bees that share a nest, and there is persistent and sometimes violent competition between them—often taking the form of egg eating (Roubik and Hanson 2004). We have never found a long-lived colony of females that clearly have a queen, and which look different among them, with neuter females. However, a female and her offspring may occupy the same nest among *Euglossa*, a trait unknown among other euglossine bees (Roubik 2012).

The consumer of entomological nomenclature must take care. Orchid bees, *Melipona* spp. and their allies, and also honeybees (*Apis*) and bumblebees (*Bombus*), are sometimes relegated to subtribes, according to proponents of a single family of bees, but only they propose a division of tribes into subtribes, to model diversity. This practice is rare in current literature. Further, Kimsey (1982) "sunk" a bee genus, *Euplusia*, still found in the older orchid literature. The *Eufriesea* species were contained in a natural species group of "short-tongued" orchid bees, the majority of which were later classified as the genus *Euplusia*. Because the name *Eufriesea* was coined in the taxonomic literature before the name *Euplusia*, the priority rules governing taxonomic practice require that *Eufriesea* replace *Euplusia, when applied to those species*. In a thorough examination of the medium-size orchid bees (with no sociality) made by Kimsey (1982), the result was that rather than treat *Eufriesea* as a small species group within the genus *Euplusia*, or as a subgenus, *Eufriesea* became the name of the entire genus. This genus has never received subgeneric classification, in contrast to the genera *Euglossa* and *Eulaema*, thus differing greatly from other categorical modifications within a relatively large group of bees, such as *Bombus, Centris, Trigona*, or *Megachile* spp., in which numerous subgenera are named and applied (see Michener 2007).

Darwin had experience with *Bombus* and *Andrena* spp. (family Andrenidae) on European wild orchid flowers, and mentioned *Euglossa* spp. on Neotropical orchids in his 1877 book. The Darwinian-European imprint on orchid-bee biological interpretation lasted at least into the 1960s, when Vogel (1963a) and van der Pijl and Dodson (1966) still brought up sexual deception and narcotic behavioral modification (thinking of *Ophrys* spp. and *Andrena* spp. [Chapter 3], and of *Euglossa* spp. falling into orchid buckets) in orchids and orchid bees. As a sign of significant growth in studies, Darwin's original 1862 volume proposed 6000 species and 433 genera among the world's orchids (quoting Lindley 1846), now approaching 30 000 and 600 taxa, respectively (Govaerts et al. 2012). The bulk of this chapter shows a similar multiplication in raw knowledge. Furthermore, I would not hesitate to say there are at least 5 times as many bee genera involved with orchid pollination in the Neotropics as were mentioned in Darwin's book. The most significant bees there, servicing Darwin's showy, hothouse orchids, are within the tribe Euglossini in toto, genera *Euglossa, Eulaema, Eufriesea, Exaerete* (and perhaps *Aglae*), the tribe Bombini (*Bombus*) in the highlands, with a few widespread Amazonian lowland species, and on deceit flowers (van der Pijl and Dodson 1966; 80–81), and the tribe Meliponini—all tribes in family Apidae, subfamily Apinae (Michener 2007).

Euglossine bees collect floral fragrances and store them in a very special manner, within a fibrous hind-leg pouch (Kimsey 1984; Eltz et al. 1999). The male bees must hover during transfer of their collected orchid fragrances (see Color Plate 14). The bees move the collected fragrance (Eltz 2005a) by combining it with a fine oil they secrete from cephalic glands into the slit on top of their enlarged hind tibia (Figure 9.8). They later extract the stored volatile odors (Eltz et al. 2005a) and let them be carried by air currents to attract bees to their perch site on the stem of a woody plant. A female bee flying to that site lands, allowing the male to copulate with her (summaries in Roubik and Hanson 2004; Zimmermann et al. 2006).

Orchid visitors are almost all male euglossine bees (*Sobralia* spp. are often nectar flowers and are visited by bees of both sexes; see Ackerman et al. 2011), and sterile workers in the tribe Meliponini—but are evidently just the females in other bee families. In Panama, we have seen *Epicharis* spp. (tribe Centridini) visit female flowers of *Cochleanthes* spp. (a deception flower; Ackerman 1983b), while bees in the family Halictidae visit *Maxillaria* spp. and some other orchid flowers (Singer and Cocucci 1999a). We are largely ignorant of any orchid chemical or nutrient value to species within the Meliponini, the family Halictidae, or the tribe Centridini, but the first have been often seen gathering pseudopollen (proteinacous trichomes, or shed epidermal cells) as food (Singer 2003; Davies 2009). It is unknown whether bees use orchid substances as building material (Roubik 2000, 2006). Deception at the level of perceived food may entail wastage by some flower visitors, and the species within the Meliponini are the ubiquitous forest visitors of tropical flowers (Roubik 1989).

Vanillin, which is a general chemical attractant for orchid bees (see below), is found naturally in flowers of *Maxillaria rufescens* (van der Pijl and Dodson 1966), and *Trigona pallens* relentlessly pursues this attractant through Suriname and French Guiana (personal observation). The crystals are carried on the hind legs (on the worker bee corbicula—normally used for pollen or resin transport) back to the nest, as was also witnessed with the large stingless bee *Melipona insularis* on Coiba Island (Roubik and Camargo 2012). What use this material has to the social bees is unknown. At the same time, the flowers of *Vanilla* spp. are said to produce no odor that attracts pollinators to them (Lubinsky et al. 2006). I find 2 species in Panama that do, and they attract *Euglossa* and *Eulaema* spp.

Baiting for Orchid Bees

Bee baiting, an application of chemical compounds from orchid fragrances in the field, came about all at once in the 1960s and revolutionized the way that orchid pollination was studied (Lopez 1963; Dodson et al. 1969; Dressler 1976; Williams 1978). Before then, flowering orchids were taken in the forest, and the arriving bee visitors or pollinators were then observed. Preliminary work using chemical baits addressed and continues to assess which bees are found in nature preserves or other habitats (Dressler 1981; Roubik and Hanson 2004), and is not reviewed here. Nor can I justifiably include the fascinating chemical details on structure, manufacture, stability, volatility, and modification of individual compounds in field conditions, or how they are utilized or learned in the brains of bees. However, our long-term study of bee abundance and species at chemical baits on Barro Colorado Island, Panama (Ackerman 1983a), coupled with a phenological study of *Spathiphyllum* spp. (Araceae) and its orchid bee male attendants through the year, demonstrated convincingly that baits can provide a window into the workings of a tropical ecosystem—and those of its euglossine bee–pollinated orchids.

When we place the baited 7-cm pads on tree trunks, we record and identify each orchid bee species that arrives, without killing the bees, and count how many there are. Counts are made every 15 minutes, on all 3 baits positioned at a baiting site, for 16 times in a morning—before the afternoon rains, and when temperatures most favor odor dispersal and bee arrival (Williams and Whitten 1983; Roubik and Ackerman 1987; Roubik and Hanson 2004). We also collect and note orchid pollinaria and pollen from other plants on the bees (Ackerman and Roubik 2012).

A 20-year study of orchid bees demonstrated little variation between sites, and a general trend could be detected in phenology and abundance of pollinators (Roubik and Ackerman 1987; Roubik 2001; and see Armbruster 1993). A 30-year study conferred a special "no" on the proposition that more specialized orchids require more stable, abundant, or specialized pollinators (Ackerman and Roubik 2012). These studies can be made, because central Panama contains the large nature reserves of the Chagres River watershed and the forest island of Barro Colorado (formerly the Canal Zone Biological Area [1923–1968], now part of the Barro Colorado Nature Monument, protected by the Smithsonian Tropical Research Institute and the Republic of Panama), along with Soberanía National Park. All the chemical fragrances known to attract the bees were tried individually (Ackerman 1983a). Pure wintergreen (methyl salicylate), eucalyptus (1–8 cineole), and skatole are the best general attractants. Vanillin, clove (eugenol), p-dimethoxybenzene, beta ionone, and a number of other attractants occurring in orchids, aroids, fungi, and other forest sources (Ramírez et al. 2011b) are also quite effective. Their use with adequate sampling protocol is sufficient to collect a wealth of data on pollen sources, nectar sources, and bees, along with orchid pollinaria.

Orchid fragrances mimic cephalic secretions of stingless bee workers (Meliponini) that coordinate colony foraging, and thus may attract bees, as would a foraging nest mate (Roubik 2000). These are not well-tested ideas. In contrast, an important literature is accumulating on chemical fragrances and pollinating euglossine bees, primarily as a means of studying populations and their evolutionary ecology.

One of our early findings was that most orchid bees, with the exception of many *Eufriesea* spp., are relatively stable in populations and active yearlong (Roubik and Ackerman 1987), but species often change their fragrance preferences in time and place (Ackerman 1989). Surprisingly, Ackerman's work also established that orchid bees are not entirely dependent on orchids, but the orchids are certainly completely dependent on orchid bees for their pollination (Ackerman 1983c). That study has since been confirmed in a variety of ways (see below), which represents the greatest single advance in understanding orchid-bee interactions since Darwin. Aside from this, we know, in great detail and for periods of over 30 years, which euglossine bees visit which orchids.

Bees and Orchids in Nature

The orchid bee chemical and fragrance story has been elaborated in fascicles that are both anatomical and behavioral (see Eltz et al. 1999, 2003, 2005a, 2006, 2007, 2008, 2011). After all this development, it is positively defamatory to find that "orchid bees do not need orchids," but this is precisely what we now believe (Ackerman 1983c; Roubik and Ackerman 1987; Pemberton and Wheeler 2006; Ramírez et al. 2011b). What we do not know is what proportion of most important bee chemicals in the hind legs came from orchids. Very few, less than a dozen among nearly 600, were exclusively from orchids (Ramírez et al. 2011a,b)—a statement that required examination of 7000 male bees of 30 species, and the fragrances extracted from all known sources of eulgossine male bee fragrances—orchids, aroids, other dicots and monocots, and fungi. In the final analysis, most of the orchid bees' leg contents were of unknown origin and also are—with this level of sophistication, unlikely to occur—in orchids. Behavioral assays with live males and females in their mating perch areas offer the solution to the specific fitness value of a chemical constituent from orchids or elsewhere—but how can one go straight to those perches and manipulate the bees for study?

The chemical odors add to a strong general attractant and may "fine-tune" the pollinator spectrum by eliminating certain visitors and reinforcing the proven attractants of those that are better pollinators (see "Orchid Floral Biology" above). Thus, all complex blends of chemicals exhibited in orchid odors, or in the legs of male orchid bees, are casting a widening net but expanding the mesh size. There may be substantial "noise" in the contents, whether by design, deceit, or random mutation and genetic drift. Such considerations were all but absent in Darwin's publications.

Orchids have apparently utilized one nonchemical option for attracting more pollinators—they flower for a longer period (Ackerman and Roubik 2012). We found no correlations between potentially fitness-enhancing characteristics, such as a predictable high abundance or predictable visitation and phenology, in pollinators and their host orchid. We did find that the longer an orchid species flowers, the less "faithful" or "specific" its pollinators are. It is an encouragement to field biologists that we obtained similar results in a 1-year, 7-year, and 30-year study in the same place. The orchid bees and orchids have an "asymmetric" relationship, as already mentioned. For instance, euglossine orchids usually depend on 3 different orchid bee species, often more than 1 genus (Roubik and Hanson 2004; Ramírez et al. 2011b; Ackerman and Roubik 2012). The orchid bees depend on an average of 6 orchids, but they may use none, and certain orchid species use over a dozen pollinating species.

Invasive orchid bees are scarcely known, but one such bee arrived in southern Florida, probably aided by inadvertent human dispersal, several years ago (Pemberton and Wheeler 2006). (I think that it is strange when I find that the 2 most abundant bees in Dade County, seen at nature reserves, and even near the American Orchid Society building in pine forest, are honeybees from Africa and an orchid bee from Mexico. Further, that the orchid bee is an undescribed species [Eltz et al. 2011]). Nonetheless, Pemberton and Wheeler (2006) determined that its hind-leg chemical constituents, remarkably, were composed from chemicals found in the leaves of allspice (*Pimenta dioica*, Myrtaceae), an Indonesian plant cultivated in Florida. The bees used no local or exotic orchids, and could obtain all their required resources, including those for mating, elsewhere.

One of the tenets of pollination ecology, often applied to crops and orchards, is that pollinator abundance determines the quality and quantity of fruit. If "quality" includes the biological trait of fitness—thus the chance of seed survival and growth to a reproductive age—we know the "pollinator abundance implies success" analogy is untrue for *Catasetum*

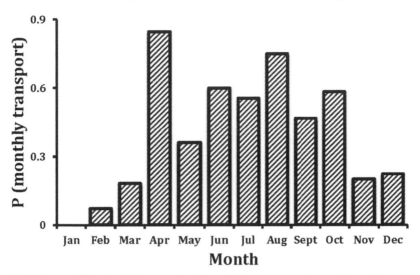

Catasetum viridiflavum transport by *Eulaema marcii* (N =164 censuses, 1979-2011)

Fig. 9.9 The probability of pollination, monitored by pollinaria presence or absence, during a 4-hour baiting study with 3 chemical attractants in the Soberanía National Park of central Panama. *Eulaema marcii* is the pollinator of *Catasetum viridiflavum* in this area (see also Zimmerman et al. 1989). David Roubik.

viridiflavum (Zimmerman et al. 1989). The study found evidence that nature and natural selection juggle a number of variables at once. When we published the paper, we summarized abundance data for 7 years on the sole pollinator, *Eulaema marcii* (then called *E. cingulata*), in central Panama. Now I can add 25 years to the pollinator abundance data, and a component that we had neglected: when, during the year, is *E. marcii* an active pollinator of the plant (Figure 9.9)? The bees go to the flower all year. Only January has never produced a bee with the pollinaria at the site, Pipeline Road, in Soberanía National Park, chosen for this example. This is the same forest and some disturbed areas as that of Barro Colorado Island Natural Monument, where several other studies on *Catasetum* spp. have been made (Ackerman 1983a,b; Murren 2002; Zimmerman and Aide 1989; Zimmerman et al. 1989).

Bee abundance is highest in April, then remains at generally lower levels during the rest of the year. The major flowering of *C. viridiflavum* takes place as far from April as it is possible to be—in October. The greatest chance of seeing *E. marcii* with pollinaria of *C. viridiflavum* is in April, but the chance is also quite good in August and October (Figure 9.10).

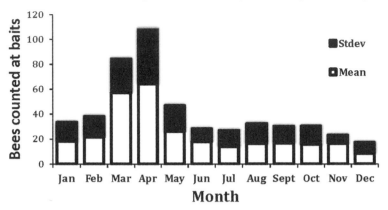

Bee abundance
Eulaema marcii, Soberania Park, Panama (1980-2011)

Fig. 9.10 The total abundance of *Eulaema marcii* at chemical baits throughout the year at Soberanía National Park, central Panama. David Roubik.

From the plant's "point of view," however, seeds that are dispersed in the dry-season trade winds—January until the rainy season in late April—are much more likely to find a favorable germination site and grow. Fruit on those plants in the wet season rot and fall to the ground. The phenology of flowering plants has responded to this selection pressure accordingly. Fewer plants are in flower outside the late wet season, although plants apparently flower throughout the year (Figure 9.9). *Catasetum viridiflavum* may respond to new selection pressures, including a new phenological schedule of a pollinator or seed mortality and fruit rot, at any time in the future. For now, this gender-dimorphic orchid seems to maintain a balanced yet elastic relationship between the forces of selection for seedling establishment and seed production. Flowering and survival are part of the same species biology, and require time and repeated study to appreciate in nature.

CONCLUSION

We find everything in Darwin's writing from minutiae to conceptual framework. There is good reason to believe that in science, the prescription is to do fieldwork, and if concepts do not keep up with field data, they had better be revised. Darwin got as far as he could, then many field and theoretical biologists continued and expanded his original work.

Why have orchids evolved to be outcrossed? It is because their life history requires merging their needs with the needs of an ant, a fungus, or sometimes a bee. This requires an extraordinary amount of trial and error, through the production of millions of seeds and pollen grains used to explore such possibilities. Under those improbable circumstances, the epidendroid orchids and their relatives hooked certain bees into sampling their displays and later, their fragrances. Over time, this meant a shift to male pollinators from females, who needed to provide pollen protein for their brood.

The link between epidendroid orchids and bees of different genders was provided by floral fragrances and perhaps resins. In contrast, pollen rewards left dangling in the rain forest seem obdurately impractical. Therefore, pollen as a reward was replaced first with visual or olfactory mechanisms of deceit. The small amount of nectar secreted was rerouted to bodyguards (e.g. ants) or to some chance pollinators. Finally, the ancestors of modern Neotropical orchids made an offer that could not be refused. They began offering odors to augment the sexual legacy of mutualist male orchid bees. If tropical stingless bees didn't affect the evolution of floral traits and orchid speciation in their role as parasites, mutualists, or commensals, we would be shocked. If bucket orchids depend on ant poop to survive, so be it. What is a better theory than a half-baked one, waiting for validation?

ADDENDA AND CODA

Darwin was a brave voice in rational biology, in his time. We perhaps need to be a brave voice of conservation and ecology, in ours. Determined to fill the gaps in our understanding of orchids and their pollination, Darwin made no pretense of having solved every riddle and seen through each apparent contradiction. If he had used a video camera, or had a university-funded molecular lab, I doubt we would have been more enriched and motivated. Darwin appreciated nature. He summarized a fondness for the pursuit of understanding orchid floral and reproductive evolution by writing that they "appear to us as if they had been modeled in the wildest caprice, but this no doubt is due to our ignorance of their requirements and conditions of life" (Darwin 1877; p. 285).

PART V

Extravagant Architecture:
The Diandrous Orchids

TEN

Pollination and Floral Evolution of Slipper Orchids
(Subfamily Cypripedioideae)

Retha Edens-Meier, Yi-bo Luo, Robert Pemberton, and Peter Bernhardt

INTRODUCTION

Four species in the genus *Paphiopedilum* (*barbatum*, *insigne*, *purpuratum*, and *venustum*) have the unique distinction of producing flowers that "fooled" Charles Darwin (see Chapter 1). In the first edition of Darwin (1862), these 4 species were placed within the genus *Cypripedium*. As usual, Darwin dissected the flowers and conjectured on the process of pollen removal and pollen deposition on the stigma by an insect vector. These flowers, like all flowers in the subfamily Cypripedioideae (sensu Cribb 1987, 1997, 1998, 1999), produce 3 openings. The large, solitary opening on the dorsum of the saclike labellum is produced by expansion and infolding of the labellum petal. In contrast, the 2 basal openings are produced by the base of the labellum contacting the staminodium of the column (Figure 10.1). The 2 fertile anthers are then positioned in the 2 gaps made by the staminodium converging with the labellum. Darwin presumed that in order for insect pollination to occur in these flowers, the insect must first enter through 1 of the 2 basal openings, contacting 1 of the 2 dehiscent anthers. If the insect then continued to push itself into the hollow labellum chamber, it would then contact the inverted, receptive surface of the stigma.

Why didn't Darwin (1862) accept the possibility that the insect entered through the larger dorsal opening? He presumed that the insect would be attracted to the large, contrastingly colored staminodium, and

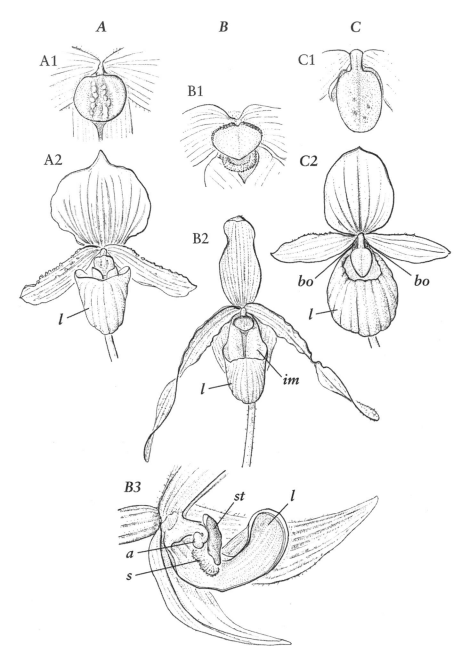

Fig. 10.1 Drawings of comparative floral morphology of (*A*) *Paphiopedilum* sp., (*B*) *Phragmipedium pearcei*, and (*C*) *Cypripedium reginae*. *A1*, *B1*, and *C1* = position and sculpturing of the staminodium. *A2*, *B2*, and *C2* = whole flower. *B3* = longitudinal section of the labellum. *a* = anther, *bo* = basal openings, *im* = infolded margins of the labellum, *l* = labellum, *st* = staminodium, *s* = receptive stigma. Artist: John Myers. Used by permission.

therefore would be drawn to the basal part of the labellum, not to the forefront. After all, the first 4 species he studied were all *Paphiopedilum* spp. with large, sculptured, and vividly colored staminodia (Darwin 1862). In fact, pollinators of *Paphiopedilum* spp. do attempt to land on the staminodium first (Bänziger 1996; Table 10.1; and see below). Furthermore, Darwin believed that the staminodium and the stigma blocked access to the anthers, because they almost closed the "medial slit." The "medial slit" is produced by the infolding of the labellum base (Darwin 1862).

Darwin did not believe wholeheartedly in his own interpretation. On 27 August 1863, he wrote to the British botanist Roland Trimen, "By the bye, I believe I have blundered on Cypripedium. Asa Gray suggested that small insects enter by the toe and crawl out by the lateral windows—I put in a small bee and it did so and came out with its back smeared with pollen."

The full story of this experiment with a bee did not appear in book form until Darwin (1877). By then, Darwin had 2 new specimens. They were the flowers of the North American species *C. pubescens* (now *C. parviflorum* var. *pubescens*) and *C. acaule*. At first, he placed some flies into the labellum of *C. pubescens*. After careful observations, he noted, "but they were either too large or too stupid, and did not crawl out properly" (Darwin 1877; p. 230). Darwin had success introducing a live specimen of *Andrena parvula* (Andrenidae). The bee could not exit through the large dorsal entrance, but escaped by crawling under the stigma and out through one of the basal openings, contacting the dehiscent anther, which smeared the "glutinous" pollen on its back. Darwin felt little pity for the bee and wrote, "I repeated the operation five times, always with the same result" (Darwin 1877; p. 231). In this way, he showed that just as in most orchids that have only 1 fertile stamen, pollination in a slipper orchid occurs when the pollen-carrying insect enters the flower, and first contacts the stigma. Pollen dispersal occurs as usual only when the bee leaves the flower (Figure 10.1; see Chapters 2–4, 6–9).

Darwin (1862) believed that the glandular hairs located within the flower's labellum were probably nectar glands that provided edible rewards for pollinators. His opinion changed by the second edition (1877; p. 229), where he stated he was unable to find nectar within these flowers. To date, no one has identified nectar within any flower in this subfamily (Bernhardt 1989b). Darwin also described the pollen of his Asian and North American species as glutinous (Darwin 1862, 1877). In fact, there are some interesting variations in the configuration of pollen masses in the Cypripedioideae. As Darwin predicted, many species in the subfamily

Table 10.1 Pollination system and fruit set in 3 genera in the subfamily Cypripedioideae

Species	Self-pollinated	Pollinator(s) (number of species)	Insect contacts staminodium	% Conversion of flower/fruit	References
Cypripedium acaule		*Bombus* spp. (gynes)	–	01.3–25.0	Stoutamire, 1967; Primack & Stacy 1998
C. arietinum	–	*Lasioglossum* (1)	–	NA	Stoutamire 1967
C. bardolphianum		*Drosophila* spp.	–	10.8–13.2	Zheng et al. 2010
C. calceolus s.s.		*Andrena* (8); *Lasioglossum* (5); *Colletes* (1); *Halictus* (1); *Nomada* (1)	–	04.0–57.0	Nilsson 1979; Kull 1998
C. candidum	–	*Augochlorella* (1); *Halictus* (4); *Lasioglossum* (3); *Andrena* (1)	–	25.0	Catling & Knerer 1980
C. fargesii	–	*Cheilosia* (1) (males & females)	–		Ren et al. 2011
C. fasciculatum		*Cinetus* spp. (females)	–	18.0–69.2	Ferguson & Donham 1999; Lipow et al. 2002
C. flavum		*Andrena* (3); *Bombus* (2)	–	07.1–09.2	Bänziger et al. 2008
C. guttatum		*Lasioglossum* (5)	+	NA	Bänziger et al. 2005
C. henryi		*Lasioglossum* (3)	–	17.0–22.2	Li et al. 2008b
C. japonicum		*Bombus* (gynes) (3)	–	05.2–07.7	Sun et al. 2009
C. leutiginosum	–	Syrphids	–	NA	Liu et al. 2008
C. macranthos		*Bombus* spp. (gynes) (1)	–	01.2–16.8	Sugiura et al. 2002
C. micranthum	–	*Drosophila* spp. (1?)	–	0.45	Li et al. in progress
C. montanum		*Andrena* (4); *Ceratina* (1); *Halictus* (1); *Lasioglossum* (9)	–	75.0–85.0	Bernhardt in preparation
C. parviflorum		MSB*	–	10.5–40.0	Case & Bradford 2009; Edens-Meier in progress
C. plectrochilum		*Lasioglossum* (3); *Ceratina* (1)	–	38.7–45.9	Li et al. 2008a

Species		Pollinators		%	Reference
C. reginae		*Apis* (1); *Hoplitis* (1); *Megachile* (1); *Anthophora* (2); *Syrphus?, Trichiotinus?* spp.	–	04.6–33.0	Vogt 1990; Edens-Meier et al. 2011
C. sichuanense	–	*Scathophaga* (1)	–	19.0–25.0	Li et al. in progress
C. smithii	–	*Bombus* (2)	–	NA	Li and Luo 2009
C. tibeticum		*Bombus* spp. (gynes) (3)	–	09.57–26.0	Li et al. 2006
C. yunnanense		*Lasioglossum* (1)	–	21.0	Bänziger et al. 2008
Paphiopedilum armeniacum	–	*Lasioglossum* (1); *Ceratina* (1); *Eristalis* (1)	+	43.3	Liu et al. 2005
P. barbigerum	–	*Allograpta* (1); *Erisyrphus* (1)	+	21.0–35.0	Shi et al. 2008
P. bellatum	–	Syrphidae	+	66.0	Bänziger 2002
P. callosum	–	Syrphidae	+	90.0	Bänziger 2002
P. charlesworthi	–	Syrphidae	+	0.0	Bänziger 2002
P. dianthum	–	*Episyrphus*	+	51.0–65.0	Shi et al. 2007
P. hirsutissimum	–	*Allobaccha* (1); *Episyrphus* (1)	+	07.0	Shi et al. 2009b
P. parishii	–	*Allographa* (1)	+	30.0	Bänziger 2002
P. purpuratum	–	*Ischiodon* (1)	+	82.5	Liu et al. 2004
P. rothschildianum	–	*Dideopsis* (1)	+	NA	Atwood 1985
P. villosum	–	*Betasyrphus* (1); *Episyrphus* (1); *Syrphus* (1)	+	9.0	Bänziger 1996
Phragmipedium beeseae	?	Wasps	–	04.3	Pemberton personal observation
Ph. caudatum	–	*Syrphus* (1)	–	NA	McCook 1989
Ph. lindenii	+	NA	NA	100.0	Pemberton 2013
Ph. longifolium	–	*Chlerogella* (1); *Syrphus* (1)	–	NA	Dodson 1966; Pemberton 2013
Ph. pearcei	–	*Ocyptamus* (1)	–	50.0	Pemberton 2013
Ph. reticulatum	+	NA	NA	100.0	Pemberton 2013

Note: Portions of this table were published previously in Pemberton 2013.

*MSB = Medium to Small Bees; NA = not assessed.

release pollen as irregular smears or globs. In other species, for example *Cypripedium fargesii*, the entire contents of each anther cell are deposited on the pollinator's body as a paired unit (double loaf) when it exits the flower (Ren et al. 2011). Some botanists refer to these masses as pollinia. However, there is no true pollinarium in these flowers, because a true pollinarium must consist of pollinia attached to a viscidium made by a rostellum (Dressler 1981; see Chapter 9). *Cypripedium* species placed in the section Sinopedilum do release the contents of each anther cell as paired pollinia complete with a sticky tip derived from the anther instead of the stigma (Li et al. 2011). In all cases studied thus far, the pollen of *Cypripedium* spp. is deposited dorsally on the thorax (Li et al. 2006, 2008a,b; Edens-Meier et al. 2011) and/or head of the pollen vector (Nilsson 1979).

A most important question is why Darwin (1877) introduced a living specimen of a small, solitary bee (*Andrena parvula*) into the labellum of *C. parviflorum* (syn. *pubescens*). Although this is a North American orchid pollinated at least in part by *Andrena* spp. (Edens-Meier in preparation), no correspondence has been found to date between Darwin and Asa Gray or any other North American naturalist to indicate that any *Andrena* spp. was observed visiting this flower before 1877. Darwin (1877; p. 230) explained that he needed a very small bee because it "seemed of about the right size." His seemingly simple insight is remarkable. Indeed, until the last decade of the 20th century, no other pollination biologist made the connection that the insect's dimensions and the floral architecture reflect the dominant evolutionary trend within the subfamily. As we will see below, the physical dimensions of the insect as well as the exterior and interior architecture of the flower of most members of the subfamily Cypripedioideae ultimately determine pollinator efficiency (Edens-Meier et al. 2011). In fact, some species in this subfamily are well visited by a variety of insects. Yet only insects having the correct dimensions actually pollinate the flower (Edens-Meier et al. 2011; Li et al. 2008a,b).

Why did Darwin (1862) mistake Asian tropical *Paphiopedilum* species s.s. for the *Cypripedium* species s.s. restricted to the temperate zones of the Northern Hemisphere? In fact, he did not. In 1862, most British botanists followed Lindley (1826) in recognizing only the solitary genus *Cypripedium*. A few British naturalists (see Riches 1876) agreed with French taxonomists and separated *Cypripedium* s.s. from the slipper orchids native to the tropical Americas. However, it was not until 1886 that the German taxonomist E. Pfitzer (1846–1906) published a revision in which he segregated the genus *Cypripedium* s.s. from tropical Asian orchids now placed in the genus *Paphiopedilum* (Bernhardt 2008). Although segregation of the genus *Paphiopedilum* from *Cypripedium* s.s. is credited to Pfitzer, we now recognize 5 genera in the subfamily: *Cypripedium, Paphiopedilum, Mexipedium, Phragmipedium,* and *Selenipedium* (Dressler 1981; Albert and Chase 1992). Therefore, to review pollination within the subfamily, we must address all 5 modern genera.

CYPRIPEDIUM

Distribution and Taxonomy

The genus *Cypripedium* consists of 47–50 species confined to the temperate–subtropical regions of the Northern Hemisphere (Cribb 1997). The majority of species are distributed through China, the Himalayas, and the temperate to subpolar islands of the northern Pacific. *Cypripedium guttatum* is trans-Pacific (Bänziger et al. 2005). The study of pollination systems within the genus *Cypripedium* remains the most complete of the 5 genera within the subfamily (Table 10.1; Bernhardt and Edens-Meier 2010). Initial studies by Edens-Meier et al. (2010) suggest that self-compatibility is common if not universal within this genus (see also Chapter 8). However, a residual and minor self-incompatibility response may be expressed by a minority of the pollen tubes entering the ovaries of *C. reginae* (Edens-Meier et al. 2011). Full documentation of mechanical self-pollination (autogamy) remains rare and appears to be an adaptation to colder, shorter seasons with a depauparate seasonal pollinator fauna, as in *Cypripedium passerinum* (Table 10.1; Catling 1990). In fact, Catling and Bennett (2007) found that there are some outbreeding floral forms extant within this species. All other species examined to date require insect-mediated pollination emphasizing cross-pollination.

Floral Presentation

Floral presentation within the genus is highly variable in 3 ways. First, the biochemistry of floral fragrances ranges from a mere 2 volatiles in *C. reginae* to at least 50 in *C. fargesii* (Barkman et al. 1997; Ren et al. 2011). The human nose detects these floral fragrances as ranging from "pleasant," resembling floral aromas of commercial, perfume flowers, to "unpleasant," resembling rotting vegetation to dung-like odors (Barkman et al. 1997; Edens-Meier et al. 2011; Ren et al. 2011).

Second, to the human eye, the color patterns of the labellum and the interlocking staminodium vary widely between species. In some cases, the staminodium has a color or color pattern that contrasts with the labellum's exterior. Based on the phylogeny produced by Li et al. (2011), labella with certain dominant colors evolved independently and repeatedly throughout the lineage. For example, yellow labella appear in species belonging to 3 out of 12 subgeneric sections. Likewise, flowers with a rich, deep-purple labellum are found in sections Acaulia and Cypripedium. To the human eye, the exteriors of many labella are blotched and/or striated. The interior often displays lines or spots (e.g. *C. reginae*; Edens-Meier et al. 2011; see photographs in Chen et al. 1999). We also note the unusual pigmentation patterns on the leaves of species restricted to section Trigonopedia (sensu Li et al. 2011). In *C. fargesii*, *C. lichiangense*, and *C. margaritiaceum*, the flower appears nestled within 2 leaves. The leaves bear discrete dark–spotted, hairy tufts. Ren et al. (2011) suggests that this is part of the visual pattern that attracts atypical fly pollinators to the drab, red-brownish flowers.

Third, the labella of *Cypripedium* spp. show a remarkable range of length, from 1.0 cm in *C. palangshanense* (Chen et al. 1999) from China to 6.3 cm in *C. irapeanum* from Mexico (Luer 1975).

There are only a few obvious correlations between these 3 floral variables (fragrance, labellum color, and labellum size). Species in sections Trigonopedia and Sinopedilum all tend to produce smaller, mottled, darker labella on short stalks that are noted for producing unpleasant odors. The 2 sister species in section Arietinum are the only *Cypripedium* spp. with keeled labella. Authorities working on the taxonomy and/or classification of species in sections Irapeana, Obtusipetala, and Cypripedium have commented on their pleasant fragrances. Their labellum colors, however, vary from white to yellow through various shades of pink to deep purple (Luer 1975; Edens-Meier et al. 2011; Bernhardt and Edens-Meier, personal observations).

There does appear to be a correlation between labellum size (floral architecture), pigmentation pattern, and possibly fragrance once the pollinators of these species are identified (Table 10.1). The papers by Luo and his associates (Bänziger et al. 2005, 2008; Li et al. 2006, 2008a,b) and Edens-Meier et al. (2011) show up to 3 interlocking aspects of floral architecture. As predicted above, the dimensions of the pollinator must fit within the dimensions of the labellum. However, fitting into the labellum is never enough. First, the pollinator must also fit between the surface of the receptive stigma and the floor of the labellum to transfer incoming pollen. Second, the pollinator's size determines whether the insect will be able to fit between the dehiscent anther and the opening made by the staminodium and the labellum base if the insect is going to disperse pollen (Figure 10.1). This explains why most insects observed visiting the flowers of *C. plectrochilum* (Li et al. 2008a) and *C. reginae* (Edens-Meier et al. 2011) failed to carry the host's pollen. *Apis cerana* was far too large to fit into the flowers of *C. plectrochilum*. Members of the genus *Bombus* entered the labellum of *C. reginae*, but always exited through the same opening by which they entered (Edens-Meier et al. 2011). Ants visiting *C. plectrochilum* and the smallest bees visiting *C. reginae* exited the flower through the rear lateral openings repeatedly, but never contacted the dehiscent anthers.

In recent years, more information on the composition of floral fragrances has become available in the study of *Cypripedium* pollination. The data are complicated. Species pollinated by bees have received the most attention (Barkman et al. 1997; Li et al. 2006, 2008a,b), but it is difficult to associate a genus of bees with a suite of volatiles at this time, as fragrance analysis of this genus is so incomplete. We note that at least 7 *Cypripedium* spp. are pollinated primarily or in part by the nearly pandemic bee genus *Lasioglossum* (Table 10.1). *Cypripedium tibeticum* is pollinated primarily by queens in the genus *Bombus* (Li et al. 2006). The fragrance of *C. fargesii* reminded Ren et al. (2011) of the smell of decaying leaves. Flowers of this species are pollinated exclusively by male and female syrphid flies (Table 10.1).

What, then, is the primary function(s) of the staminode in the genus *Cypripedium* in the attraction of potential pollinators (see also Chapter 7)? In almost all the *Cypripedium* spp. native to North America (Luer 1975), and in many of the species native to China (Chen et al. 1999), the staminodium is in a color that contrasts with the labellum exterior and/or is marked with contrasting spots and/or splotches like the interior of the labellum (see photos in Chen et al. 1999; Edens-Meier et al. 2011). Bänziger et al. (2005) observed the pollination of *C. guttatum*, and noted that the bees attempted to land on the bright-yellow staminode before they slipped off and fell into the labellum. However, this is the only *Cypripedium* spp. studied thus far in which bees were observed to land repeatedly on the staminode (Table 10.1). Past observers described insects either flying directly into the labellum or landing on the surface of the labellum and entering the large dorsal opening once they crawled too close to the edge.

Edens-Meier (in progress) observed bees landing on the labellum of *C. parviflorum* and then crawling directly into the pouch via the dorsal opening. Bernhardt et al. (in progress) observed over 100 bee visits to *C. montanum* at 2 different sites in 2003, 2004, and 2006, but never witnessed a true pollinator landing on the staminodium. When Bernhardt intentionally removed the labellum of 7 flowers of *C. montanum*, he observed bees landing on the staminodium. In fact, contrasting blotches or lines on the staminodium of both *C. parviflorum* and *C. montanum* tend to match much the same patterns found on the interior floor of the labellum directly under the dorsal opening (Bernhardt and Edens-Meier, personal observation). A few cases (e.g. see photographs of *C. tibeticum* and *C. yunnanense* in Chen et al. 1999) suggest that the pigmentation pattern on the staminode appears to blend with the pigmentation pattern around the rim of the dorsal opening on the labellum (Li et al. 2006; Bänziger et al. 2008). Therefore, we suggest that in most *Cypripedium* spp. (except *C. guttatum*), staminodium pigmentation serves as part of an interlocking, irregular, 2-dimensional landing cue instead of acting as a central landing platform. Dafni and Kevan (1996) referred to this pattern in other flowers with bilateral symmetry as the "irregular blotch."

What is (are) the function(s) of the lateral petals? We suggest that lateral petals may also serve as part of the visual display. For example, the twisted lateral petals of *C. parviflorum* flutter, flap, and twirl in the wind, movements that may serve to attract potential pollinators (Edens-Meier, personal observation). In addition, what is (are) possible function(s) of the dorsal sepal? We propose that the dorsal sepal may function like an umbrella, serving on one hand to prevent excessive amounts of water from entering the labellum during rain showers (Edens-Meier, personal observation).

Pollination Ecology

Interest in the pollination ecology of *Cypripedium* spp. revived in the late 1970s, thanks primarily to the work of Nilsson (1979), who showed that the deposition of pollen smears vary on the bodies of different pollen vectors of *C. calceolus*. The studies that followed this seminal paper revealed further variation and trends in floral evolution within this lineage (Table 10.1). Yes, all insect-pollinated *Cypripedium* spp. studied to date are pollinated by deceit, but interpretations of floral presentation differ. Most species studied (Table 10.1) now appear to be pollinated by female polylectic bees, but as noted above, the physical size of bees effective as agents of pollen dispersal/pollination depends specifically on floral dimensions. Therefore, bee pollination in *Cypripedium* spp. breaks down into 3 modes that may overlap.

The most common mode (*n* = 7 *Cypripedium* species) depends primarily on small, solitary-parasocial bees under 11 mm in length. This mode occurs repeatedly over the natural distribution of the genus. The second, far less common (*n* = 1) mode depends on medium-large, solitary-eusocial bees 12–15 mm in length and is expressed by North American *C. reginae*

(Edens-Meier et al. 2011). The third mode (n = 4) is dependent primarily on gynes (queens) of *Bombus* spp., and has also evolved independently throughout the modern range of the genus excluding Western Europe, where only one species, *C. calceolus*, exists. This solitary species depends on small-bodied, solitary bees (Table 10.1).

While we recognize 3 bee size modes (small, medium, and large), we are not suggesting that each bee-pollinated *Cypripedium* species is always pollinated exclusively by bees belonging to the same size mode. As regional variation occurs, we should expect that both bee diversity and labellum size within the same species may vary. For example, we note that Li et al. (2011) segregated *C. flavum* into 3 variants. This probably explains why some populations of *C. flavum* are pollinated by small *Lasioglossum* spp. (<10 mm in length), while others are pollinated by *Bombus* spp. (>10 mm; see Table 10.1). We must also consider the Mexican *C. irapeanum*. It has the largest labellum in the genus (Luer 1975), but without pollination studies we do not know whether its flowers are dependent on large, medium, or small bees. We predict, based on the relative size of the labellum, that these flowers rely on large bees such as *Bombus* spp. or even larger carpenter bees (Xylocopina; Apidae) to complete the pollination process.

While pollination by small-bodied, solitary-parasocial bees is common within this genus, the diversity of bee pollinators varies between *Cypripedium* species. Bees carrying pollen smears after visiting the flowers of *C. montanum* (Table 10.1) at 2 sites over 4 seasons represented 12 taxa divided between the families Andrenidae, Apidae, and Halictidae (Table 10.1). Conversely, it is possible that *C. arietinum*, *C. guttatum*, *C. henryi*, *C. plectrochilum*, and *C. yunnanense* are pollinated primarily or exclusively by bees in the genus *Lasioglossum* (Halictidae; Table 10.1).

The reason why bees, of any size, visit these flowers largely depends on their foraging preferences. Female bees identified as pollinators of *Cypripedium* spp. share the same foraging habits regardless of physical size or family. All appear to be polylectic and/or polyphagic, visiting a variety of co-blooming flowers in situ (Li et al. 2006, 2008a; Edens-Meier et al. 2011). At some sites, there are only a few co-blooming species, so the diversity of pollen grains found on bees carrying the pollen of the orchid is low (Li et al. 2006, 2008a). In other cases, *Cypripedium* spp. bloom at the same time as a wide variety of unrelated but generalist, bee-pollinated species. As a result, the sheer diversity of pollen grains on bees exiting the *Cypripedium* flower increases (Edens-Meier et al. 2011). Bernhardt (in progress) compared pollen load analyses of bees collected as they exited the flowers of *C. montanum* at 2 isolated sites in Oregon over 4 years. Bees bearing the pollen smears of *C. montanum* carried pollen grains from an estimated 12 co-blooming flowering taxa (see also Chapter 7).

Not only does the degree of attractiveness of *Cypripedium* flowers versus other co-blooming species remain unclear, but we also don't understand the degree of floral mimesis within this genus. For example, is *C. macranthos* var. *rebunense* a specific mimic of the flowers of *Pedicularis schistostegia* (Sugiura et al. 2002)? Some *Cypripedium* spp. probably do not mimic other flowers within their habitats. Does *C. tibeticum* mimic the odors of co-blooming guilds of *Bombus*-pollinated species, does the entrance to the dorsal opening on the labellum mimic the entry holes of *Bombus* nesting sites (Li et al. 2006), or does a combination of both occur? Likewise, evidence that *Cypripedium* spp. are pollinated by females of parasitoid wasps (Hymenoptera: Diapriidae) is restricted currently to *C. fasciculatum* (Table 10.1). The mode of mimesis is not understood here either, as foraging habits of winged *Cinetus* spp. are underexplored. Do these females feed on nectar of plants with similar colors and odors, or does the labellum mimic an oviposition site? These wasps lay their eggs on maggots of mushroom gnats (Diptera: Sciaridae). Does the flower mimic a maggot-infected mushroom (Ferguson and Donham 1999)?

Species in subgeneric sections Trigonopedia and Sinopedilum (sensu Li et al. 2011) show a completely different mode of pollination that appears dependent on true flies (Diptera) as vectors. Here, floral presentation probably mimics putrefaction or decay. Almost all these exclusively Asian species produce small, dark-mottled flowers on reduced scapes or peduncles (Chen et al. 1999). *Cypripedium bardolphianum* and *C. micranthum* (sect. *Sinopedilum*) have 2 features unique to this genus. First, they are the only species so far known to be pollinated by *Drosophila* spp. (Diptera: Drosophilidae; Table 10.1). Second, they appear to be the only species in which the contents of each chamber in the same anther are removed as a single unit, because these 2 pollinia are interconnected by a sticky disc (Perner 2008; see above). This disc is not a true viscidium, because it is made by the anther instead of by a rostellum. Individuals who have smelled the flowers report a musty-fermented fragrance. Fly pollinators are more diverse in section Trigonopedia, and there is a greater range of pollinator size. *Cypripedium sichuanense* depends on a *Sarcophagus* sp. (Sarcophagidae; Table 10.1). Liu Z-J et al. (2008) have suggested that pollination of *C. lentiginosum* is accomplished by hoverflies (Syrphidae). Photographs taken of the pollinators show that they remove the entire contents of an anther when they exit the flower, but the paired, loaf-shaped pollinia adhere directly to the fly's thorax without a modified sticky disc (Liu Z-J et al. 2008).

Hybridization

One would conclude that diversification in pollination systems within the genus *Cypripedium* promotes interspecific isolation where and when species are sympatric and co-blooming. The most convincing evidence for this was presented by Bänziger et al. (2008). When *C. yunnanense* and *C. flavum* flowered together in the same site in southwestern China, no hybrids were found. Field observations and bee collections showed that *Lasioglossum* spp. restricted their visits to *C. yunnanense*, while *Andrena* spp. were pollinators of *C. flavum*. When *Cypripedium* spp. share the same distribution as well as flowering season and pollinators, we might expect hybridization and a history of introgression in some instances (Hu et al. 2011). This is the interpretation of the North American species *C. parviflorum* and *C. candidum*, as they share the same small bee pollinators from the bee families Andrenidae and Halictidae (Table 10.1). They have a history of introgression in Manitoba (Worley et al. 2009) throughout parts of their respective ranges (see also Chapter 7). When *C. tibeticum* and *C. smithii* grow and bloom together, they are pollinated by the same *Bombus* spp. and produce such a broad range of intermediate forms that Li and Luo (2009) concluded that *C. smithii* s.s. was best submerged within *C. tibeticum* s.s. However, different pollinators may not always result in interspecific isolation when the species is known to produce different forms or variants. *Cypripedium tibeticum* pollinated by *Bombus* gynes (see above) has a history of asymmetric introgression with *C. yunnanense* (Hu et al. 2011), which is pollinated by much smaller *Lasioglossum* spp. (Halictidae).

Reproductive Success

As *Cypripedium* spp. are classified variously as threatened, rare, and/or endangered, fecundity remains a major issue in the reproductive ecology of surviving populations (Bernhardt and Edens-Meier 2010). Low conversion rates of flowers into fruit characterize orchid species with food-mimic flowers (Tremblay et al. 2005; see Chapter 2). Therefore, low conversion rates in. *Cypripedium* spp. should not surprise us (Table 10.1). What should surprise us is that although all *Cypripedium* spp. studied to date lack rewards, some species have populations that produce capsules at rates competitive or superior to other species of flowering plants that only set seed when they are cross-pollinated (Bernhardt and Edens-Meier 2010).

Reproductive success within this genus must be considered on both a species-to-species basis and a population-to-population basis. Fruit set rate varied significantly between isolated populations of *C. calceolus*, *C. fasciculatum*, *C. parviflorum*, and *C. reginae*. Fruit production also varied from year to year at the same site within the same populations of *C. acaule*, *C. calceolus*, *C. reginae*, and *C. tibeticum* (Bernhardt and Edens-Meier 2010). Populations of some, but not all, endangered species show extremely low conversion rates of flowers into fruit (e.g. *C. fargesii* and *C. micranthum*; Ren et al. 2011; Li et al. in progress). It would be tempting to suggest that *Cypripedium* species with the most diverse spectrum of pollinators have the highest fruit set, as species pollinated by a wide variety of small bees (e.g. *C. calceolus*, *C. montanum*, and *C. parviflorum*) show fruit rates as high as 45–85%, but this cannot be used to explain the 69% conversion ratio in an Oregon population of *C. fasciculatum*. This species, after all, appears to be pollinated exclusively by 2 or 3 *Cinetus* spp. (Ferguson and Donham 1999) and shows no regional trend toward mechanical self-pollination (Lipow et al. 2002). We do note that fruit set may increase in some populations of *Cypripedium* spp. if their flowering periods overlap with a diverse range of pollen- and/or nectar-rich flowers (see above), but once again the correlation is not universal (see also Chapter 6). The nearly 100% conversion ratio of flowers into fruit in some populations of *C. passerinum* is due obviously to obligate self-pollination (Catling and Bennett 2007).

PAPHIOPEDILUM

Distribution and Taxonomy

While we recognize more than 60 species in the genus *Paphiopedilum*, earlier reports of more than 80 species may reflect an overestimate, as some rare species are often reclassified as recurrent hybrids (see Cribb 1987). The genus is distributed from India through the Philippines and as far south as the Solomon Islands. Unlike *Cypripedium* spp., *Paphiopedilum* spp. occupy a much wider range of microhabitats. Some are terrestrial, preferring dense deposits of humus and leaf litter. Others are true epiphytes, while many are true lithophytes, growing on eroded boulders varying in pH. While we remind the reader that Darwin (1862, 1877) examined *Paphiopedilum* spp. before he examined *Cypripedium* spp., field studies on the pollination ecology of *Paphiopedilum* spp. are recent, and only 11 species have received significant investigation (Table 10.1). The first successful field study to capture the pollinator of a *Paphiopedilum* spp. was accomplished by Atwood (1985) and was based on a population of the now rare *P. rothschildianum*. Currently, there are no records of spontaneous self-pollination in this genus (Table 10.1).

Floral Presentation

Floral attractants should include both odor and pigmentation pattern, but biochemical analyses of floral fragrances in *Paphiopedilum* spp. remain seriously understudied. Cribb (1998) reported a range of fragrances within the genus from spicy odors to those reminiscent of urine. However, Shi et al. (2009a) were unable to detect fragrance compounds in *P. barbigerum*, but noted that pigmentation patterns varied within this species. There are no species with true red flowers, but Cribb (1998) discriminated between con-colored flowers and bi-colored flowers. In con-colored flowers, the sepals, petals, and labellum have the same color (to the human eye). Bi-colored flowers show a labellum colored differently from the exaggerated sepals and petals. *Paphiopedilum* spp. with dull- to somber-colored perianth segments often bear sculptured and/or hairy protuberances.

Paphiopedilum spp. are known for their large, highly sculptured staminodia that are often contrasting in color from each of the perianth segments (Cribb 1998). In many if not all *Paphiopedilum* spp., the center of the staminodium has a large protuberance or "wart" (sensu Bänziger 1996). Staminodium sculpture and presentation (Figure 10.1) within this genus and in *C. guttatum* (Bänziger et al. 2005) offer a unique contrivance (sensu Darwin 1862) unknown in monandrous genera within the family Orchidaceae. The pollinator is attracted specifically to the staminodium of a *Paphiopedilum* sp., but when the pollinator attempts to clutch the wart it slips and falls into the labellum. Once the pollinator is inside the labellum, the act of pollen removal and pollen deposition proceeds as in *Cypripedium* spp. (see above).

Pollination Ecology

Pollinators of *Paphiopedilum* spp. include bees and flies, as in the genus *Cypripedium* (Table 10.1). However, staminodium presentation combined with perianth sculptures in some *Paphiopedilum* spp. show different modes of floral mimeses (Bänziger 1994). While *Paphiopedilum arminiacum* appears to attract a wide range of bees, wasps, and flies (Liu KW et al. 2005), *Eristalis tenax* (Syrphidae), *Ceratina morawiz* (Apidae), and *Lasioglossum pronotale* (Halictidae) appear to be the dominant pollinators. This species is probably best interpreted as a generalist food mimic in which the staminodium represents a flower or a giant yellow anther. Bumblebee pollination has been observed in *P. micranthum* (Luo in progress). The large insect attempts to clutch sculptures on the yellow staminode (see below) before it falls into the labellum. In all other *Paphiopedilum* spp., hoverflies (Syrphidae) predominate.

Pollinators of 11 *Paphiopedilum* spp. represent members of 7 genera in the family Syrphidae (Table 10.1). However, syrphid fly pollination within the genus *Paphiopedilum* falls into 2 possibly overlapping modes of deceit. Some species (e.g. *P. armeniacum*) appear to be generalist food mimics. Their flies prefer the color yellow (Weiss 2001), so yellow staminodia predominate. In other species (e.g. *P. rothschildianum*), female hoverflies appear attracted to the dark and contrastingly colored warts and bumps. These gravid females are believed to mistake these sculptures for the colonies of insect hosts on which they oviposit (Atwood 1985; Bänziger 1994, 1996, 2002). Eggs of the duped females are found occasionally on the flower and/or in the labellum (Atwood 1985; Bänziger 1994). Labellum size in *Paphiopedilum* spp. correlates with pollinator dimensions, as in *Cypripedium* spp. (see above).

Hybridization

As mentioned above, interspecific hybridization in *Paphiopedilum* has been proposed by Cribb (1998), but the breakdown of interspecific isolation mechanisms in situ has not been studied to date.

Reproductive Success

As in *Cypripedium* spp., the conversion rate of flowers into fruit in *Paphiopedilum* is highly variable at the interspecific level (Table 10.1). Bänziger (2002) was unable to find fruit set in a population of *P. charlesworthii*, but 90% of the flowers of *P. callosum* set fruit. Shi et al. (2007, 2008, 2009a) found that fruit set was higher in *P. dianthum* (51–65%) compared with *P. barbigerum* (21–35%).

PHRAGMIPEDIUM, MEXIPEDIUM, AND *SELENIPEDIUM*

Distribution, Taxonomy, and Floral Presentation in *Mexipedium*

The 3 remaining genera are exclusive to the Neotropics. *Mexipedium xerophyticum* is found only in southeastern Mexico and was described originally as *Phragmipedium xerophyticum* (Soto et al. 1990). Albert and Chase (1992) segregated the species and established the monotypic genus

Mexipedium. Two or more flowers may be produced on the same scape. These flowers are white tinged with pink, but unscented to the human nose. The staminodium is often tinged with pink and appears large and prominent in relation to the relatively small, white labellum.

Pollination Ecology and Reproductive Success in *Mexipedium*

Although Soto et al. (1997) observed many capsules on flowering stems, they saw no pollinators. Soto et al. (1997) insisted that these flowers failed to self-pollinate, but did not include information on how data were collected.

Distribution, Taxonomy, and Floral Presentation in *Selenipedium*

Pridgeon et al. (1999) recognized 5 to 6 species in the genus *Selenipedium*, distributed from Panama through northern South America eastward to Trinidad. These rhizomatous herbs produce tall, woody, secondary stems terminating in multiflowered racemes. The flowers appear to be short lived, and Crüger (1864) found that flowers of *S. palmifolium* smelled like vanilla. Online color photographs suggest that unlike so many *Paphiopedilum* spp. (see above), the staminodium is not exaggerated, highly ornamented, or vividly pigmented, and its color pattern may merge with the infolded margins at the base of the labellum. We do note that in some species (e.g. *S. aequinoctiale* and *S. chica*), the infolded rim of the large dorsal entrance produces a contrasting color to the inflated labellum exterior.

Pollination Ecology in *Selenipedium*

An unpublished observation by Calaway Dodson noted an *Anthophora* sp. (Apidae) visiting the flower of *S. aequinoctiale*, but did not note whether it contacted the staminodium (see van der Pijl and Dodson 1966).

Distribution, Taxonomy, and Floral Presentation in *Phragmipedium*

Phragmipedium consists of about 15 species found from southern Mexico through southern Brazil (Cribb 1999), where they colonize tree limbs and/or boulders. The flowering stem ends in one to many short-lived flowers. Species like *Ph. longifolium* produce flowers that have long, ribbonlike, often spiraling lateral petals. As they are often hairy organs, we propose they may function as scent glands (see illustrations in Vogel 1990) and/or to attract pollinators as visual cues. As in *Selenipedium* (see above), the infolded margins forming the rim and entrance to the dorsal opening on the labellum may show pigmentation patterns that contrast with the overall colors of the labellum surfaces. In particular, the infolded margins that connect to the staminodium in flowers of *Ph. reticulatum*, *Ph. longifolium*, and *Ph. pearcei* are broad and intensely spotted (see Pemberton 2011). Additional sculpturing was noticed on the staminodium of *Ph. pearcei*, with 2 discontinuous rows of dark bristles along its margins where each fertile anther emerges and faces the lateral exit holes. However, in *Ph. caudatum*, these margins appear upraised and winged (see photographs in Pemberton 2011).

Pollination Ecology and Reproductive Success in *Phragmipedium*

Insect pollination in *Ph. pearcei* and *Ph. longifolium* appears restricted to ovipositing flies in the family Syrphidae, as in *Paphiopedilum* spp. (Table 10.1). Pemberton (2011) did not note female hoverflies landing on the staminodia of either species. Those insects lay their eggs on aphids. The female flies visited the spots on the wide, infolded margins of the labellum before falling into the pouch. Some female flies died before they could escape, trapped between the lateral opening and the dehiscent anther. However, Pemberton (2011) and Koopowitz (2008) noted that at least 3 species in the genus self-pollinated in the absence of insects (see also Chapters 7 and 8), so the conversion rate of flowers into fruit in individual plants of *Ph. lindenii* and *Ph. reticulatum* was 100% (Table 10.1). Curiously, each flower of *Ph. lindenii* contained 3 anthers, with the central (middle) anther attached to the receptive stigma surface guaranteeing mechanical self-pollination (Pemberton 2011). Koopowitz (2008) proposed interspecific hybridization within this genus, but did not provide data supporting breakdown of interspecific isolation mechanisms.

MATTERS FOR DISCUSSION

The various contrivances by which species within the subfamily Cypripedioideae are fertilized by insects appear identical when we consider only the adaptive anatomy within the flower's interior. In all slipper-orchid flowers studied to date, the pollinator enters the flower through a wide opening on the inflated labellum, passes under the receptive stigma, and contacts the anther only when it exits via one of the 2 narrow passages at the labellum base. These 2 passages are made of the interlocking labellum and column. Interior floral architecture, regardless of the slipper orchid species and the pollinator species, remains remarkably conservative throughout the subfamily. In all cases, the labellum must be large enough to accommodate the body of the pollinator. In addition, the distance of the receptive stigma to the floor of the labellum must be narrow enough to contact the dorsum of the insect. Likewise, the dimensions of the dehiscent surface of the anther and the exit hole must be narrow enough to smear or release pollen/pollinia onto the insect's dorsum. Consequently, entrance and release processes in a slipper-orchid flower are identical whether the pollinator is a tiny drosophilid fly or a large bumblebee queen (Table 10.1). Regardless of flower size, the appropriate pollinator enters the flower with ease, but often exits the flower with some difficulty. In a few cases, some insect fatalities occur (e.g. *C. montanum*, N. Vance, personal observation; *C. reginae*, Edens-Meier et al. 2011; *Phragmipedium* spp., Pemberton 2011).

Consequently, macroevolution of the slipper-orchid flower shows the exact same pattern regardless of genus. The floral architecture of the labellum and the column entrance and exit must have evolved long before this lineage diversified into what taxonomists now recognize as separate genera. After personally examining floral organs of *Paphiopedilum* and *Cypripedium* spp., and reviewing the literature on *Selenipedium*, Darwin (1877; p. 262) came to the conclusion that these orchids represent a remnant within the family "in a simpler or more generalised condition." Darwin's initial hypothesis has been validated repeatedly by recent molecular analyses (see review in Li et al. 2011).

However, when we consider the evolution of species within this subfamily, we see distinct variations of floral characters within and between genera. Looking at Table 10.1, the photographs presented in Li et al. (2011), and the color plates in Cribb (1998) and Koopowitz (2008), it is obvious that the diversification of pollination systems within this subfamily also plays a significant role in the evolution and diversification of species. Tucker (1997) argued convincingly for a hierarchy in floral development that related to the evolution of the lineage and the individual species. As modes of floral presentation and primary attractants relate to specific pollinators of flowers, these traits should always be the last to develop, just before or after the flower opens. Such late changes in floral development within the subfamily Cypripedioideae have resulted in the exploitation of bees and/or flies in rather different ways.

Merging the pigmentation pattern of the staminodium and labellum in *Cypripedium* suggests that the 2 organs work together to attract pollinators. In contrast, expansion of the size, lobing, sculpturing, and position of the staminodium produced the system in the genus *Paphiopedilum*.

Here the insect falls into the labellum after it attempts to grasp the "mesmerizing" wart (Bänziger 1996, 2002). For *Phragmipedium* spp., increasing the width of the infolded margins of the labellum and adding spots to the margin's surface account for the attraction of pregnant flies. What do these 3 different modes of floral development and presentation have in common? Each time an appropriate pollinator is attracted successfully, the insect ends up inside the labellum and must exit through the smaller, lateral openings.

Isn't it interesting that some *Paphiopedilum* spp. and some *Phragmipedium* spp. are both pollinated by flies from the family Syrphidae? These flies usually lay their eggs on the bodies of smaller insects living in colonies. However, the flowers in each genus exhibit their fake infestations on completely different floral organs. Within a genus in this subfamily, pollinator(s) size and foraging behavior correlate ultimately with floral size and presentation (Edens-Meier et al. 2011). At the simplest level, no *Bombus* queen will ever fit into the labellum pouch of *C. bardolphianum*. Likewise, floral presentation and odor in *C. fargesii* are unlikely to attract female bees foraging for pollen and nectar to provision their offspring (Ren et al. 2011). *Cypripedium* is the best-studied genus in the subfamily for the time being. Therefore, it currently shows the most diverse modes of pollination. When we look closely at this genus, we cannot discuss bee pollination without subdividing labellum presentation into 3 distinct modes that exploit 3 distinct bee sizes. Fly pollination in this genus also subdivides into at least 3 different modes (small—drosophilid, medium—syrphid, and medium—sarcophagid), based on respective foraging preferences (fruit, fungi, feces) by the primary pollinators.

Are there any other floral novelties aside from variation in color, pigmentation pattern, and the physical size of certain organs within the Cypripedioideae? We argue there are 4. First, there is the unique third anther in the self-pollinating *Phragmipedium lindenii*, which "fits" onto the stigmatic surface. This is probably the result of a recurrent but successful mutation, as 3 anthers have been found infrequently in other species (e.g. *C. reginae*; see Edens-Meier et al. 2011). Second, there is both the central wart on the staminodium of *Paphiopedilum* spp. and the trend toward decorative bumps and protrusions on the flower as possible mimics of host insects.

Third, there are the sticky tips that both unite 2 pollinia in each anther and glue them to the bodies of some of the smallest pollinators in *Cypripedium* sect. Sinopedilum (sensu Li et al. 2011). As this "pseudoviscidium" attaches 2 whole pollinia to the much-reduced space on the small fly's back, we wonder if it ultimately prevents the feeble pollinator from sticking to and dying on the stigma before it can leave the flower with a second pair of pollinia. Remember, in all other members of this subfamily, pollen is released as smears or irregular globs, or the entire contents of each anther cell adhere directly to the insect's dorsum. By uniting the contents of the anther with a sticky plug, a tiny fly could carry far more pollen to each stigma.

Finally, there is the distinctly keeled labellum in *Cypripedium* s.s. restricted to 2 species in section *Arietinum* (sensu Li et al. 2011), both pollinated by *Lasioglossum* spp. (Table 10.1). At this time we do not understand the function of this architecture.

However, as medium to large eusocial apid bees (*Apis* and *Bombus*) as well as flower flies (*Eristalis*; Syrphidae) have been used extensively as model systems in animal behavior, we should ask what the cross-pollinated slipper orchid tells us about foraging theory (Stephens and Krebs 1986) or even older concepts in ethology. At the time they bloom, the flowers of slipper orchids tend to be among the largest flowers in their microhabitats. Presuming that visual acuity is low in insect pollinators, we conclude that the much-enlarged, sculptured, often-spotted staminodium in *Paphiopedilum* spp., the spotted pattern in *Phragmipedium* spp., and the inflated, contrastingly pigmented patterns on and within the labellum of *Cypripedium* spp. have the same effect on appropriate foragers. All 3 are supernormal stimuli (sensu Manning 1967). These "lures" interrupt the standard foraging behavior of insects with "promises" of a larger supply of rewards and/or oviposition hosts, as in sexual mimics (see Chapter 3).

How well does this system work? The only way to measure it is by the seasonal production of cross-pollinated fruit. Here reproductive success, based on the canalization of pollinator behavior, is far less obvious, as fruit set is highly variable between and within species in *Cypripedium* and *Paphiopedilum* (Table 10.1). Fruit set varied significantly by site in *C. fasciculatum* (Lipow et al. 2002) and *C. calceolus* (Kull 1998) and from year to year in the same populations of *C. acaule* (Primack and Stacy 1998), *C. reginae* (Edens-Meier et al. 2011), *C. macranthos* (Sugiura et al. 2002), *P. barbigerum* (Shi et al. 2009a), and *P. dianthum* (Shi et al. 2007, 2008). Let's compare fruit set via insect-mediated cross-pollination at the level of genus. It may be as low as 1.3% in *C. acaule* and as high as 85% in *C. montanum* (Bernhardt and Edens-Meier 2010). In *Paphiopedilum* the range is from 0% in a population of *P. charlesworthii* to 90% in *P. callosum* (Table 10.1). Why this variation occurs must remain unanswered in this chapter and remains open to greater experimentation.

We speculate that some slipper orchid species may be superior to others in luring appropriate pollinators. Some may now be pollinator limited (sensu Committee on the Status of Pollinators in North America 2007) because of long-term environmental changes/degradation over time. For example, 6 of the species studies in Table 10.1 failed to include fruit-set rates in their controls. Ten of the species studied (cross- and self-pollinated) reflected work on only 1 population. Based on records taken over more than one season and/or at more than one site, at least 12 insect-pollinated species usually converted fewer than half their flowers into fruit. Of those 12 species, 8 did not convert more than 26% of their flowers into fruit (Table 10.1). This suggests that some species are pollinator-limited regardless of site or season.

We suggest that the staminodium serves a variety of functions depending on genus and species within this subfamily. In all cases, it covers and conceals the stigma and forms a barrier separating the 2 dehiscent anthers. It is also part of the pollinator escape route, and in some species it obviously helps to attract potential pollinators. In *Paphiopedilum* spp. and *C. guttatum*, the staminodium provides a false landing cue. It remains to be tested whether any slipper orchid has scent glands on its staminodium. However, when considered collectively, staminodium architecture is pivotal in the evolution of specialist pollination systems within this subfamily.

CONCLUSION

What do we need to do in the future? Although pollination ecology research on *Cypripedium* spp. has increased since the 1970s, we strongly encourage additional research in order to extend and expand existing knowledge on the evolution of reproductive mechanisms. In particular, sites where populations of 2 or more species overlap are particularly useful (Bänziger et al. 2008), as they allow scientists to test hypotheses on more than one interrelated taxon. We need to emphasize, however, that any ongoing studies must consider more than one population at more than one site for more than one season. This is necessary because so many slipper orchids now survive in small, isolated populations (see also Chapter 6). After all, field scientists and naturalists have reported, and will continue to report, entire seasons in which no visitations to the flowers are observed. All the tropical genera still require intense research, as only a fraction of *Paphiopedilum* spp. have been studied, and virtually nothing is known to date about either *Mexipedium* or *Selenipedium* spp. In all cases, morphometric analyses of pollinator dimensions and floral architecture must continue. Finally, we encourage more fragrance analyses within each genus. The existing literature for fragrances within the subfamily covers just a few *Cypripedium* spp. Results of these analyses are so intriguing that they must be expanded to include all species within the subfamily if we are ever to understand the phylogenetic bases for so many variations in biochemical expressions (see also Chapter 9).

ACKNOWLEDGMENTS

We thank the entire library staff and the horticultural staff of the Missouri Botanical Garden for their professional and courteous assistance. Our thanks also to Dr. Andrew Huber of the G.R.O.W.I.S.E.R. Reserve for access to his populations of *C. montanum*. We are grateful to Dr. Nan Vance for locating populations of *Cypripedium* spp. in Oregon and Idaho. We have benefited from the insect identifications provided by Dr. Michael Arduser of the Missouri Department of Conservation. Finally, we thank Jared and Jake Edens for their assistance and encouragement.

PART VI

Overview:
The Influence of Color Perception and Climate Change

Color and Sexual Deception in Orchids: Progress toward Understanding the Functions and Pollinator Perception of Floral Color

A. C. Gaskett

INTRODUCTION

Why are many orchids so colorful? Plant color and pigments have many biotic and abiotic functions, including defense against herbivores and protection from stressors such as UV and extreme temperatures (Dixon et al. 2001; Schaefer and Rolshausen 2006). Some studies demonstrate that expression of floral color can also be affected by drift, pleiotropy, mutation, or complex biosynthetic pathways linked to other traits, such as scent (Aragón and Ackerman 2004; Majetic et al. 2007; Rausher 2008; Delle-Vedove et al. 2011). However, pollinators and their preferences, sensory systems, cognition, and behavior, and how these factors interact with and select for floral color, are what have attracted the greatest research attention (Kevan and Baker 1983; Chittka 1996b; Kevan and Backhaus 1998).

Pollinators may have innate color preferences, or they learn to associate colors with floral rewards (Giurfa et al. 1995; Lunau and Maier 1995; Giurfa 2004; Dyer et al. 2006). Shifts in floral color can lead to a sometimes dramatic shift in pollinator type (Paige and Whitham 1985; Bradshaw and Schemske 2003; see Chapter 8). When flowers provide their pollinators with rewards such as nectar, color functions in pollinator attraction, enhances floral detectability, and reinforces floral constancy (e.g. Chittka et al. 1999; Hempel de Ibarra et al. 2000; Gegear and Laverty 2001; Wertlen et al. 2008). More sophisticated signaling strategies involve colors that attract or exclude specific species or guilds of pollinators via learning, private sensory channels, or pollinator syndromes (Fenster et al. 2004; Johnson et al. 2006). Colors can also signal changes in the presence, abundance, or quality of floral rewards (Weiss 1995; Lunau 1996). When pollination systems are deceptive or nectarless, color is likely to have additional or different functions than rewarding plant-pollinator interactions (see Chapters 2–4, 6, 7, and 10). Orchids are renowned for their extraordinary diversity and unusual deceptive pollinating systems, providing broad scope for considering whether or not floral color does indeed influence plant fitness via pollination, and how color influences pollinator behavior.

In his marvelous book on orchids and their pollinator-driven "contrivances," Charles Darwin supported K. C. Sprengel's then-controversial hypothesis that showy floral colors are pollinator attractants and that floral trait combinations reflect pollinator types (Darwin 1877). Consequently, Darwin noted that he had few explanations for why pollinators were attracted to orchids with drab green or inconspicuous flowers, many of which had hinged labella. These unusual floral features are now strongly associated with sexually deceptive pollination systems in which orchids mimic female insects, and pollination occurs when male insects are fooled into sexual behavior with flowers (Chapters 3 and 6). (For a list of sexually deceptive orchids and their pollinators, see Gaskett 2011; and see Chapters 3, 5, and 6.) In these sexually deceptive pollination systems, pollinators are primarily attracted by olfactory signals mimicking female sex pheromones (e.g. Peakall et al. 2010; Vereecken et al. 2010a), but floral color can still play a role.

In this chapter, I summarize progress in understanding the functions of color in the pollination of sexually deceptive orchids. I briefly define color and some nonpollinator functions of color. The known functions of color in rewarding pollination systems are compared with the possible functions of color in sexually deceptive orchids, with a focus on the role of pollinator cognition and behavior. To assist with attempts to measure, model, and understand pollinator perception of orchid color, I describe the vision systems of 2 common orchid pollinator types, trichromatic species of Hymenoptera and tetrachromatic species of Diptera, in some detail, including their receptor types and spectral sensitivities. Competing methods of modeling floral colors into hymenopteran and dipteran visual spaces are evaluated, incorporating some discussion of quantum catch, color contrast, and the role of achromatic versus chromatic signals and brightness.

COLOR

Color, or chromatic contrast, involves both hue and saturation (Schaefer 2010). Hue describes the wavelengths absorbed and reflected by an object, typically relating to a color category such as red, yellow, green, or blue, or a mixture of any 2 of these, for example orange or purple. Saturation, or chroma, refers to the intensity of a color or how much achromatic gray is present. Highly saturated (high-chroma) colors have little or no gray, whereas low-saturation (low-chroma) colors are diluted by large amounts of gray (Schaefer 2010). Achromatic elements such as the absence of hue (e.g. white, gray, or black), brightness, and luminance also influence the perception of color (Pokorny et al. 1991). In practice, the influence of these factors on perception depends on lighting conditions and backgrounds.

Before considering the roles of color in plant-pollinator communication, the functions of flower and leaf color that are not pollinator driven must be acknowledged. Color pigments have several physiological functions for plant tissue, including photosynthesis, scavenging of free radicals, and protection from cold as well as from excessive light, UV, and other forms of radiation (Schaefer and Rolshausen 2006). Anthocyanins and associated compounds are also implicated in the deterrence of herbivores and pathogens, although this may sometimes be a secondary function correlating with the production of pigments for primarily physiological purposes (Frey 2004; Schaefer and Wilkinson 2004; Schaefer and Rolshausen 2006). Regardless of the proximal trigger of pigment production, color changes can be an honest signal of a plant's capacity for antiherbivore defense (Hamilton and Brown 2001; Archetti and Brown 2004). Multicolored leaves may also decrease the effectiveness of herbivore crypsis against standard leaf colors, thereby facilitating attack on herbivores by their predators (Lev-Yadun and Gould 2007; but see Schaefer and Rolshausen 2006).

Color and Pollinators

In rewarding pollination systems in which pollinator and plant receive mutual benefits, floral color has multiple functions. First, if there is sufficient contrast, color may enhance the detectability of a flower against a background (Hempel de Ibarra et al. 2000; Spaethe et al. 2001). Color also attracts pollinators (see Chapter 9). It is a key signal in the evolution of pollinator syndromes or "private channels" that are "tuned" to attract particular types or functional groups of pollinators (Chittka et al. 2001; Fenster et al. 2004). Many pollinators have innate preferences for certain floral colors (Giurfa et al. 1995; Lunau and Maier 1995; Pohl et al. 2011). Pollinator syndromes may arise via convergent evolution if these innate preferences select for similar colors across plants that share the same pollinators or pollinator functional groups. For example, flowers pollinated by moths are often white, suggesting that moths innately prefer white flowers (Kelber et al. 2003a; Goyret and Raguso 2008). However, floral syndromes need not involve innate color preferences, because pollinators can learn to favor other colors in response to rewards. Experiments show that although hawkmoths (*Manduca sexta*) innately prefer blue flowers, they learn to preferentially visit white flowers when they are rewarding (Goyret and Raguso 2008; see Chapter 8).

Color facilitates floral constancy, the tendency of a pollinator to continue to visit a particular species of rewarding flower in a foraging bout, regardless of the presence of other types of flowers (Kevan and Backhaus 1998; Chittka et al. 1999). This fidelity leads to more efficient and reliable choices for pollinators when foraging, and less pollen wastage and stigma clogging for flowers during pollination (Chittka et al. 2001). Floral color can also indicate the quality, presence, or value of rewards. For example, postpollination color change may be a strategy to maintain a large floral display yet reduce pollen wastage and geitonogamy by directing pollinators away from senescent and rewardless flowers (Weiss 1995; Lunau 1996; Kudo et al. 2007). Selection for this type of floral color change is also likely to be facilitated by pollinator preferences for efficient foraging and ability to associate each color with its reward.

Pollinators' visual attraction to flowers often involves symmetry (Dafni and Kevan 1996; Neal et al. 1998; Giurfa et al. 1999). Bilateral/zygomorphic symmetry (which is common in orchids) may restrict pollinators to a more limited approach direction than radial/actinomorphic symmetry that allows a wider range of pollinator access points and orientations (Neal et al. 1998; Sargent 2004; see Chapter 7). It is unclear how symmetry interacts with floral color, although color-contrasting nectar guides typically reproduce the same symmetry as the overall flower shape (Dafni and Kevan 1996). Symmetry is perhaps likely to be perceived at very close range, after color has already been perceived (Neal et al. 1998). Flower-naïve bumblebees (*Bombus terrestris*) innately prioritize symmetry over floral color in rewarding arrays (Rodríguez et al. 2004), but the hierarchy or function of color versus bilateral symmetry in orchids appears unstudied.

Functions of Color in Pollination by Deception

The deceptive pollination systems of orchids rely on different aspects of insect foraging and reproductive behavior. Accordingly, color is likely to function differently depending on the behavioral context involved. In food-deceptive or nectarless pollination systems in which pollinators receive no rewards, flowers are nevertheless pollinated by foraging insects (Jersáková et al. 2009). Since both food-deceptive and rewarding flowers rely on the same types of insects (Chapter 7) and foraging behaviors, floral color in food-deceptive systems is likely to share many of the functions of color in rewarding systems. Foraging pollinators regularly encounter nectarless flowers, either because a flower has recently been emptied by another animal or because many ostensibly rewarding plants often have a few nectarless flowers (Renner 2006). Therefore, the experience of foraging on a nectarless flower and the consequent associated behaviors likely involve a similar behavioral repertoire.

In deceptive pollination systems involving sexual or other reproductive behaviors, color is likely to function quite differently than in rewarding or food-deceptive systems. In sexually deceptive pollination systems that exploit male insect mating behavior, color will interact with innate and learned behaviors associated with mate choice, courtship, and copulation. In brood site–deceptive pollination systems that exploit female oviposition behavior, color will involve female host preferences, and male and female attraction to mating or rendezvous sites.

Mimicry

Mimicry is regularly suggested as a function of floral color for a range of deceptive pollination systems including food, sexual, and brood-site deception (Schiestl 2005; Jersáková et al. 2006a; Paulus 2006; Jersáková et al. 2009; Gaskett 2011; see Chapter 5). The first experiments on the role of color in sexual deception were performed by Kullenberg (1961), who made some preliminary spectral measurements of *Ophrys* flowers and conducted experiments with artificial paper and fabric flowers, as well as colored hoods concealing real flowers (Chapter 3). After experimenting with *Ophrys insectifera*, Kullenberg concluded that dark colors, UV, gloss, and a velvety appearance mimic female insects and are important releasers of pollinator sexual behavior. Paulus (2006) identified 3 pollinator-driven categories of coloration in *Ophrys*: (1) strong resemblance to female insects in *Ophrys* species pollinated by insects that use visual signals in mate choice (e.g. *O. vernixia* [or *O. speculum*], pollinated by *Dasyscolia ciliata*, Scoliidae), (2) distantly related *Ophrys* that share the same pollinator and have similar color patches or patterns, presumably due to convergent evolution (e.g. *O. atlantica*, *Ophrys bertolloni*, and *O. ferrum-equinum*, all pollinated by *Chalicodoma parietina*; Megachilidae), and (3) *Ophrys* with dark-colored labella, mostly pollinated by andrenid and anthophorine bees (e.g. species from the *O. fusca* and *O. sphegodes* aggregates; Paulus 2006). In andrenid and anthophorine (previously Anthophoridae, now Apidae) bees, mating systems are dominated by chemical rather than visual signals (Spaethe et al. 2010). Correspondingly, studies of *Ophrys* from this final group find little effect of floral color; for example, experiments with *O. arachnites* show that color does not influence pollinator attraction (Vereecken and Schiestl 2009; see also Chapter 3).

Sexually deceptive orchids may indeed mimic the colors of female insects' bodies and the UV-reflectance of their wings, but this has been explored in only a few studies (Kullenberg 1961; Paulus 2006; Gaskett and Herberstein 2010). To human eyes, the flowers of *Ophrys vernixia* strongly resemble females of its pollinating species (*Dasyscolia ciliata*), while flowers of *Ophrys regis-ferdinandii* resemble the colors of male hoverflies (Syrphidae) and are pollinated by aggressively territorial males (*Merodon velox*; Paulus 2006). Modeling with a Hymenoptera-type visual system indicates that the colors of 4 *Cryptostylis* species that share a single pollinator (male *Lissopimpla excelsa*) are likely indistinguishable from those of female *L. excelsa* (Gaskett and Herberstein 2010; see Chapter 5). A series of behavioral assays (reported by Paulus 2006) show that manipulation of UV reflectance from sexually deceptive *Ophrys* flowers resulted in modified pollinator attraction to *O. heldreichii* (pollinated by *Eucera rufa* [previously *E. berlandi* or *Tetralonia berlandi*]) and *O. vernixia* (pollinated by *Dasyscolia ciliata*). However, it is unclear whether these situations indicate mimicry, or a form of sensory trap (see below).

As well as mimicking mates, floral colors and visual patterns have been suggested to imitate a range of other resources. For example, the glistening staminodes of food-deceptive *Paphiopedilum villosum* may mimic drops of nectar (Bänziger 1996). Contrasting dots on the petals of brood site–deceptive *Paphiopedilum* species may mimic a colony of aphids and thus attract pollinators such as hoverflies that typically oviposit into aphid colonies (Bänziger 1996; Shi et al. 2007; Shi et al. 2009a; see Chapter 10). The dark, featureless labellum of *Ophrys helenae* may mimic a dark hole into which its duped pollinator, male *Eucera* spp., attempts to crawl, seeking overnight shelter (Paulus 2006). By comparison, fewer rewarding flowers reflect UV (Chittka et al. 1994), and the trait is correspondingly rare in food-deceptive orchids (e.g. Johnson 2000; Johnson et al. 2003a).

In many of the cases above, it is difficult to conclude whether signals are truly mimetic, or are instead the result of convergent evolution or the exploitation of perceptual biases (Grim 2005; Ruxton and Schaefer 2011). Cases of similar signals via convergent evolution are perhaps most likely to occur among food-deceptive orchids. When the models (rewarding flowers) and putative mimics (food-deceptive orchids) grow sympatrically and share the same environmental conditions, pollinators, and pollinator behaviors, it is unsurprising if convergent evolution resulted in similar floral signals. Alternatively, similarity in signals between a deceptive orchid and an important resource such as food, mates, or oviposition sites may evolve when an orchid exploits the sensory preferences or innate biases of its pollinator via a sensory trap (see below). Exploitation of perceptual biases may be widespread in rewarding and deceptive pollination systems, and in some cases it may even represent an early stage in the evolution of more specific mimicry (Schaefer and Ruxton 2010; Schiestl et al. 2010).

Sensory Drive and Sensory Traps

Deceivers may take advantage of a target receiver's sensory drive or innate sensory biases via a "sensory trap" (Schaefer 2010). Sensory drive is the tendency for signal formats used by an animal to be similar across several different contexts in their behavioral repertoire (Schaefer et al. 2004; Raguso 2008). For example, an insect's vision system may be tuned toward certain colors that allow successful foraging. If the insect is sensitive or preadapted to the detection of these colors, they also become important signals in other behaviors, such as mating and territoriality (Schaefer et al. 2004). Therefore, such insects are vulnerable to exploitation by orchids that can produce similar signals.

Yellow or UV wavelengths are good candidate signals for use in sensory traps. Many insects (and other animals) are strongly attracted to yellow or UV (Goldsmith 1994; Tovée 1995; Kevan and Backhaus 1998; Goulson et al. 2007; Lunau 2007; Kelber and Osorio 2010). These wavelengths can be "releasers" of specific wavelength-dependent behaviors, from simple phototaxis to more complex routines associated with foraging or reproduction (Goldsmith 1994). Furthermore, many deceptive orchids also reflect yellow and/or UV as blocks of continuous color, or as spots, patches, or raised bumps, as in food- and/or brood site–deceptive

Paphiopedilum species (Bänziger 1996; Shi et al. 2009a) and sexually deceptive *Cryptostylis* and *Ophrys* species (Kullenberg 1961; Paulus 2006; Gaskett and Herberstein 2010). The UV receptor of winged members of the Hymenoptera is considerably more sensitive than the blue or green receptors (Chittka et al. 1994; Lunau and Maier 1995), suggesting that a small amount of UV reflectance can provoke a strong sensory and behavioral reaction. Universally important and attractive signals such as UV may be vulnerable to exploitation by deceptive orchids, but the fitness benefits of attending to such signals outside the realm of pollination are likely to strongly reinforce the behavioral response and override any costs of deception by orchids.

Enhancing Detectability

Much as in rewarding or food-deceptive systems, color can attract pollinators by enhancing the detectability of a flower against a background, which may manifest in increased attraction or reduced search time by the pollinator (e.g. Chittka et al. 2001; Spaethe et al. 2010; see Chapter 6). Kullenberg (1961) demonstrated that in the presence of scents from a hidden *Ophrys insectifera* flower, any dark-colored object against a contrasting background attracts pollinators. A series of experiments with *Ophrys heldreichii* show that pollinator attraction is enhanced by a pink or white perianth (or artificial floral border) that contrasts with the labellum and background (Spaethe et al. 2007; Streinzer et al. 2009; Spaethe et al. 2010; Streinzer et al. 2010). Kullenberg preempted this result by suggesting that colored perianths resemble rewarding flowers that are "conspicuous in the ordinary way" (Kullenberg 1950a,b,c).

The presence and function of a contrasting perianth appear to depend on the typical sexual behavior of the pollinator involved. In their literature survey, Spaethe et al. (2010) noted that a contrasting perianth was present in ~80% of *Ophrys* species pollinated by male insects from the tribe Eucerini (Apidae; Hymenoptera). These males rely strongly on vision when searching for a mate, unlike male *Andrena* spp., which rely largely on chemical signals, and pollinate many of the other *Ophrys* species that lack the color-contrasting perianth (Spaethe et al. 2010; see Gaskett 2011 for a list of pollinators).

Do some sexually deceptive orchids use color to hide their appearance via camouflage? Many species have greenish and often inconspicuous flowers, for example South and Central American *Geoblasta pennicillata*, *Lepanthes glicensteinii*, *Mormolyca ringens*, *Stellilabium*, and *Telipogon* spp., and Australian and New Zealand *Chiloglottis* and *Pterostylis* spp. Green flowers could indicate camouflage, or alternatively, relaxed selection on floral color because it is unimportant in pollinator attraction. Pale cream and white flowers are also common among sexually deceptive orchids (e.g. Australian *Caladenia syn. Arachnorchis* and synonymous genera; see Chapter 6). Some vision models and experiments suggest that white flowers may not be detectable (or learnable) against many typical backgrounds, including soil and sand (Chittka et al. 1994; Kevan et al. 1996; Chittka 1999; Spaethe et al. 2001). However, this remains controversial, because other models and tests suggest this may not be the case (Hempel de Ibarra et al. 2000). Important factors in detectability appear to be the distance and visual angle between the viewer and the target, and the brightness and stimulation of the LW (green) receptor by the target in relation to the background (Giurfa et al. 1996). The spatial resolution of eyes of unique Hymenoptera also suggests that pollinators lack the sensitivity necessary to use fine color patterns when orienting themselves on a flower (Hempel de Ibarra et al. 2000; Wertlen et al. 2008).

Cognitive Interactions

It has been often suggested that intraspecific variation in the appearance of deceptive orchids functions to hinder pollinator recognition of orchids (Juillet and Scopece 2010). Like most insects, pollinators are highly capable of learning, recognizing, and consequently avoiding signals associated with aversive experiences (Burns et al. 2011; Gaskett 2011; Kevan and Menzel 2012). Ingenious experiments by Paulus and colleagues showed that pollinators do rapidly recognize and subsequently avoid individual *Ophrys* flowers (Paulus 2006). When presented with halved flowers, pollinators willingly visited all the first halves of each flower, but then ignored the second halves of any flower they had already encountered. Intraspecific variation in floral color or scent in deceptive orchids may challenge this aversive learning (or "reverse floral constancy") and ensure that pollinators continue to visit and transfer pollinia between flowers of an orchid species (Juillet and Scopece 2010; Chapters 2, 6, 7). Streinzer et al. (2010) argued that this explains the variable fine color patterning on the labella of *Ophrys heldreichii* flowers. Manipulations or the presence or absence of the patterns do not appear to influence initial pollinator attraction (indicating no innate preferences), and these patterns vary between but not within species (Streinzer et al. 2010). Experiments reported by Paulus (2006), in which scent was excluded, confirmed that pollinators preferentially avoided individual *O. heldreichii* flowers if they had already encountered flowers with the same patterns. However, in their review, Juillet and Scopece (2010) found no consistent fitness benefit associated with floral trait variability.

An alternative and untested hypothesis for how deceptive orchids can combat pollinator aversive learning involves sensory traps. Orchids could distract pollinators from remembering or avoiding their flowers by producing irresistible yellow or UV signals that exploit pollinators' innate preferences. Studies of other flowers suggest this may be a possibility in orchids. For example, in plants that provide pollen as a reward and have distinct male and female phases, pollinators should learn to avoid the unrewarding female flowers, and flowers should evolve counteradaptations that prevent pollinator avoidance (Pohl et al. 2008). In experiments with artificial flowers, bumblebees did have difficulty distinguishing rewarding and unrewarding flowers when all flowers had bright-yellow markings resembling pollen-loaded anthers. The specific color and location of the markings were vital, and the authors argued that pollinator attraction to bright yellow was able to override their aversion to deceptive flowers (Pohl et al. 2008). Perhaps instead of focusing solely on floral trait variability in deceptive orchids, we should consider "distractors" that might deliver similar cognitive effects.

FUTURE DIRECTIONS IN UNDERSTANDING FUNCTIONS OF FLORAL COLOR

Floral color may evolve in response to multiple target and nontarget receivers, including pollinators and herbivores (Schaefer 2010). Floral color may also have multiple, simultaneous functions in pollinator attraction. For example, the dark spots on *Paphiopedilum* species may mimic aphid colonies at the same time that their bright-yellow staminodes create a sensory trap (Shi et al. 2007; Shi et al. 2009a; see also Chapter 10). Similarly, UV-reflecting patches on the labella of sexually deceptive *Cryptostylis* orchids may both mimic the wings of female insects and exploit innate preferences for UV wavelengths (Gaskett and Herberstein 2010). The most persuasive deceivers may elicit the strongest sexual behavior from their pollinators by incorporating several strategies across multiple sensory modes (Gaskett et al. 2008).

At present, many aspects of color function remain unexplored. Does the green coloration of many sexually deceptive orchids indicate some form of camouflage or even crypsis of surrounding vegetation? Do color patterns such as blotches and stripes function as disruptive coloration that hides the true shape of an orchid against a background? Finally, what can explain the extraordinary frequency of red, white, and green color-ation among sexually deceptive orchids in all 3 of their centers of diversity: Australia, South and Central America, and Mediterranean Europe? Perhaps these aspects of color function involve some orchid exploitation of the pollinators' sensory vulnerabilities in color regions that are difficult for these pollinators to detect or differentiate (Gaskett and Herberstein 2010). Further investigations integrating cognition and sensory perception with deception and plant fitness are required.

THE VISION OF ORCHID POLLINATORS

Any attempt to understand the function or evolution of floral color in pollination should consider the visual capacities of the pollinator. Information about color vision, photoreceptors, and detected wavelengths is available for a surprisingly wide variety of animals, including insects, spiders, mollusks, birds, reptiles, fish, and mammals, in terrestrial, aquatic, diurnal, and nocturnal environments (Briscoe and Chittka 2001; Kelber et al. 2003b; Warrant and Locket 2004; Warrant 2008; Théry et al. 2010). Models exist for a wide range of orchid-pollinating taxa, such as members of insect orders Hymenoptera, Diptera, and Lepidoptera, as well as birds. Along with basic data on spectral sensitivities, understanding pollinator perception of orchid colors requires information about other aspects of vision. For key model species such as bees (*Apis* and *Bombus* spp.) and flies (*Drosophila* spp.), a wealth of information about optical anatomy, visual resolution, neural processing of color, and visual cognition and learning is available for swift application to orchid research (Kevan and Backhaus 1998; Spaethe et al. 2001; Kelber and Osorio 2010; Avarguès-Weber et al. 2011; Dyer et al. 2011). Here I summarize some current knowledge about the visual system and color perception of 2 important deceptive orchid–pollinating groups, Hymenoptera and Diptera.

The Trichromatic Vision of Hymenoptera

Members of the order Hymenoptera are trichromats, with 3 types of photoreceptor that detect wavelengths across 3 spectral ranges (Daumer 1956). In the honeybee, *Apis mellifera*, the spectral sensitivities of the 3 ranges peak at 344 nm (UV or short wavelengths [SW]), 436 nm (blue- or mid-wavelength [MW] region), and 556 nm (green- or longer-wavelength [LW] region) (Menzel and Blakers 1976; Peitsch et al. 1992). These spectral sensitivity values are generated by intracellular recordings of spectral stimulation of single photoreceptors (Menzel and Backhaus 1991). Comparison of spectral sensitivities across a wide range of species in this order indicates very little variation in sensitivity ranges or peaks, and any minor differences do not correlate with phylogenetic relationships (Peitsch et al. 1992). Thus, honeybee spectral sensitivities have become a model system applied to represent bee and wasp vision in a range of study systems, including studies of pollination, predator avoidance, and camouflage (Théry and Casas 2002; Heiling et al. 2003; Bruce et al. 2005; Arnold et al. 2009; Defrize et al. 2010).

In trichromatic Hymenoptera, the 3 photoreceptor types vary in their relative sensitivity, with the UV/SW receptor being most sensitive; a small amount of UV light provokes a comparatively strong photoreceptor response (Daumer 1956; Chittka et al. 1994; Lunau and Maier 1995). Although species of Hymenoptera do not generally have a receptor focused on red wavelengths, red objects are not invisible to these insects (Chittka et al. 1994; Reisenman and Giurfa 2008). This is because the green/LW receptor detects wavelengths across a wide range, including up to ~650 nm, which extends into and overlaps with some wavelengths considered "red" (~630–740 nm). However, since red is at the periphery of detection for most Hymenoptera, there is therefore less sensitivity to color differences in this region (Gaskett and Herberstein 2010), although differentiation improves with practice (Reisenman and Giurfa 2008). A few exceptional members of the Hymenoptera species do have red receptors (Peitsch et al. 1992), so investigations of orchids pollinated by poorly known bees or wasps should take this possibility into account.

Across a honeybee's eye, the distribution and abundances of the 3 UV, blue, and green photoreceptor types are not homogenous (Briscoe and Chittka 2001). The abundance of each receptor type affects the total quantum catch of each wavelength entering the viewer's sensory system, and therefore the perception of color (Defrize et al. 2010). Three different combinations of photoreceptors are found in ommatidia: 1 UV, 1 blue, and 6 green (type I); 2 UV, 1 blue, and 6 green (type II); and 1 UV, 2 blue, and 6 green (type III; Wakakuwa et al. 2005). The ratios of the 3 types of ommatidia vary slightly between Hymenopteran species and between studies, but Defrize et al. (2010) have proposed an average ratio of 1:0.471:4.412, calculated from existing literature.

Dipteran Tetrachromatic Vision

The order Diptera (true flies) are important pollinators of several sexually deceptive orchids, including South and Central American *Lepanthes, Stel-*

lilabium, Telipogon, and *Trichoceros* species and Australian and New Zealand *Pterostylis* s.l. spp. (see review by Gaskett 2011). Flies are known or implicated in the pollination of a wide range of food-deceptive orchids, including South African *Disa* and *Satyrium* spp. (Anderson and Johnson 2006; Jersáková and Johnson 2006; van der Niet et al. 2011; Chapter 4), Asian *Paphiopedilum* (Bänziger 1996; Shi et al. 2009a), and Australian and New Zealand *Thelymitra* (Christensen 1994). Brood site–deceptive orchids pollinated by flies include Australian and New Zealand *Corybas* s.l. and *Nematoceras* species (and synonymous genera; Gaskett 2011), Mediterranean *Epipactis consimilis* (Christensen 1994), and Neotropical *Dracula* spp. (Endara et al. 2010).

Much is known about the visual systems, neural processing, and photoreceptors of flies (Burns et al. 2011). There are also data on the spectral sensitivities of fly photoreceptors (Bernard and Stavenga 1979; Hardie and Kirschfeld 1983). Perhaps surprisingly, then, little is known about how color is processed and perceived, and how this influences fly behavior. This may be because, unlike honeybees, flies are difficult to train to color stimuli. In a key paper attempting to quantify fly vision with behavioral assays, Troje (1993) suggests a categorical rather than a continuous color vision system. Blowflies (*Lucilia* sp.) appear to be able to distinguish between colors in 4 categories: UV, blue, yellow, and purple, but not between colors or shades within a category (Troje 1993). These categories coincide with 4 photoreceptors and their peak regions of sensitivity: R7p (UV) at 341 nm, R7y (blue) at 362 nm, R8p (yellow) at 465 nm, and R8y (purple) at 537 nm (Hardie and Kirschfeld 1983).

MEASURING AND EXPERIMENTING WITH FLORAL COLOR

Using UV-sensitive cameras to photograph flowers provides a rapid method for demonstrating some of the biases of human vision (Kropf and Renner 2005; Indsto et al. 2006; Indsto et al. 2007). To quantify color, reflectance across the entire UV and visible light spectrum is required. A wide variety of spectral equipment is available from Ocean Optics (Florida, USA), only some of which are suitable for measuring and modeling color reflectance. Orchid pollination studies utilizing spectral reflectance measurements include Johnson (2000), Johnson and Andersson (2002), Johnson et al. (2003a), Anderson et al. (2005), Jersáková et al. (2006a), Johnson and Morita (2006), Shi et al. (2009a), Streinzer et al. (2009), Gaskett and Herberstein (2010), and Streinzer et al. (2010). (For procedural advice on making measurements, see Chittka and Kevan 2005.

Methods for quantifying the area of each color patch are reviewed in Théry et al. 2010.)

During experiments, color can be manipulated with colored filters, sunscreen, or UV-absorbing chemicals to remove UV (Johnson and Andersson 2002; Göth and Evans 2004; Heiling et al. 2005; Peter and Johnson 2008). Paulus (2006) described behavioral assays in which *Ophrys* flowers were manipulated to cover their UV-reflecting parts, or were sealed inside transparent boxes that either allowed or filtered out UV wavelengths. In the broader field of behavioral ecology, Plasticine modeling clay and paper are sometimes used to make models that accurately match natural objects' spectral reflectances, including UV (Fan et al. 2009; McLean et al. 2010).

OTHER IMPORTANT ASPECTS OF FLORAL COLOR

Along with chromatic or hue-based color, there are several other influential properties, including gloss or matte, transparency, brilliance, and speculance or sparkle (Pokorny et al. 1991). Flowers and insects such as orchid-pollinating euglossine bees may also fluoresce under UV light (Gandía-Herrero et al. 2005; Nemésio 2005; Chapter 9).

Achromaticity

Achromatic elements lack hue (e.g. white, gray, or black; Pokorny et al. 1991). Across a range of animals (including humans), achromatic cues are used for motion detection and for determining distances, textures, and edges (Kevan et al. 1996; Giurfa et al. 1997; Osorio and Vorobyev 2005). For species of Hymenoptera, the role of chromatic and achromatic contrast depends on the distance and therefore the visual angle between themselves and the object they are viewing (Giurfa and Lehrer 2001). When scent is primary in long-range pollinator attraction, visual assessment of color is likely to operate at close range, when chromatic contrast with all 3 receptors would be involved (Spaethe et al. 2001; Gaskett and Herberstein 2010). At longer distances or when looking at small flowers (resulting in a visual angle of 5–15 degrees), achromatic contrast becomes primary (Giurfa et al. 1997; Spaethe et al. 2001; Dyer et al. 2008). Both honeybees and bumblebees detect achromatic contrast with their long-wavelength (green or LW) receptors (Giurfa and Lehrer 2001), so the degree of contrast at a distance can be calculated by considering excitation or quantum catch of this receptor alone, as in Théry et al. (2005). Mammals are also thought to similarly use their long-wavelength receptor for both chromatic and achromatic vision, but flies and birds probably have separate types of receptors for chromatic and achromatic purposes (Osorio and Vorobyev 2005).

Brightness

The role of the achromatic brightness of an object in relation to its background is still somewhat unclear (Menzel and Backhaus 1991; Hempel de Ibarra et al. 2000). Some experiments suggest honeybees did not use brightness when trained to chromatic signals (Menzel and Backhaus 1991). Although some visual models predict that bright targets are easier to detect against dim backgrounds (e.g. Brandt and Vorobyev 1997), others predict the reverse, (e.g. Hempel de Ibarra et al. 2000]). The hexagon model of Chittka 1992 could be interpreted to predict that bright targets are difficult to detect because the viewer's receptors become saturated when the target is more intense than the background (Chittka 1992; Hempel de Ibarra et al. 2000). However, behavioral experiments demonstrate that honeybees can detect both bright targets against dim backgrounds and dim targets against bright backgrounds, and that the degree of color contrast is key (Hempel de Ibarra et al. 2000). In his 1999 rebuttal, Chittka describes studies where there were difficulties in training bees to bright white signals, and makes the point that even if an animal can detect a signal, this does not necessarily mean it can learn and use it (Chittka 1999).

MODELING COLOR IN INSECT VISUAL SYSTEMS

Methods for modeling insect color perception and discriminability involve the wavelengths reflected by an object and its background, the ambient light, the quantum catch of each photoreceptor, and the differential sensitivities of receptors to their target wavelengths. All models for trichromats propose that they code chromatic color by directing signals from their 3 receptor types into 2 independent color-opponent mechanisms which define a 2-dimensional color space (Hempel de Ibarra et al. 2000). When mapped into the color space, the distance between the loci of different objects indicates their perceptual difference in color to the viewer (Backhaus 1991; Menzel and Backhaus 1991; Backhaus 1992). This is often called the color contrast, ΔS. The threshold at which colors become so close in the color space that they are indistinguishable for the viewer depends on the modeling method used (see "Color Contrast" below).

There are 2 commonly used methods for modeling trichromatic vision of Hymenoptera: the hexagonal color space developed by Lars Chittka, Werner Backhaus, Randolf Menzel, and colleagues (Kevan 1972; Backhaus and Menzel 1987; Chittka 1992; Chittka et al. 1994; Chittka 1996a) and the Maxwell triangle color space, first applied by Christa Neumeyer and then further developed and characterized by Misha Vorobyev, Robert Brandt, Natalie Hempel de Ibarra, and colleagues (Neumeyer 1980; Brandt and Vorobyev 1997; Vorobyev and Osorio 1998). (See Hempel de Ibarra et al. 2000 for an excellent comparison of the theory and calculations involved in these models and their variants.)

The hexagon is a generalized color opponent space based on how color is typically processed and incorporating a brightness dimension absent from other models (Chittka 1992). The hexagon color space has been widely applied to a range of studies assessing bee and wasp perception of flowers, predators, and mates (Chittka et al. 2003; Heiling et al. 2003; Tso et al. 2004; Bruce et al. 2005; Dyer and Neumeyer 2005; Wignall et al. 2006; Dyer et al. 2008; Arnold et al. 2009; Chiao et al. 2009; Defrize et al. 2010; Gaskett and Herberstein 2010; Llandres et al. 2011). However, some experiments with controlled colors and artificial stimuli do not appear to support all the predictions of the hexagon model, which has generated controversy in the literature, especially regarding the detection of white flowers against achromatic, green leaf, or soil backgrounds (Kevan et al. 1996; Chittka 1999; Vorobyev 1999; Hempel de Ibarra et al. 2000). The most recent triangle models address how color discrimination is affected by both internal receptor noise and photon noise, and how these are affected by variation in light intensity (Hempel de Ibarra et al. 2000; Vorobyev et al. 2001). These models were devised to explain results drawn from electrophysiological measurements of receptors and behavioral experiments performed with color stimuli.

Both the hexagon and the triangle models involve calculating the quantum catch (or photon flux) of the photoreceptors, that is, the total amount of light received by each type of photoreceptor (UV, blue, and green) across all wavelengths within the range of wavelengths visible to Hymenoptera (300–700 nm). Quantum catch (Q_i) is the product of ambient light (e.g. standard function D65), the spectral sensitivity of the receptor (e.g. from Menzel and Backhaus 1991), and the wavelengths reflected from a specimen (e.g. as recorded by a spectrometer; Backhaus and Menzel 1987). In the hexagon model, the loci of specimens within the color space (i.e. their x and y coordinates in the hexagon) are mapped by calculating receptor excitations for the UV, blue, and green receptors (based on quantum catch), and then using these to calculate the coordinates with 2 straightforward color opponent formulae (for equations, see Chittka 1992). The Euclidean distance between the loci in a color space represents the color contrast (ΔS), or degree of color difference, between specimens. In the triangle model, the loci coordinates are calculated directly from the quantum catches of each of the UV, blue, and green receptors (for equations, see Brandt and Vorobyev 1997). The vertices of the frame of the hexagon or triangle on a graph represent points at which 1 of the 3 receptors is maximally stimulated (excitation = 1) and the other 2 are unstimulated (excitation = 0), so these values can be used in the loci equations to calculate where to place the vertices of the triangle when constructing a graph.

COLOR CONTRAST

Color contrast (ΔS) is perhaps the most important calculation for orchid color research. Color contrast values indicate the degree of color difference between 2 objects as seen by the viewer (Hempel de Ibarra et al. 2000). Calculating the color contrast may reveal what parts of an orchid flower are most visible, if an orchid is visible or camouflaged against a background, or if it resembles the color of another flower or a female insect when seen by a viewer in the order Hymenoptera. Calculating color contrast adds considerable power to any study on floral color.

The threshold value at which the viewer can detect a difference in color (ΔS^t) varies according to the modeling method used, but is generally based on or validated by behavioral experiments. These typically involve training insects to respond to a certain color, then testing for discriminability of the target and similar nontarget colors in pairwise assays (e.g. Hempel de Ibarra et al. 2000; Dyer and Chittka 2004b). For the hexagon model, behavioral experiments with bumblebees (*Bombus terrestris*) suggest a conservative threshold of $\Delta S^t = 0.062$ hexagon units, although this depends on some prior experience and learning (Dyer and Chittka 2004b). After 10 experiences, approximately 55% of bumblebees could distinguish between objects with a contrast as low as 0.062 hexagon units, but even after 100 experiences, 20% of bumblebees still could not distinguish between colors with a contrast of 0.012 hexagon units (Dyer and Chittka 2004a,b). If the pollinators were naïve, a threshold as high as $\Delta S^t = 0.1$ is not unreasonable (A. G. Dyer, personal communication). When there are negative consequences or punishments for failing to discern between target and similar deceptive colors, as could occur in a sexually deceptive pollination system, insects learn more quickly and achieve finer color discrimination (Chittka et al. 2003; Dyer and Chittka 2004a,b). Bumblebees can discriminate between rewarding and punishing colors with a color contrast of only 0.185 hexagon units after a single experience (Dyer and Chittka 2004a,b). In Vorobyev and Osorio's triangle model, ΔS is already presented in units of JND (Just Noticeable Differences), and a value of less than 1 indicates no detectable color difference to the viewer (Defrize et al. 2010). The triangle model takes into account the effects of receptor noise and how it is affected by increasing light intensity.

MODELING FLY VISION

The few studies addressing the color vision of flies suggest a categorical rather than a color opponency system (Hardie and Kirschfeld 1983; Troje 1993; Lunau andKnüttel 1995). The relative quantum catch across 2 pairs of photoreceptors (R7p vs. R8p and R7y vs. R8y) generates 4 possible color categories: +,+ or "fly UV"; +,– or "fly purple"; –,+ or "fly blue"; and –,– or "fly yellow or green." Quantum catch is calculated with the same equation as for honeybees, as in Backhaus and Menzel (1987), but using the spectral sensitivities of *Lucilia* sp. available from Hardie and Kirschfeld (1983). Although these techniques have been applied in a few published studies (Arnold et al. 2009; Defrize et al. 2010), more behavioral experiments are required to corroborate the observations that led Troje (1993) to conclude that flies have a categorical color vision system. Spectral sensitivity measurements for a wider range of flies and wavelengths are urgently required.

CONCLUSION

Considerable progress has been made toward understanding color, insect perception of color, color processing, and color-based learning since Darwin's time. This wealth of information can be readily applied to orchid pollination studies. Scent may be the primary attractant for sexually deceptive orchids, but even the earliest experiments by Bertil Kullenberg indicate that color also plays a role (Kullenberg 1950a; Kullenberg 1961). This is particularly so for pollinators whose natural sexual and mate choice behaviors typically involve some aspect of visual signaling or assessment. Furthermore, while previously many studies focused on color in mimicry, it is now clear that color can interact with many cognitive arenas, including learning and, perhaps most important, sensory drive or the exploitation of innate biases. As the field of vision science progresses, we will have access to considerably more information about the visual systems of an even greater array of insect taxa. The more data we have on spectral sensitivities, color processing mechanisms, and the psychophysics of color, the better we can understand the extraordinary colors and fascinating natural history of orchids.

TWELVE

Impacts of Extreme Weather Spells on Flowering Phenology
of Wild Orchids in Guangxi, Southwestern China

Hong Liu, Chang-Lin Feng, Xiao-Qing Xie, Wuying Lin,

Zheng-Hai Deng, Xin-Lian Wei, Shi-Yong Liu, and Yi-Bo Luo

INTRODUCTION

Climate Change and Phenological Studies

According to Charles Darwin (1877; p. 281), "In other cases the paucity
of the flowers that are impregnated may be due to proper insects having
become rare under the incessant changes in which the world is subject;
or to other plants which are more highly attractive to the proper insects
having increased in number." This remark indicates Darwin's forward-
thinking approach to the possible impacts of environmental changes—
including ones currently affecting us—on the reproductive success of orchids. Darwin mentions the impact of unusual weather events on orchid
reproduction in his book *The Various Contrivances by Which Orchids Are
Fertilised by Insects* (Darwin 1877). However, he does not emphasize extreme weather spells as drivers of temporal variations in orchid reproduction. Annual changes in ambient temperature and/or rainfall may trigger
changes in plant phenology (Zhu and Wan 1983; Wan 1986, 1987; Chen
X-Q 2003; Schwartz 2003; Inouye 2008), which can have consequences
for fruit set and other processes related to population dynamics. This
is especially true if the species in question are animal pollinated or are
otherwise dependent on other organisms, whose phenologies may also be
influenced by annual temperature and rainfall changes.

Phenological studies in the past decade have documented the association of phenological events with global warming (Parmesan and

Yohe 2003; Menzel et al. 2006; Schwartz et al. 2006; Cleland et al. 2007; Miller-Rushing and Primack 2008). Most current phenological studies address mean responses to average weather trends. Only recently have a few studies focused on the effects of naturally occurring or manipulated weather extremes on phenology (e.g. Menzel 2005; Jentsch et al. 2007, 2009; Luterbacher et al. 2007; Rutishauser et al. 2008; Menzel et al. 2011).

Extreme weather spells are not single events. Instead, they are associated with periods of unusual temperature or precipitation that last for some predefined duration (Klein Tank et al. 2009). Extreme warm and cold spells triggered prominent changes in plant phenology in Europe, where phenology of most species is temperature controlled (Menzel et al. 2011).

Early initiation of flowers and other spring events due to the current global warming have been reported for many temperate species (Fitter and Fitter 2002; Menzel et al. 2006; Miller-Rushing and Primack 2008). However, few data are available on the response of subtropical species to global climate change, although extreme weather spells also occur in subtropical regions. These events can provide excellent opportunities to assess which species' phenology is influenced by temperature and/or precipitation changes in subtropical regions, where phenology may be driven by either factor, or others (Corlett and Lafrankie 1998).

Most modern phenological monitoring focuses primarily on common tree species and crops, because they are easy to observe and can be compared across a wide range of locations (Zhu and Wan 1983; Wan 1986, 1987; Chen X-Q 2003; Schwartz 2003; Chuine et al. 2004). The responses of herbaceous understory species to climate change are, however, largely unknown. These plants include orchids, many of which are rare and threatened. Most orchid species have specialized insect pollination systems, relying on one to a few pollinator species (van der Cingel 2001; Tremblay et al. 2005; see Chapters 3, 6, 8–10). The potential mismatches in phenology between orchids and their pollinators due to the current changes in climate present a significant challenge for orchid conservation (Liu et al. 2010).

The Yachang Orchid Nature Reserve is situated between 24°44′16″ and 24°53′58″N, and between 106°11′31″ and 106°27′04″E; it has a subtropical climate, as does most of southwestern China (Corlett 2009). In this region, there are pronounced seasonal variations in both rainfall and temperature (Corlett and Lafrankie 1998; Huang C-B et al. 2008), with nearly 60% of the rainfall occurring in the hot summer months and less than 10% in the cold winter months (Huang C-B et al. 2008). The phenology of woody plants in southwestern China is characterized by regular annual cycles at the individual, population, and community levels (Wan 1986, 1987; Corlett and Lafrankie 1998), probably triggered by temperature and/or water availability (Corlett and Lafrankie 1998).

Many of the orchid species in Yachang have specialized insect pollination systems. Pollination systems of 10 orchid species in this reserve have been studied; these orchids are pollinated by a single species of pollinator (Cheng et al. 2007, 2009; Shi et al. 2007, 2008, 2009b; Shangguan et al. 2008; Lin et al. in revision). One species (*Geodorum densiflorum*) can also self-pollinate (Lin et al. in revision). Species with these pollination patterns may be vulnerable to climate change because of the impacts of potential asynchrony in phenology between the orchids and their pollinators.

Average Climate and Extreme Weather Events in Southwestern China

Records indicate that in the past 100 years, warming in southwestern China has been around 0.5°, slightly lower than the global average of 0.7° (IPCC 2007; Huang X-S et al. 2005). While in other parts of this area warming has largely been caused by an increase in winter, spring, and fall temperatures (He et al. 2007; Chen Y-G et al. 2008; Wang et al. 2008), in the Yachang region it has been reflected largely in summer and fall and, to a much lesser extent, spring temperatures (Figure 12.1). Total annual rainfall, on the other hand, has remained the same or increased slightly for the region (Huang X-S et al. 2005; Bates et al. 2008). It is projected that the warming trend will continue during the next 2 centuries in southwestern China (Jiang et al. 2005; IPCC 2007; Xu et al. 2009). Precipitation, though, is projected to increase only slightly (Bates et al. 2008; Jiang et al. 2005; Xu et al. 2009), and will not keep pace with the increase in evaporation rates due to warming (Bates et al. 2008). A slight decrease in soil moisture is also predicted (Bates et al. 2008).

Against the backdrop of this average climate trend, several extreme weather events have occurred in the region. In 2008, there was an unusually long cold period, that is, an extreme cold spell, which started in early January and lasted through early February (Stone 2008; Zhou et al. 2011). This was followed by an extreme drought in the dry season of 2009–10 (Figure 12.2B; Stone 2010), from October 2009 to April 2010. The purpose of this study was to examine the relationship between changes in flowering dates of selected wild orchid species and these extreme weather events, and to determine if and how certain species' flowering phenologies may be driven by temperature and/or precipitation.

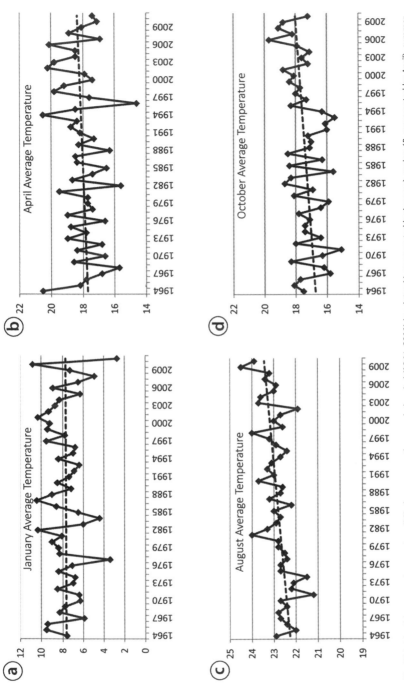

Fig. 12.1 Monthly average temperature (°C) changes over the recorded period (1964–2011) in winter (*A*, represented by January), spring (*B*, represented by April), summer (*C*, represented by August), and fall (*D*, represented by October) in Leye County at 972 m a.s.l. in the Guangxi Zhuang Autonomous Region, southwestern China (location of the Yachang National Orchid Nature Reserve). The linear trend lines (*dashed*) indicate significant temperature increase in summer and fall, slight increase in spring, and no changes in winter.

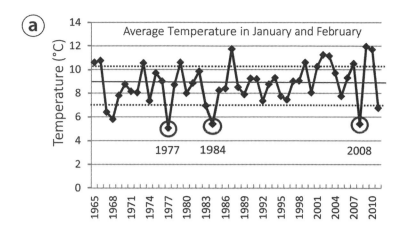

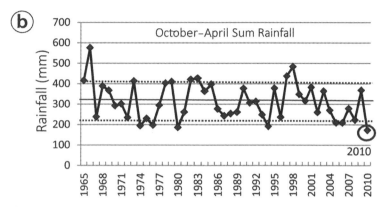

Fig. 12.2 Mean and standard deviations of January and February (*A*) average monthly temperature (°C) (1964–2011) in Leye County at 972 m a.s.l. in the Guangxi Zhuang Autonomous Region, southwestern China, where the Yachang Orchid National Nature Reserve is located. Both the January and February temperatures in 2008 were lower than one standard deviation (*dashed lines*) below the 47-year average (*solid line*), indicating an exceptionally long cold spell. The annual total rainfall during the dry season (October–April) (*B*) in 2010 is the lowest in the recorded history of Leye County, indicating the severity of the drought. Circles indicate rainfalls that fell below one standard deviation.

MATERIAL AND METHODS

Study Site

We made the phenological survey before and after the extreme cold event in 2008 on wild orchids at Fengyandong, a small, limestone mountain site 970–1000 m a.s.l. (N 24°50.89′, E106°24.44′). The vegetation there is characterized as mixed evergreen and deciduous broad-leafed forests (The Comprehensive Scientific Investigation Team of Guangxi Yachang

Table 12.1 Flower phenology of 23 orchid species in a single limestone mountain site in the Yachang Orchid National Nature Preserve, south-western China

Species	N	Flowering time (month/day)			Flower time change[1]	Habit[2]	Narrow endemic[3]	Spring flower[4]	Small local population[5]
		2006	2007	2008					
Bulbophyllum andersonii (Hook.f.) J.J.Smith 梳帽卷瓣兰	13	—	9/19	9/8	-11	L	No	No	No
Cleisostoma nangongense Z.H.Tsi 南贡隔距兰*	17	—	8/7	8/16	9	E	Yes	No	Yes
Cleisostoma williamsonii (Rchb.f.) Garay 红花隔距兰*	3	5/13	—	6/3	21	E	No	Yes	No
Coelogyne fimbriata Lindl. 流苏贝母兰	12	8/5	8/18	8/29	11	L	No	No	No
Cymbidium aloifolium (L.) Sw. 纹瓣兰*	24	4/24	5/26	4/29	5	L	No	Yes	No
Cymbidium cyperifolium Wall.ex Lindl. 莎叶兰	36	—	10/2	10/6	4	GT	No	No	No
Cymbidium floribundum Lindl. 多花兰*	7	3/18	—	3/25	7	L	No	Yes	No
Cymbidium kanran Makino 寒兰*	9	—	11/2	11/16	4	GT	No	No	Yes
Cymbidium lancifolium Hook. 兔耳兰	31	10/31	10/11	10/5	-6	L	No	No	No
Cymbidium tracyanum L.Castle 西藏虎头兰*	11	—	10/25	10/21	-4	E	No	No	No
Dendrobium aduncum Lindl. 钩状石斛*	2	5/25	—	5/25	0	E	No	No	No
Dendrobium loddigesii Rolfe 美花石斛*	54	4/25	—	5/16	21	E	No	Yes	No
Dendrobium officinale Kimura et Migo 铁皮石斛*	4	6/1	7/2	5/23	-39	E	No	No	Yes

Species	N				[1]	[2]	[3]	[4]	[5]
Dendrobium williamsonii Day et Rchb.f. 黑毛石斛*	30	3/15	3/15	4/14	30	E	No	Yes	Yes
Eria corneri Rchb.f. 半柱毛兰*	19	8/2	7/30	8/7	8	L	No	No	No
Eria coromaria (Lindl.) Rchb.f. 足茎毛兰	29	10/28	10/20	10/13	-7	L	No	No	No
Habenaria dentata (Sw.) Schltr 鹅毛玉凤花	33	—	8/20	8/28	8	GT	No	No	No
Kingidium braceanum (Hook.f.) Seidenf. 尖囊兰	4	5/15	5/15	5/18	3	E	Yes	Yes	Yes
Liparis esquirolii Schltr. 贵州羊耳蒜	30	5/1	—	6/1	31	L	Yes	Yes	Yes
Liparis viridiflora (Bl.) Lindl. 长茎羊耳蒜*	2	4/21	10/12	10/15	3	L	No	No	No
Malaxis purpurea (Lindl.) Kuntze 深裂沼兰	3	—	7/1	7/5	4	GT	No	No	No
Paphiopedilum dianthum T.Tang et F.T.Wang 长瓣兜兰	24	8/5	7/31	8/9	9	L	Yes	No	Yes
Paphiopedilum hirsutissimum (Lindl.ex.Hook) Stein 带叶兜兰	30	4/9	4/15	4/23	8	L	No	Yes	No

Notes: The phenological observations were carried out in 2008, a year with an unusually long cold spell (in January and February) as well as in 1 or 2 previous years with relatively normal winter temperatures. *N* is the number of individuals observed. Species in bold are likely pollinated by the Chinese honeybee (*Apis cerana* L.). An * next to the species name indicates species rescued from a lower-elevation site. A — indicates no observation made.

[1]The difference in the number of days between first flower starting date in 2008 and in 2007 (or 2006 if 2007 data are not available). A positive number means delay in the flowering date in 2008 compared with the previous year.

[2]E = epiphyte; GT = geophytic terrestrial; L = lithophyte.

[3]Yes = species' current range includes only Guangxi, Guizhou, Yunnan, and northern Vietnam or fewer areas, because these areas are adjacent to one another and share similar limestone and climatic characteristics.

[4]Yes = species flowered on or before 15 May.

[5]Yes = species found in fewer than 3 locations within the Yachang Reserve (each location contains less than 50 reproducing plants).

Orchid Nature Reserve 2007). The canopy species include *Carpinus purpurinervis* Hu, *Platycarya longipes* Wu, *Cyclobalanopsis glauca* (Thunb.) Oerst., *Quercus phillyraeoides* A. Gray, and *Lysidice rhodostegia* Hance. The shrub layer includes *Myrsine africana* L., *Rapanea neriifolia* (Sieb. et Zucc.) Mez, and *Raphiolepis indica* Lindl. The 17-ha site is super rich in orchid species, with 35 occurring naturally, 11 of which were monitored for this study (Table 12.1). In addition, more than 20 species had been transplanted into the same general area from their former location at a hydrological project area (Longtan Hydropower Station) that was slated to be flooded (Liu et al. 2012). We monitored 12 of these species (Table 12.1).

Phenological monitoring around the time of the extreme drought in 2010 was carried out on a small hill located near Dingshu Village, along the Hongshui riverbank at about 500 m a.s.l. The site is in a highly disturbed state and was severely affected by farming between 2008 and 2010; the area had been fenced to prevent such activities in 2010 (Liu 2010, 2011). It contained sporadic Yunnan thin-leafed pine trees (*Pinus yunnanensis* var. *tenuifolia* Cheng et Law), a canopy species endemic to this narrow stretch of the Hongshui River. Small trees of various *Eucalyptus* spp. planted in early 2008 dominated half the area, while *Chromolaena odoratum* (L.) R. King et H. Rob., an introduced invasive species, dominated the shrubby layer at the entire site. Native plants that were present included *Callicarpa macrophylla* Vahl., *Urena lobata* L., *Lygodium japonicum* (Thunb.) Sw., *Desmodium pulchellum* (L.) Benth., *Rosa rubus* Lévl. et Vant., and *Rhus chinensis* Mill.

Responses to Extreme Cold Event of 2008

We monitored the flowering phenology of 54 orchid species in 3 years (26 species in 2006, 30 species in 2007, and 31 species in 2008). Twenty-three of these species were monitored for 2 years or more, including the year 2008. Eleven of the 23 species were transplanted from a lower-elevation site as a part of an endangered-species rescue mission in spring 2006. For each species, 2–54 individuals were tagged and monitored, depending on availability. A trained ranger carried out the observation daily when possible, or once every 2 to 3 days.

The following parameters were monitored: first inflorescence emergence, first flower bud emergence, first flower open/initiation of flowering season, flowering peak (when more than half the monitored plants were in bloom), first flower to wilt, and last flower to wilt. The first-flowering date was used in our study because this parameter has the most complete record relative to other parameters. In addition, it has minimal human error associated with it, because it is easy to observe in field conditions. We excluded species that flower more than once a year (e.g. *Panisea cavalerei* Schltr.), as well as those that get damaged directly by the cold, like *Calanthe argentro-striata* C. Z. Tang et S. S. Cheng.

Responses to Extreme Drought in 2010

Flowering phenology of 3 co-occurring *Geodorum* species (*G. eulophioides*, *G. recurvum*, and *G. densiflorum*) were monitored during late May until August, the flowering period of the 3 species. Similarly to the protocol mentioned above, we recorded the first-flower-open date and the beginning of peak flowering for 3 years (2009, 2010, 2011). *Geodorum eulophioides* is a narrowly endemic species whose distribution within China is restricted to one section of the Hongshui riverbank. The other 2 co-occurring congeners have wider geographic distributions. Populations of *Geodorum recurvum* and *G. densiflorum*, however, are small at the study site. Flowering data were absent for *G. recurvum* in 2011, because we could not locate any flowering plants for the species at our study site. Nevertheless, flowering periods of *G. eulophioides* and *G. recurvum* overlapped almost completely in other years. We did not collect the peak flowering time for *G. densiflorum* in 2011, due to logistical difficulties.

Data Analyses

We created a binary variable to note whether or not a species responded to the extreme weather events with flowering delay. A species received a Yes if its date of first flowering was delayed for at least 5 days following the 2008 extreme cold event. The difference in flowering time is between 2008 and 2007, unless the latter is not available (Table 12.1). The 5-day criterion is arbitrary to allow for human recording errors, as sometimes the ranger surveyed the route only once every 2 days. We used multiple binary logistic regressions to explore which of the following variables was or was not associated with flowering delay: flowering time (spring flowering or not), habit (terrestrial or epiphytic), and global distribution range (narrow endemic or not). Species flowering on or before 15 May were considered spring-flowering species. A species was considered a narrow endemic if its current range included only Guangxi, Guizhou, Yunnan, and northern Vietnam or fewer areas, because these areas are adjacent to one another and share similar geological (limestone) and climate characteristics.

Local population size was also known for each of these species but was not included in the logistic regression analysis, because it was co-related with statistical significance to whether or not a species is a narrow endemic to the region (Pearson correlation = 0.694, P < 0.001). Narrow endemic species tend to have small local populations in the Yachang National Orchid Nature Reserve. Any species found in fewer than 3 locations within that reserve and with fewer than 50 reproducing plants was classified as having a small local population.

To determine if species pollinated by the same pollinators responded to the cold event in the same way, we compared the phenological responses of species that were or were likely to be pollinated by the Chinese honeybee (*Apis cerana* L.), a major native pollinator in the region. The pollination information was derived from the literature (Shangguan et al. 2008; Shi et al. 2008, 2009b; Yu et al. 2008; Chen I-CH et al. 2009) and unpublished personal observations. To determine whether responses to a cold spell were phylogenetically conservative, we compared the responses of congeneric species, especially those in the genera *Cymbidium* (6 species) and *Dendrobium* (4 species).

RESULTS

Extreme Cold-Spell Impacts

Overall, 54.5% of the study species responded to the 2008 extreme cold spell with a delay in flowering initiation of 5 days or more (Table 12.1). Specifically, those species flowering early in the year were more likely to be affected by the cold (77.8%) than those flowering later in the year (38.5%) (Figure 12.3A); a larger proportion of narrowly endemic species (75%) responded than did those with wide distribution ranges (47.4%) (Figure 12.3B). A larger proportion of epiphytic orchids (57.9%) delayed flowering than was true of terrestrial orchids (25%) (Figure 12.3C). The binary logistic regression indicated that the flowering season was a significant predictor of whether or not a species would respond to the extreme cold event (Wald = 4.011, df = 1, $p = 0.045$).

Among the 13 species that are or are likely pollinated by the Chinese honeybee (Table 12.1), 5 species responded to the 2008 cold spell with a delay. All but 1 (*Eria corneri*) of the 5 were spring-flowering species. Only 1 honeybee-pollinated, spring-flowering species (*Kingidium braceamum*) did not respond with a flowering delay of more than 5 days. It delayed flowering for only 3 days.

Among the 6 sets of congeners, only 2 (*Cleisostoma* and *Paphiopedilum*) showed consistent responses to the 2008 cold spell, but for these

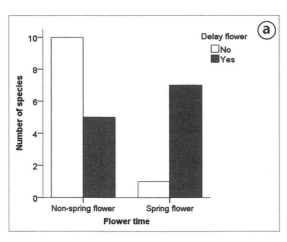

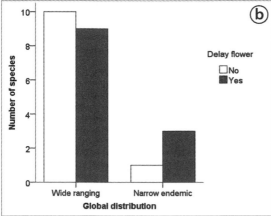

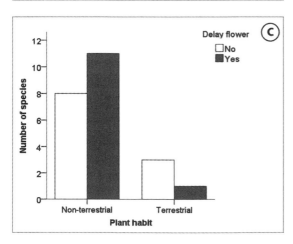

Fig. 12.3 Responses to the 2008 extreme cold spell of 23 selected species of wild orchids in the Yachang Orchid National Nature Reserve in southwestern China on the first flowering date. Contrasts in responses are given between orchid flowers in the spring season (before mid-May) versus other seasons (*A*), between species with narrow versus wide global distributions (*B*), and between ground versus epiphytic species (*C*).

we had a very small sample size of only 2 species apiece (Table 12.1). In contrast, a mixed reaction was recorded among the 6 *Cymbidium* species and 4 *Dendrobium* species. Two *Cymbidium* species were spring flowering, and both delayed flowering following the cold spell. Among the 4 non-spring-flowering *Cymbidium* species, 2 advanced their flowering schedule in 2008, while the other 2 were delayed by 4 days (Table 12.1). Similarly, the 2 spring-flowering *Dendrobium* species delayed their flowering in 2008, while the other 2 non-spring-flowering *Dendrobium* species did not. One *Dendrobium* species actually advanced its flowering by 39 days.

Drought Impacts

Compared with the flowering time in the relatively normal year of 2009, in the drought year of 2010 all 3 *Geodorum* species delayed flowering by up to 5 days (*G. densiflorum*) (Table 12.2). More important, the flowering durations, as suggested by the period between flower initiation and flowering peak, were shorter in the drought year than in 2009. Specifically, the flowering durations in 2010 were only half those in 2009 (11 vs. 21 days for *G. eulophioides* and *G. recurvum*, and 11 vs. 20 for *G. densiflorum*). Flowering was delayed in 2011 compared with 2009 or 2010, and flowering duration in 2011 was similar to that in 2010.

DISCUSSION

Temperature-Influenced Phenology in the Subtropical Region

More than half the orchid species studied responded to the 2008 extreme cold spell with a delay in flowering initiation, suggesting that temperature could be a controlling factor in flowering for these species in this subtropical region. This finding contrasts with phenological observations made in Hong Kong, also in a subtropical region of southern China (Corlett 1993). The reproductive phenology of Hong Kong's mostly woody plants is highly seasonal but varies little between years, despite annual fluctuations of winter temperatures (Corlett and Lafrankie 1998).

Those differences in phenological patterns in relation to winter temperatures may have been caused by several factors. First, Hong Kong is a coastal island, but our study site is 800 km away from the coast, and Hong Kong is 2 degrees of latitude south of Yahcang. Both these characteristics encourage the occurrence of a more tropical flora, which are less responsive to temperature changes (Corlett and Lafrankie 1998). Second, the variation in winter temperatures during the study periods in Hong

Table 12.2 Flowering phenology of natural populations of 3 sympatric ground orchid species, Yachang Orchid Nature Reserve, southwestern China, 2009–2011 flowering seasons

Species	Year	Number of plants monitored	First flower date	Peak flower date	Flower season duration
G. eulophioides	2009	33	May 24	June 15	43
	2010	12	May 27	June 11	32
	2011	12	June 5	June 17	25
G. recurvum	2009	8	May 29	June 21	38
	2010	7	June 1	June 15	27
G. densiflorum	2009	11	June 20	July 10	35
	2010	9	June 25	July 6	28
	2011	6	June 30	?	

Note: Geodorum recurvum flowering data were absent in 2011 because no flowering plants for the species could be found at the study site. Nevertheless, flowering periods of G. eulophioides and G. recurvum overlapped almost completely in other years.

Kong (1988–1993) was unlikely to have been as great as that encountered during the wild orchid study period we report here. Finally, Corlett (1993) had carried out weekly phenological surveys, and did not follow the same individuals of various species between surveys. Unlike our work, those survey intervals and methods could not have detected phenological changes of less than one week.

Our results also contrast with what is known about the true tropical regions, where phenology is thought to be controlled entirely by endogenous developmental factors, as reported in the Neotropics (see review by Reich 1995) and paleotropical rainforests (Corlett 1990), or by annual precipitation patterns, as reported in tropical wet and dry forests (Reich 1995; Morellato 2003; Sanchez-Azofeifa et al. 2003). Our study area lies in the transition zone between the tropics and the temperate zones and has pronounced wet/hot and dry/cold seasons. Perhaps some species have temperature-driven phenologies, as temperate species do, while others do not. The woody canopy layer at the study site is composed of a mixture of evergreen broad-leafed and deciduous broad-leafed trees. The latter species are mostly found in the temperate regions, and are likely to have their phenology driven in part or entirely by temperature, as is the case for most tree species in the temperate regions (Walther et al. 2002; Chen X-Q 2003; Parmesan and Yohe 2003; Cleland et al. 2006, 2007; Schwartz et al. 2006). In the temperate European region, extreme warm and cold spells have led to greater than usual changes in natural or agricultural phenology (Menzel et al. 2011). More research is needed to determine the extent of the phenological changes we observed in relation to those occurring during nonextreme weather periods.

The extreme drought in 2010 triggered a small delay in flower initiation, and significantly shortened flowering duration for populations of 3 *Geodorum* species. Studies of plant phenology in the tropical region found that the timing of onset and length of the wet versus dry seasons determined phenological events such as leafing out and flowering (Corlett 1990; Reich 1995; Morellato 2003; Sanchez-Azofeifa et al. 2003). The 2010 drought in southwestern China occurred within the regular dry season. Therefore, the shortened flowering season observed in this study was likely caused by stress induced by the intensity of the drought, not the timing. Based on anecdotal observations, shortened flowering duration likely occurred in several other orchid species in the Yachang Reserve as well. Temperature also seems to be an important factor to the phenology of the 3 *Geodorum* species: their delayed flowering coincided with a late cold snap (a brief, 3-day cold event) in March 2011, and/or an unusually cold January that year (the coldest on record).

By contrast, many of the summer- and fall-flowering species advanced their flowering phenology significantly in 2008, instead of delaying it. Differences in summer and fall temperatures among these monitoring years were not large and were not likely to contribute to the advanced phenology in these species. Perhaps their phenology is influenced mostly by internal developmental factors, as is the case with many true tropical species (Reich 1995).

Ecological and Evolutionary Significance

Most (80%) of the spring-flowering species responded to the 2008 cold spell. However, some of the summer-flowering species may also respond to cold spells or snaps. For example, the 3 *Geodorum* spp., not included in the 2008 cold-spell monitoring, were observed to delay flowering in 2011 following a cold snap. Whether or not pollinators of these orchids would respond similarly to these weather events needs more research.

Our knowledge of insect responses to current climate change is just beginning to accumulate. There is evidence that some butterflies and moths have migrated poleward or up to higher elevations within the past 5 decades as their response to global warming (Parmesan et al. 1999; Chen I-CH et al. 2009). Among the known insect pollinators of orchids in the Yachang Reserve, the Chinese honeybee (*Apis cerana* var. *cerana*) is the best known. The biology/ecology of this species is similar to that of the European honeybee (*Apis mellifera*) in many ways, including its overwintering behavior (Oldroyd and Wongsiri 2006). In general, the honeybees stop flying and stay in their hives when ambient temperatures dip down into the teens (in Celsius) or below. During overwintering, they maintain their brood temperature, up to 30°C above the ambient temperature (Oldroyd and Wongsiri 2006).

In a long-term phenological study conducted in temperate China, a honeybee species (Chinese or European *A. mellifera*) was reported to respond to spring temperature fluctuations by flying early during warm springs and late during cool springs (Wan 1986, 1987). This evidence suggests that these honeybees may have responded to the 2008 cold spell by emerging later than usual. Among the 14 orchid species that are or are likely pollinated by the Chinese honeybee, 5 species flower in the spring; 1 of these did not respond to the extreme cold event, presenting the likelihood of a mismatch between initial flower and initial honeybee fly times.

Not all species in the same genus responded consistently, indicating that if a species has a temperature-controlled flowering schedule, that trait is not always phylogenetically conservative. As with many other ecological traits, how a species times its flowering phenology has potentially large consequences for reproductive fitness, and is likely to be subjected to natural selection (Cavender-Bares et al. 2004; Silvertown et al. 2006; Losos 2008). Therefore, it is not surprising that this trait is not always phylogenetically conservative.

A slightly greater proportion of epiphytic orchids responded with a delay in flowering than did ground orchids, suggesting that epiphytic orchids may be more sensitive to environmental changes. Epiphytes, living on tree trunks, branches, or exposed rocks, grow without the benefit of a belowground temperature buffer. They are known to be exceptionally sensitive to climate fluctuations (see references in Benzing 1998; Zotz and Bader 2009). Epiphyte survival may require a greater responsiveness to temperature and moisture changes compared with terrestrial, soil-dwelling plants. In our study, the proportional difference of orchids responding to the 2008 cold spell was not large between the epiphytic and terrestrial orchids, possibly because many of the epiphytic orchids are true lithophytes, growing on bare rocks or rocks wearing very shallow soil. This could merely be an artifact because of the small number of terrestrial soil orchids (only 4 species) sampled in our study.

Orchid species that have narrow distribution ranges responded proportionately more to the 2008 cold spell than did those with wider distributions. This is somewhat unexpected, because species with narrow distributions may have narrower ecological niches and show lower plasticity in their response to environmental changes (Sultan 2001). However, there are exceptions to this possibility. For example, *Acacia* spp. (Fabaceae) with narrow geographic distributions have equal or greater phenotypic plasticity compared with those with wider distributions (Pohlman et al. 2005). Nevertheless, the difference shown in this study between the 2 groups was not significant statistically, probably because of the relatively small sample sizes in narrow endemic species.

CONCLUSIONS

The ability of species to respond to global warming varies (Miller-Rushing and Primack 2008). Whether the change or lack of change in plants' flowering phenology can synchronize with their pollinator's activities is a conservation issue (Memmott et al. 2007; Liu et al. 2010). It is logical to expect that specialized pollination relationships, such as the ones presented by orchids and their pollinators, would be more vulnerable to such mismatches than in more generalist interactions (Ashworth et al. 2004; Dixon 2009). Darwin's concern regarding "incessant changes in which the world is subject" (1877; p. 281) may be our present reality regarding the fertilization of orchid flowers. This vulnerability is due in part to the skewed relationships between orchids and pollinators, as orchid species are much more dependent on the pollinators than vice versa (Dixon 2009; Pemberton 2010; Vereecken et al. 2010b; Chapters 2–4, 6, 7, 9, 10). We demonstrated here that a large portion if not more of the wild orchid species in a subtropical region can respond to climate changes, at least in the very short term. More research is needed to determine the consequence for fitness when an orchid species can or cannot respond to environmental changes, and place these results in the perspective of their long-term survival.

ACKNOWLEDGMENTS

We thank the Vice Governor of Guangxi, Dr. Chen Zhangliang, for his vision for biodiversity conservation and his unwavering support of this endeavor. We thank Drs. Cheng Jin and Shi Jun, who provided supplemental data for a few species. We are in debt to the staff of Yachang, especially Mr. Wu Tiangui, for their logistical support of the conservation research in Yachang Reserve. This research is supported by the following grants: Guangxi Science and Technology Bureau (Chairman's Foundation grant # 09203–04) to H. L. and Y. B. L.; the Mohamed bin Zayed species conservation fund (0905324) to H. L. and Y. B. L.; and the Social Welfare Research Project (2005DIB6J144) of the Ministry of Science and Technology of the People's Republic of China to C. L. F.

Retha Edens-Meier and Peter Bernhardt

The preceding chapters have shown how Charles Darwin (1862, 1877) gave us a viable and novel subdiscipline within plant evolutionary biology. Like Darwin, we have traveled through these pages from the English countryside into the Eurasian continent, down to the temperate and Southern Hemisphere, then off to the tropical forests, taking the time to pause there to examine the inflated blooms of the slipper orchids. Darwin completed both editions of his book with detailed reviews on how different flowers in the same family share or express adaptations favoring cross-pollination. Our Chapters 11 and 12 take Darwin's concern with cross-pollination a step further by relating it to the importance of pigmentation and climatic variation. As readers may surmise, we are still asking some of the same questions Darwin first posed a century and a half earlier. The main difference is that we have access to advanced technologies illuminating, and contributing to, this 150-year-old branch of scientific literature.

Let's return to the two most general but important questions Darwin presented. The first question is one that he addressed in the title (slightly modified) of the second edition (1877) of his book—Just what *are* the various contrivances by which orchids are "pollinated" by insects? The second question Darwin (1877; p 288) posed was "Why do the Orchideae exhibit so many perfect contrivances for their fertilisation?"

Most of the "contrivances" that Darwin identified were based on the adaptive floral morphology derived from descent by modification

(Chapter 1). Darwin showed that these "contrivances" promote cross-pollination. He saw the relationship between the column architecture and the labellum in conjunction with the structure and construction of the pollinarium. Earlier plant anatomists dissected and illustrated these structures over and over again, but Darwin showed how they work together to exploit the foraging behavior of a prospective pollen vector.

By the second edition, Darwin had even more evidence that the deposition of a pollinarium on an insect is often self-consistent for both the insect and the orchid species. Butterflies and moths wear pollinaria toward the bases of their proboscides (Chapter 8). Wasps carrying the pollinaria of *Epipactis* wear them on their heads (but see Chapter 6). The morphology of the orchid column does not merely disperse the pollinarium onto the insect—it usually disperses the pollinarium onto the same part of the same insect. Cross-pollination is a matter of precision. How precise are the dimensions of the pollinator and the parameters of the orchid flower? In the case of *Cypripedium* spp. (Chapter 10), this difference is recorded in millimeters, and determines whether the insect carries pollinia and deposits pollen smears on receptive stigmas.

Of course, Darwin did not catch all the structural and biochemical novelties found in the orchid family, because he did not receive fresh flowers from all known genera and he did not have the technology we take for granted today. For example, his interpretation of pollination in the flowers of *Thelymitra* spp. (Chapter 7) began and ended with his correspondence with Fitzgerald. Darwin never speculated on the function of the lobed, colorful, and ornamental hood (mitra) as a structure that could entice foraging insects. Likewise, he never saw euglossine bees carrying the pollinaria of an orchid with the anther cap still clinging to the pollinia (Chapter 9). Had he done so, he might have been the first to show that this is an effective but temporary "contrivance" lowering the frequency of self-pollination via geitonogamy.

Decades after Darwin's death, scientists made field observations that led to the second great advance in knowledge about the adaptive morphology of orchid flowers. Darwin had been unable to accept the idea that many orchid flowers lack edible rewards and instead mimic co-blooming flowers that do have them. From South Africa (Chapter 4) to Australia (Chapters 5–7) to Eurasia (Chapters 2 and 3) and back to northern Africa, we now recognize and acknowledge that even the smallest floral sculptures may offer false promises. These "contrivances" go beyond the empty (nectarless) spur in Chapter 2. Some pretend to be tufts of dehiscent anthers, while others masquerade as brood sites or the bodies of receptive females (Chapters 3–8, 11). In fact, the scanning electron microscope offers new prospects of viewing microadaptations in these false females (Chapter 3). How tactile perception of these false females reinforces the behavior of the "rutting" male insects may be as important for seed set as the physical appearance and odors of their flowers.

Darwin never elaborated on the roles of pigmentation patterns and fragrances of orchid flowers in either edition of his book. Yet thanks to 20th- and 21st-century advances in photographic technology and scent analyses, scientists have been able to dissect and interpret pigment patterns and individual floral odors much as Darwin first dissected and interpreted the functions of floral organs. However, color and scent analyses of orchid flowers still would be speculative were it not for ongoing experimental procedures using living insects to respond to color and scent models (Chapters 2, 3, 6, 11). As in the morphological "contrivances" described by Darwin, Chapters 3, 6, and 11 show that color and odor in sexually deceptive orchids vary at the intra- and interspecific levels. These chapters also show us that we can't understand the role and evolution of sexual deception, so common in orchid species, without a better understanding of how insects perceive color patterns. Within sexually deceptive orchids, fragrance often works in conjunction with color patterns to ensure greater reproductive success (Chapters 3, 6).

As for the second question Darwin (1877; p 288) posed—"Why do the Orchideae exhibit so many perfect contrivances for their fertilisation?"— the answer is twofold. First, the sciences of genetics and plate tectonics were embryonic in Darwin's time. Second, the "contrivances" are not perfect (rigidly identical in every member of the same population), because they are products of sexual selection. Variation in every suite of adaptive characters is the rule, not the exception, in evolution. There is no evidence that floral color, odor, nectar chemistry, and/or floral morphology are based on a single gene (Chapter 5). Localized natural selection clearly plays an important role in the floral evolution of orchid species.

Note that as previous chapters (Chapters 2–10) compared the pollination of orchid floras in different parts of the world, our coauthors found that pollination systems are dependent entirely on different lineages of animals. Therefore, referring to an orchid lineage as bee pollinated, or moth pollinated, or fly pollinated, and so on, now seems too vague. The selection pressures leading to the exploitation of Neotropical bees in the Euglossini (Chapter 9) resulted in the evolution of different sets of primary attractants and rewards compared with the rewards (and false rewards) offered by orchids of southern Africa pollinated by long-proboscid flies (Chapter 4). The flowers of *Ophrys* spp. are sexual mimics like so many of the orchid species of southern Australia (Chapters 5–6). However, the Australian lineage evolved in the presence of thynnine wasps, a group of flower-visiting insects less diverse in Eurasia (Chapter 6).

Why, then, are there so many "contrivances"? Because orchids are by and large pandemic, whereas so many of the animals they exploit as pollen vectors are localized and have discrete centers of species diversity. Orchid "contrivances" may look conservative and fixed, but Chapters 2–10 strongly suggest that natural selection has probably caused significant pollinator shifts over time. Otherwise, why would a lineage of tropical orchids contain species pollinated by long-tongued moths, small perching birds, and a raspy cricket (Chapter 8)? Here's the paradox. The "contrivance" expressed by orchid flowers indicates that pollination systems are unusually plastic within the orchid family, but do not always represent examples of long-term coevolution (Chapter 9). Yes, floral evolution within the orchid family is unusually opportunistic and labile. Why else are orchid flowers such spectacular models for the ongoing study of evolution?

In fact, a new evolutionary model has emerged, and we are delighted that some orchid flowers are prime examples suggesting a new era of lab and field studies. We are also delighted that some of our coauthors have addressed this model in their chapters. Specifically, coevolution between orchids and animals, while very important, is insufficient to explain the vast, and often rapid, diversification of orchid pollination systems. Many orchid species did not and do not merely coevolve with their pollinators. Many orchid lineages now appear to be the result of floral evolution "co-opting" prospective pollinators. It looks as though many orchid species evolved to exploit animals as pollinaria carriers long after those animals evolved specialized life histories, behaviors, and physiological requirements. Many orchid species evolved by co-opting preexisting insect sexual behavior (Chapters 3, 4, 6, 7, and 9). In other words, while orchid populations responded to the selective pressures imposed on them by the foraging of their pollinators, that selection was not reciprocal. Consequently, although many orchid species may have only 1 to 3 pollinator species, almost all orchid-pollinating insects never depend on simply 1 to 3 orchid species to complete their life cycles (Chapters 2–6, 8–10). The flowers of *Ophrys* and *Catasetum* spp. evolved responding to the ways in which different groups of male bees and wasps responded to scent molecules and color patterns. The insects, on the other hand, did not evolve modified equipment or behavioral patterns to exploit the orchids. Descent through modification, in many orchid-pollinator interactions, is one-sided, with adaptive "concessions" made by the flowers.

We support the idea that the evolution of orchid pollination deserves greater study, particularly as we pass into a period in which human- and climate-based changes influence the survival of orchid species. If such issues as climate change alter the flowering seasons of some orchids (Chapter 12), what impact does this have on their pollinators? As Darwin (1877, pp. 281–282) wrote, "In other cases the paucity of the flowers that are impregnated may be due to the proper insects having become rare under the incessant changes to which the world is subject; or to other plants which are more highly attractive to the proper insects having increased in number." Remember, Darwin studied orchids in a time when there was no need to write grants, request research permits, submit receipts, or justify every mile of his research trip to receive reimbursement for travel and field expenses. As a self-financed country gentleman, he had both the necessary funding and the time to rigorously and meticulously study orchids. Times have definitely changed in the past 150 years! Therefore, all of us who elect to study orchid pollination must do so in the names of conservation and extramural funding. In conclusion, we must impress upon all prospective grant-giving institutions that the study of orchid pollination represents a singularly important segment of conservation. It's the only way that future generations will see natural populations of long purples, blue ladies, yellow slippers, and Darwin's comet orchids with their prospective pollinators.

Orchids are diverse—and so are the people who study them. We share with one another, as with Darwin, an intense interest in orchid ecology and evolution. Just as Darwin established a worldwide network, so we have continued and expanded this global collaborative effort to investigate scientifically and protect our beloved orchids.

Aceto S, P Caputo, S Cozzolino, L Gaudio, A Moretti. 1999. Phylogeny and evolution of *Orchis* and allied genera based on ITS DNA variation: morphological gaps and molecular continuity. Molecular Phylogenetics and Evolution 13:67–76.

Ackerman J. 1981. Pollination biology of *Calypso bulbosa* var. *occidentalis* (Orchidaceae): a food deception scheme. Madrono 28:101–110.

———. 1983a. Diversity and seasonality of male euglossine bees (Hymenoptera: Apidae) in central Panama. Ecology 64:274–283.

———. 1983b. Euglossine bee pollination of the orchid *Cochleanthes lipscombiae*: a food source mimic. American Journal of Botany 70:830–834.

———. 1983c. Specificity and mutual dependency of the orchid-euglossine bee interaction. Biological Journal of the Linnean Society 20:301–314.

———. 1984. Pollination of tropical and temperate orchids. Pages 98–101 *in* KW Tan, editor. Proceedings of the 11th World Orchid Conference. American Orchid Society, Miami, FL, USA.

———. 1985. Euglossine bees and their nectar hosts. Pages 225–233 *in* WG D'Arcy, MD Correa, editors. The Botany and Natural History of Panama. Missouri Botanical Garden, St. Louis, MO, USA.

———. 1986. Mechanisms and evolution of food-deceptive pollination systems in orchids. Lindleyana 1:108–113.

———. 1989. Geographic and seasonal variation in fragrance choices and preferences of male euglossine bees. Biotropica 21:340–347.

Ackerman J, AA Cuevas, D Hof. 2011. Are deception-pollinated species more variable than those offering a reward? Plant Systematics and Evolution 293:91–99.

Ackerman J, EJ Meléndez-Ackerman, J Salguero-Faria. 1997. Variation in pollinator abundance and selection on fragrance phenotypes in an epiphytic orchid. American Journal of Botany 84:1383–1390.

Ackerman J, MR Mesler, M Lu, AM Montalvo. 1982. Food foraging behavior of male euglossine bees (Hymenoptera: Apidae): vagabonds or trapliners? Biotropica 14: 241–248.

Ackerman J, AM Montalvo. 1990. Short- and long-term limitations to fruit production in a tropical orchid. Ecology 71:263–272.

Ackerman J, DW Roubik. 2012. Can extinction risk help explain plant-pollinator specificity among euglossine bee pollinated plants? Oikos 121:1821–1827.

Adams PB, T Bartareau, KLWalker. 1992. Pollination of Australian orchids by *Trigona* (Tetragona) Jurine bees (Hymenoptera: Apidae). Australian Entomological Magazine 19:97–101.

Adams PB, SD Lawson. 1993. Pollination in Australian orchids: a critical assessment of the literature 1882–1992. Australian Journal of Botany 41:553–575.

Ågren L, A-K Borg-Karlson. 1984. Responses of *Argogorytes* (Hymenoptera: Sphecidae) males to odor signals from *Ophrys insectifera* (Orchidaceae). Preliminary EAG and chemical investigation. Nova Acta Regiae Societatis Scientiarum Upsaliensis, Serie V:C 3:111–117.

Ågren L, B Kullenberg, T Sensenbaugh. 1984. Congruences in pilosity between three species of *Ophrys* (Orchidaceae) and their hymenopteran pollinators. Nova Acta Regiae Societatis Scientiarum Upsaliensis, Serie V:C 3:15–25.

Albert VA, MW Chase. 1992. *Mexipedium*: a new genus of slipper orchid (Cypripedioideae: Orchidaceae). Lindleyana 7:172–6.

Alcock J. 1981. Notes on the reproductive behavior of some Australian thynnine wasps (Hymenoptera: Tiphiidae). Journal of the Kansas Entomological Society 54: 681–693.

———. 2000. Interactions between the sexually deceptive orchid *Spiculaea ciliata* and its wasp pollinator *Thynnoturneria* sp. (Hymenoptera: Thynninae). Journal of Natural History 34:629–636.

Alcock J, DT Gwynne. 1987. Courtship feeding and mate choice in thynnine wasps (Hymenoptera: Tiphiidae). Australian Journal of Zoology 35:451–458.

Aldridge G, DW Inouye, JRK Forrest, WA Barr, AJ Miller-Rushing. 2011. Emergence of a mid-season period of low floral resources in a montane meadow ecosystem associated with climate change. Journal of Ecology 99:905–913.

Alexandersson R, J Ågren. 1996. Population size, pollinator visitation and fruit production in the deceptive orchid *Calypso bulbosa*. Oecologia (Berlin) 107:533–540.

Alexandersson R, SD Johnson. 2002. Pollinator mediated selection on flower-tube length in a hawkmoth-pollinated *Gladiolus* (Iridaceae). Proceedings of the Royal Society of London B Biological Sciences 269:631–636.

Allen M. 1977. Darwin and His Flowers: The Keys to Natural Selection. Taplinger Publishing Co., New York, USA.

Ames O. 1937. Pollination of orchids through pseudocopulation. Botanical Museum Leaflets, Harvard University 5:1–29.

Anderson B, R Alexandersson, SD Johnson. 2008. The role of hawkmoth pollinators in mediating divergence and maintaining species boundaries after secondary contact in Gladiolus longiicollis. South African Journal of Botany 74:360–360.

Anderson B, SD Johnson. 2006. The effects of floral mimics and models on each others' fitness. Proceedings of the Royal Society of London B Biological Sciences 271: 969–974.

———. 2008. The geographical mosaic of coevolution in a plant-pollinator mutualism. Evolution 62:220–225.

————. 2009. Geographical covariation and local convergence of flower depth in a guild of fly-pollinated plants. New Phytologist 182:533–540.

Anderson B, SD Johnson, C Carbutt. 2005. Exploitation of a specialized mutualism by a deceptive orchid. American Journal of Botany 92:1342–1349.

Anderson GJ, G Bernardello, TF Stuessy, DJ Crawford. 2001. Breeding system and pollination of selected plants endemic to Juan Fernández Islands. American Journal of Botany 88:220–233.

Andersson S, LA Nilsson, I Groth, G Bergström. 2002. Floral scents in butterfly-pollinated plants: possible convergence in chemical composition. Botanical Journal of the Linnean Society 140:129–153.

Aragón S, JD Ackerman. 2004. Does flower color variation matter in deception pollinated Psychilis monensis (Orchidaceae)? Oecologia (Berlin) 138:405–413.

Archetti M, SP Brown. 2004. The coevolution theory of Autumn colours. Proceedings of the Royal Society of London B Biological Sciences 271:1219–1223.

Arditti J, J Elliott, IJ Kitching, LT Wasserthal. 2012. Good heavens what insect can suck it—Charles Darwin, Angraecum sesquipedale and Xanthopan morganii praedicta. Botanical Journal of the Linnean Society 169:403–432.

Arends JC. 1986. Cytotaxonomy of the Vandeae. Lindleyana 1:33–41.

Arends JC, FM van der Laan. 1983. Cytotaxonomy of the monopodial orchids of the African and Malagasy regions. Genetica (Dordrecht) 62:81–94.

Armbruster WS. 1993. Within-habitat heterogeneity in baiting samples of male euglossine bees: possible causes and implications. Biotropica 25:122–128.

Armstrong JA. 1979. Biotic pollination mechanisms in the Australian flora—a review. New Zealand Journal of Botany 17:467–508.

Arnold S, V Savolainen, L Chittka. 2009. Flower colours along an alpine altitude gradient, seen through the eyes of fly and bee pollinators. Arthropod-Plant Interactions 3:27–43.

Ascensão L, A Francisco, H Cotrim, MS Pais. 2005. Comparative structure of the labellum in Ophrys fusca and O. lutea (Orchidaceae). American Journal of Botany 92:1059–1067.

Ashman TL, M Bradburn, DH Cole, BH Blaney, RA Raguso. 2005. The scent of a male: the role of floral volatiles in pollination of a gender dimorphic plant. Ecology 86:2009–2105.

Ashman TL, DJ Schoen. 1994. How long should flowers live? Nature 371:788–791.

Ashworth L, R Aguilar, L Galetto, MA Aizen. 2004. Why do pollination generalist and specialist plant species show similar reproductive susceptibility to habitat fragmentation? Journal of Ecology 92:717–719.

Atlas of Living Australia. 2014. Website at http://www.ala.org.au. Accessed January 30, 2014.

Atwood JT. 1985. Pollination of Paphiopedilum rothschildianum: Brood-site imitation. National Geographic Research 1:247–254.

Austerlitz F, CW Dick, C Dutech, EK Klein, S Oddou-Muratorio, PE Smouse, VL Sork. 2004. Using genetic markers to estimate the pollen dispersal curve. Molecular Ecology 13:937–954.

Avarguès-Weber A, N Deisig, M Giurfa. 2011. Visual cognition in social insects. Annual Review of Entomology 56:423–43.

Ayasse M, J Gögler, J Stökl. 2010. Pollinator-driven speciation in sexually deceptive orchids of the genus Ophrys. Pages 101–118 in M Glaubrecht, editor. Evolution in Action—Case Studies in Adaptive Radiation, Speciation and the Origin of Bio-diversity. Springer-Verlag, Berlin.

Ayasse M, RJ Paxton, J Tengö. 2001. Mating behavior and chemical communication in the order Hymenoptera. Annual Review of Entomology 46:31–78.

Ayasse M, FP Schiestl, HF Paulus, F Ibarra, W Francke. 2003. Pollinator attraction in a sexually deceptive orchid by means of unconventional chemicals. Proceedings of the Royal Society of London B Biological Sciences 270:517–522.

Ayasse M, FP Schiestl, HF Paulus, C Löfstedt, B Hansson, F Ibarra, W Francke. 2000. Evolution of reproductive strategies in the sexually deceptive orchid *Ophrys sphegodes*: how does flower-specific variation of odor signals influence reproductive success? Evolution 54:1995–2006.

Ayasse M, J Stökl, W Francke. 2011. Chemical ecology and pollinator-driven speciation in sexually deceptive orchids. Phytochemistry 72:1667–1677.

Backhaus W. 1991. Color oppononent coding in the visual system of the honeybee. Vision Research 31:1381–1397.

———. 1992. Color vision in honeybees. Neuroscience and Biobehavioral Reviews 16: 1–12.

Backhaus W, R Menzel. 1987. Color distance derived from a receptor model of color vision in the honeybee. Biological Cybernetics 55:321–331.

Baker HG. 1955. Self-compatibility and establishment after "long-distance" dispersal. Evolution 9:347–348.

———. 1967. Support for Baker's law—as a rule. Evolution 21:853–856.

Bänziger H. 1994. Studies on the natural pollination of three species of wild lady-slipper orchids (*Paphiopedilum*) in Southeast Asia. Pages 201–202 *in* A Pridgeon, editor. Proceedings of the 14th World Orchid Conference. Edinburgh (Scotland): HMSO.

———. 1996. The mesmerizing wart: the pollination strategy of epiphytic lady slipper orchid *Paphiopedilum villosum* (Lindl.) Stein (Orchidaceae). Botanical Journal of the Linnean Society 121:59–90.

———. 2002. Smart alecks and dumb flies: natural pollination of some wild lady slipper orchids (*Paphiopedilum* spp., Orchidaceae). Pages 165–169, plates 45–57, *in* J Clark, WM Elliott, G Tingley, J Biro, editors. Proceedings of the 16th World Orchid Conference. Vancouver Orchid Society, Vancouver, BC.

Bänziger H, H Sun, Y Luo. 2005. Pollination of a slippery lady slipper orchid in south-west China: *Cypripedium guttatum* (Orchidaceae). Botanical Journal of the Linnean Society 148:251–264.

———. 2008. Pollination of wild lady slipper orchids *Cypripedium yunnanense* and *Cypripedium flavum* (Orchidaceae) in south-west China: why are there no hybrids? Botanical Journal of the Linnean Society 156:51–64.

Barau A, N Barré, C Jouanin. 2005. Oiseaux de La Réunion. Editions Orphie, Paris.

Barber ME. 1869. On the structure and fertilization of *Liparis bowkeri*. Journal of the Linnean Society of London Botany 5:455–458.

Barkman TJ. 2001. Character coding of secondary chemical variation for use in phylo-genetic analyses. Biochemical Systematics and Ecology 29:1–20.

Barkman TJ, JH Beaman, DA Gage. 1997. Floral fragrance variation in *Cypripedium*: Implications for evolutionary and ecological studies. Phytochemistry 5:875–882.

Barlow N, editor. 1967. Darwin and Henslow: The Growth of an Idea; Letters 1830–1860. John Murray, London, England.

Barone-Lumaga MR, G Scopece, S Cozzolino. 2006. Ecological constraints on ovule development in Mediterranean orchids. Congresso Nazionale della Società Botanica Italiana. Milano (I), Atti del 105°.

Barrett SCH. 1985. Floral trimorphism and monomorphism in continental and island populations of *Eichhornia paniculata* (Speng.) Solms. (Pontederiaceae). Biological Journal of the Linnean Society 25:41–60.

———. 1996. The reproductive biology and genetics of island plants. Philosophical Transactions of the Royal Society of London B Biological Sciences 351:725–733.

Bateman J. 1837–1843. The Orchidaceae of Mexico and Guatemala. J. Ridgway and Sons, London, England.

Bateman RM, E Bradshaw, DS Devey, BJ Glover, S Malmgren, G Sramkó, M Murphy Thomas, PJ Rudall. 2011. Species arguments: clarifying competing concepts of species delimitation in the pseudo-copulatory orchid genus *Ophrys*. Botanical Journal of the Linnean Society 165:336–347.

Bateman RM, PM Hollingsworth, J Preston, L Yi-Bo, AM Pridgeon, MW Chase. 2003. Molecular phylogenetics and evolution of Orchidinae and selected Habenariinae (Orchidaceae). Botanical Journal of the Linnean Society 142:1–40.

Bateman RM, A Pridgeon, MW Chase. 1997. Phylogenetics of subtribe Orchidinae (Orchidoideae, Orchidaceae) based on nuclear ITS sequences. 2. Infrageneric relationships and taxonomic revision to achieve monopoly of *Orchis sensu stricto*. Lindelyana 12:113–141.

Bates BC, ZW Kundzewics, S Wu, JP Palutikof, editors. 2008. Climate Change and Water. IPCC Secretariat, Geneva, Switzerland.

Bates R. 1977. Pollination of South Australian orchids—Part 6. The genus *Diuris*. Journal of the Native Orchid Society of South Australia 1:12.

———. 1982. Observations of pollen vectors on *Caladenia congesta* R. Br. Journal of the Native Orchid Society of South Australia 6:37–38.

———. 1983. Pollination of *Caladenia*: an overview. Native Orchid Society Journal 7: 65–68.

———. 1984a. The ecology and biology of *Caladenia rigida* (Orchidaceae). South Australian Naturalist 58:65–68.

———. 1984b. Pollination of *Prasophyllum elatum* R. Br. (with notes on associated biology). Orchadian 8:14–17.

———. 1987. Observations on the pollination of *Eriochilus cucullatus* in the Adelaide Hills. Journal of the Native Orchid Society of South Australia 11:5–7.

———. 1989. Observations on the pollination of *Caleana major* R. Br. by male sawflies (*Pterygophorus* sp.). Orchadian 9:208–210.

Bates R, JZ Weber. 1990. Orchids of South Australia. A. B. Caudell, Government Printer, South Australia, Australia.

Bates RJ. 2010. The *Thelymitra pauciflora* R. Br. complex (Orchidaceae) in South Australia with the description of seven new taxa. Journal of the Adelaide Botanic Gardens 24:17–32.

Batty A, K Dixon, M Brundrett, K Sivasithamparam. 2001. Constraints to symbiotic germination of terrestrial orchid seed in a Mediterranean bushland. New Phytologist 152:511–20.

Batty A, K Dixon, K Sivasithamparam. 2000. Soil seed-bank dynamics of terrestrial orchids.Lindleyana 15:227–36.

Beardsell DV, MA Clements, JF Hutchinson, EG Williams. 1986. Pollination of *Diuris maculata* R. Br. (Orchidaceae) by floral mimicry of the native legumes *Daviesia* spp. and *Pultenaea scabra* R. Br. Australian Journal of Botany 34:165–174.

Bell AK, DL Roberts, JA Hawkins, PJ Rudall, MS Box, RM Bateman. 2009. Comparative micromorphology of nectariferous and nectarless labellar spur in selected clades of subtribe Orchidinae (Orchidaceae). Botanical Journal of the Linnaean Society 160: 369–387.

Bell G. 1986. The evolution of empty flowers. Journal of Theoretical Biology 118: 253–258.

Bellusci F, A Musacchio, R Stabile, G Pellegrino. 2010. Differences in pollen viability in relation to different deceptive pollination strategies in Mediterranean orchids. Annals of Botany 106:769–774.

Benitez-Vieyra S, AM Medina, AA Cocucci. 2009. Variable selection patterns on the labellum shape of *Geoblasta pennicillata*, a sexually deceptive orchid. Journal of Evolutionary Biology 22:2354–2362.

Bentley BL. 1977. Extrafloral nectaries and protection by pugnacious bodyguards. Annual Review of Ecology and Systematics 8:407–427.

Benzing DH. 1998. Vulnerabilities of tropical forests to climate change: the significance of resident epiphytes. Climatic Change 39:519–540.

Bergström G. 1974. Macrocyclic lactones in the Dufour gland secretion of the solitary bees *Colletes cunicularius* L. and *Halictus calceatus* Scop. (Hymenoptera, Apidae). Chemica Scripta 5:39–46.

Bergström G, J Tengö. 1974. Studies on natural odoriferous compounds. IX. Farnesyl- and geranyl-esters as main volatile constituents of the secretion from Dufour's gland in 6 species of *Andrena* (Hymenoptera, Apidae). Chemica Scripta 5:28–38.

———. 1978. Linalool in mandibular gland secretion of *Colletes* bees (Hymenoptera: Apoidea). Journal of Chemical Ecology 4:437–449.

Bernard GD, DG Stavenga. 1979. Spectral sensitivities of retinular cells measured in intact, living flies by an optical method. Journal of Comparative Physiology A 134: 95–107.

Bernhardt P. 1989a. The floral ecology of Australian *Acacia*. Pages 263–282 *in* CH Stirton, JL Zarucchi, editors. Advances in Legume Biology. Monographs in Systematic Botany from the Missouri Botanical Garden 29. St. Louis, MO, USA.

———. 1989b. Wily Violets and Underground Orchids: Revelations of a Botanist. William Morrow and Co., New York, USA.

———. 1993. *Thelymitra*. Pages 146–152 *in* GJ Hardin, editor. Flora of New South Wales, vol. 4. Royal Botanic Gardens and Domain Trust. New South Wales Press, Kensington, NSW, Australia.

———. 1995a. Biogeography and floral evolution in the Geoblasteae (Orchidaceae). Pages 116–134 *in* MT Kalin Arroyo, PH Zedler, MD Fox, editors. Ecology and Biogeography of Mediterranean Ecosystems in Chile, California and Australia. Springer-Verlag, New York, USA.

———. 1995b. The floral ecology of *Dianella caerulea* var. *assera* (Phormiaceae). Cunninghamia 4:1–17.

———. 1995c. Notes on the anthecology of *Pterostylis curta* (Orchidaceae). Cunninghamia 5:1–8.

———. 1996. Anther adaptation in animal pollination. Pages 192–220 *in* WG D'Arcy, RC Keating, editors. The Anther: Form, Function and Phylogeny. Cambridge University Press, Cambridge, England.

———. 2008. Gods and Goddesses in the Garden: Greco-Roman Mythology and the Scientific Names of Plants. Rutgers University Press, New Brunswick, NJ, USA.

Bernhardt P, P Burns-Balogh. 1983. Pollination and pollinarium of *Dipodium punctatum* (Sm) R. Br. Victorian Naturalist 100:197–199.

———. 1986a. Floral mimesis of *Thelymitra nuda* (Orchidaceae). Plant Systematics and Evolution 151:187–202.

———. 1986b. Observations of the floral biology of *Prasophyllum odoratum* (Orchidaceae, Spiranthoideae). Plant Systematics and Evolution 153:65–76.

Bernhardt P, A Dafni. 2000. Breeding system and pollination biology of *Mandragora officinarum* L. (Solanaceae), in northern Israel. Pages 215–224 *in* O Totland, editor. The Scandinavian Association for Pollination Ecology Honours Knut Faegri. Det Norske Videnskaps—Adademi, Series 39, Oslo, Norway.

Bernhardt P, R Edens-Meier. 2010. What we think we know *vs.* what we need to know about orchid pollination and conservation: *Cypripedium* L. as a model lineage. Botanical Review 76:204–219.

Bierzychudek P. 1981. Asclepias, lantana, and epidendrum—a floral mimicry complex. Biotropica 13(Suppl. S): 54–58.

Bino RJ, A Dafni, ADJ Meeuse. 1982. The pollination ecology of *Orchis galilaea* (Bronm. et Schulze) Schltr. (Orchidaceae). New Phytologist 90:315–319.

Bishop T. 1996. Field Guide to the Orchids of New South Wales and Victoria. University of New South Wales Press, Sydney, Australia.

Blanco MA, G Barboza. 2005. Pseudocopulatory pollination in *Lepanthes* (Orchidaceae: Pleurothallidinae) by fungus gnats. Annals of Botany 95:763–772.

Borg-Karlson A-K. 1990. Chemical and ethological studies of pollination in the genus *Ophrys* (Orchidaceae). Phytochemistry 29:1359–1387.

Borg-Karlson A-K, I Groth, L Ägren, B Kullenberg. 1993. Form-specific fragrances from *Ophrys insectifera* L. (Orchidaceae) attract species of different pollinator genera: evidence of sympatric speciation? Chemoecology 4:39–45.

Borg-Karlson A-K, Tengö J. 1986. Odor mimetism? Key substances in *Ophrys lutea–Andrena* pollination relationship (Orchidaceae: Andrenidae). Journal of Chemical Ecology 12:1927–1942.

Borrell BJ. 2007. Scaling of nectar foraging in orchid bees. American Naturalist 169: 569–580.

Bosser J. 1987. Contribution à l'étude des Orchidaceae de Madagascar et des Mascareignes. 22. Adansonia 3:249–254.

Boulter M. 2009. Darwin's Garden: Down House and the Origin of Species. Constable and Robinson, London, England.

Bower CC. 1992. The use of pollinators in the taxonomy of sexually deceptive orchids in the subtribe Caladeniinae (Orchidaceae). Orchadian 10:331–338.

———. 1993. Observations on the pollination of *Calochilus campestris* R. Br. Orchadian 11:68–71.

———. 1996. Demonstration of pollinator-mediated reproductive isolation in sexually-deceptive species of *Chiloglottis* (Orchidaceae: Caladeniinae). Australian Journal of Botany 44:15–33.

———. 2001. Pollination [of genera of tribe Diurideae]. Pages 69–70, 74–75, 76–78, 79, 83, 87, 89, 93–97, 98–101, 103–104, 106–107, 110, 112–113, 115, 118, 121–124, 128–130, 133, 138–139, 141–143, 146–148, 149–152, 154–155, 157, 161–162, 164, 165–167, 169–170, 172, 175, 179–181, 184–185, 189–192, 196–197, 201–202, 205,

208–213 *in* AM Pridgeon, PJ Cribb, MW Chase, FN Rasmussen, editors. Genera Orchidacearum, vol 2. Oxford University Press, Oxford,England.

———. 2006. Specific pollinators reveal a cryptic taxon in the bird orchid, *Chiloglottis valida sensu lato* (Orchidaceae), in south-eastern Australia. Australian Journal of Botany 54:53–64.

———. 2007. The wasp, ant and bird orchids of the *Chiloglottis* alliance—a general overview. Orchadian 15:402–416.

———. 2008. Serendipity and pollinators reveal a new cryptic spider orchid species related to *Caladenia concinna* (Rupp) D Jones et M Clements. Orchadian 16:60–69.

Bower CC, GR Brown. 1997. Hidden biodiversity: detection of cryptic thynnine wasp species using sexually deceptive, female-mimicking orchids. Memoirs of the Museum of Victoria 56:461–466.

———. 2009. Pollinator specificity, cryptic species and geographical patterns in pollinator responses to sexually deceptive orchids in the genus *Chiloglottis*: the *Chiloglottis gunnii* complex. Australian Journal of Botany 57:37–55.

Boyden, TC. 1980. The pollination biology of *Calypso bulbosa* var. *americana* (Orchidaceae): initial deception of bumble visitors. Oecologia (Berlin) 55:178–184.

Boyle F. 1893. About Orchids—A Chat. Chapman and Hall, London, England.

Bradshaw E, PJ Rudall, DS Devey, MM Thomas, BJ Glover, RM Bateman. 2010. Comparative labellum micromorphology of the sexually deceptive temperate orchid genus *Ophrys*: diverse epidermal cell types and multiple origins of structural colour. Botanical Journal of the Linnean Society 162:504–540.

Bradshaw HD, DW Schemske. 2003. Allele substitution at a flower colour locus produces a pollinator shift in monkeyflowers. Nature 426:176–178.

Brandt R, M Vorobyev. 1997. Metric analysis of threshold spectral sensitivity in the honeybee. Vision Research 37:425–439.

Briscoe AD, L Chittka. 2001. The evolution of color vision in insects. Annual Review of Entomology 46:471–510.

Brown A, P Dundas, K Dixon, S Hopper. 2008. Orchids of Western Australia. University of Western Australia Press, Perth, WA.

Brown GR. 1996. *Arthrothynnus*, a new genus of orchid-pollinating Thynninae (Hymenoptera: Tiphiidae). Beagle, Records of the Museums and Art Galleries of the Northern Territory 13:73–82.

Brown JH, A Kodric-Brown. 1979. Convergence, competition and mimicry in a temperate community of hummingbird pollinated plants. Ecology 60:1022–1035.

Brown R. 1810. Prodromus Flora Nova Hollandia et Insula Van-Diemen. J. Johnson and Co., London, England.

———. 1833. On the organs and mode of fecundation in Orchideae and Asclepiadeae. Transactions of the Linnean Society of London 16:685–746.

Bruce MJ, AM Heiling, ME Herberstein. 2005. Spider signals: are web decorations visible to birds and bees? Biology Letters 1:299–302.

Brundrett MC, A Scade, AL Batty, KW Dixon, K Sivasithamparam. 2003. Development of *in situ* and *ex situ* seed baiting techniques to detect mycorrhizal fungi from terrestrial orchid habitats. Mycological Research 107:1210–1220.

Buchmann SL. 1985. Bees use vibration to aid pollen collection from non-poricidal flowers. Journal of the Kansas Entomological Society 58:517–525.

Burns J, G Julien Foucaud, F Mery. 2011. Costs of memory: lessons from "mini" brains. Proceedings of the Royal Society of London B Biological Sciences 278:923–929.

Burns-Balogh P, P Bernhardt. 1988. Floral evolution and phylogeny in the tribe Thely-mitreae (Orchidaceae: Neottioideae). Plant Systematics and Evolution 159:19–47.

Burns-Balogh P, VA Funk. 1986. A Phylogenetic Analysis of the Orchidaceae. Smithsonian Contributions to Botany, no. 61. Washington, DC, USA.

Burrell RW. 1935. Notes on the habits of certain Australian Thynnidae. Journal of the New York Entomological Society 43:19–29.

Butenandt A. 1955. Über Wirkstoffe des Insektenreiches. 2 Zur Kenntnis der Sexual-Lockstoffe. Naturwissenschaftliche Rundschau 8:12.

Butenandt A, R Beckmann, D Stamm, F Hecker. 1959. Über den Sexuallockstoff des Seidenspinners Bombyx mori Reidarstellung und Konstitution. Zeitschrift für Naturforschung B 14:283–284.

Byrne M, AD Steane, L Joseph, DK Yeates, GJ Jordan, D Crayn, K Aplin, DJ Cantrill, LG Cook, M Crisp, DJS Keogh, J Melville, C Moritz, N Porch, JM Kale Sniderman, P Sunnucks, PH Weston. 2011. Decline of a biome: evolution, contraction, fragmentation, extinction and invasion of the Australian mesic zone biota. Journal of Biogeography 38:1635–1656.

Cady L. 1965. Notes on the pollination of Caleana major R. Br. Orchadian 2:34–35.

Cady L, ER Rotherham. 1970. Australian Native Orchids in Colour. Charles E. Tuttle Co., Rutland, VT, USA.

Calvo RN. 1993. Evolutionary demography of orchids: intensity and frequency of pollination and the cost of fruiting. Ecology 74:1033–1042.

Cameron KM. 2004. Utility of plastid psaB gene sequences for investigating intrafamilial relationships within Orchidaceae. Molecular Phylogenetics and Evolution 31: 1157–1180.

Cameron KM, MW Chase, M Whitten, PJ Kores, DC Jarrell, VA Albert, T Yukawa, HG Hills, DH Goldman. 1999. A phylogenetic analysis of the Orchidaceae: evidence from rbcL nucleotide sequences. American Journal of Botany 86:208–224.

Camus E-G. 1885. Iconographie des Orchidées des environs de Paris. Paris, France.

Camus E-G, A Camus, P Bergon. 1908. Monographie des Orchidées de l'Europe, de l'Afrique septentrionale, de l'Asie Mineure et des Provinces Russes Transcaspiennes. Librairie Jacques Chevalier, Paris.

Cane JH, J Tengö. 1981. Pheromonal cues direct mate-seeking behavior of male Colletes cunicularius (Hymenoptera: Colletidae). Journal of Chemical Ecology 7:427–436.

Carlquist S. 1974. Island Biology. Columbia University Press, New York, USA.

Carlsward BS, MW Whitten, NH Williams. 2003. Molecular phylogenetics of Neotropical leafless Angraecinae (Orchidaceae): reevaluation of generic concepts. International Journal of Plant Sciences 164:43–51.

Carlsward BS, MW Whitten, NH Williams, B Bytebier. 2006. Molecular phylogenetics of Vandeae (Orchidaceae) and the evolution of leaflessness. American Journal of Botany 93:770–786.

Case MA, ZR Bradford. 2009. Enhancing the trap of lady's slippers: a new technique for discovering pollinators yields new data for Cypripedium parviflorum (Orchidaceae). Botanical Journal of the Linnean Society 160:1–10.

Catling PM 1990. Auto-pollination in Orchidaceae. Pages 123–158 in J. Arditti, editor. Orchid Biology—Reviews and Perspectives, vol. 5. Timber Press, Portland, OR, USA.

Catling PM, BA Bennett. 2007. Discovery of a possibly relict outbreeding morphotype of sparrow's-egg lady's-slipper orchid, Cypripedium passerinum, in southwestern Yukon. Canadian Field-Naturalist 121:295–298.

Catling PM, G Knerer. 1980. Pollination of the small white Lady's slipper (*Cypripedium candidum*) in Lambton County, Southern Ontario. Canadian Field-Naturalist 94: 435–438.

Cavender-Bares J, DA Ackerly, D Baum, F Bazzaz. 2004. Phylogenetic overdispersion in Floridian oak communities. American Naturalist 163:823–843.

Chase MW, K Cameron, RL Barrett, JV Freudenstein. 2003. DNA data and Orchidaceae systematic: a new phylogenetic classification. Pages 69–89 *in* KW Dixon, RL Barrett, P Cribb, editors. Orchid Conservation. Natural History Publications (Borneo). Kota Kinabalu, Indonesia.

Chase MW, HG Hills. 1991. Silica gel: an ideal desiccant for preserving field collected leaves for use in molecular studies. Taxon 40:215–220.

———. 1992. Orchid phylogeny, flower sexuality, and fragrance-seeking. Bioscience 42: 43–49.

Cheeseman T. 1880. On the fertilization of *Thelymitra*. Transactions and Proceedings of the New Zealand Institute 13:291–296.

Chen I-CH, J Shiu, S Benedick, JD Holloway, VK Chey. 2009. Elevation increases in moth assemblages over 42 years on a tropical mountain. Proceedings of the National Academy of Sciences of the United States of America 106:1479–1483.

Chen S, Z Tsi, Y Luo. 1999. Native orchids of China in colour. Science Press, Beijing, China.

Chen X-Q. 2003. East Asia. Pages 11–26 *in* MD Schwartz, editor. Phenology: An Integrative Environmental Science. Kluwer Academic Publishers, Dordrecht, the Netherlands.

Chen Y-G, D-Y He, M-S Nong. 2008. Change of cold wave in the Nanning area under global warming. Advances in Climate Change Research 4:245–249.

Cheng JS, Y Liu, R He, X-L Wei, Y-B Luo. 2007. Food-deceptive pollination in *Cymbidium lancifolium* (Orchidaceae) in Guangxi, China. Biodiversity Science 15: 608–617.

Cheng JS, F-Z Shangguan, Y-B Luo, ZH Deng. 2009. The pollination of a self-incompatible, food-mimic orchid, *Coelogyne fimbriata* (Orchidaceae), by female *Vespula* wasps. Annals of Botany doi:10.1093/aob/mcp029.

Chiao C-C, W-Y Wu, S-H Chen, E-C Yang. 2009. Visualization of the spatial and spectral signals of orb-weaving spiders, *Nephila pilipes*, through the eyes of a honeybee. Journal of Experimental Biology 212:2269–2278.

Chittka L. 1992. The colour hexagon: a chromaticity diagram based on photoreceptor excitations as a generalized representation of colour opponency. Journal of Comparative Physiology A Neurethology Sensory Neural and Behavioral Physiology 170:533–543.

———. 1996a. Optimal sets of colour receptors and opponent processes for coding of natural objects in insect vision. Journal of Theoretical Biology 181:179–196.

———. 1996b. Does bee color vision predate the evolution of flower color? Naturwissenschaften 83:136–138.

———. 1999. Bees, white flowers, and the color hexagon—a reassessment? No, not yet. Naturwissenschaften 86:595–597.

Chittka L, AG Dyer, F Bock, A Dornhaus. 2003. Bees trade off foraging speed for accuracy. Nature 424:388.

Chittka L, A Gumbert, J Kunze. 1997. Foraging dynamics of bumblebees: correlates of movements within and between plant species. Behavioral Ecology 8:239–249.

Chittka L, PG Kevan. 2005. Flower colour as advertisement.pages 157–196 *in* A Dafni, PG Kevan, BC Husband, editors. Practical Pollination Biology. Enviroquest Ltd. Cambridge, Ontario, Canada.

Chittka L, NE Raine. 2006. Recognition of flowers by pollinators. Current Opinions in Plant Biology 9:428–435.

Chittka L, A Shmida, N Troje, R Menzel. 1994. Ultraviolet as a component of flower reflections, and the colour perception of Hymenoptera. Vision Research 34:1489–1508.

Chittka L, J Spaethe, A Schmidt, A Hickelsberger. 2001. Adaption, constraint, and chance in the evolution of flower color and pollinator color vision. Pages 106–126 *in* L Chittka, JD Thompson, editors. Cognitive Ecology of Pollination: Animal Behaviour and Floral Evolution. Cambridge University Press, Cambridge, England.

Chittka L, JD Thomson, NM Waser. 1999. Flower constancy, insect psychology, and plant evolution. Naturwissenschaften 86:361–377.

Christensen DE. 1994. Fly pollination in the Orchidaceae. Pages 415–454 *in* J Arditti, editor. Orchid Biology: Reviews and Perspectives 6. John Wiley and Sons, Inc., New York, USA.

Chuine I, P Yiou, N Viovy, B Sequin, V Daux, EL LeRoy. 2004. Historical phenology: grape ripening as a past climate indicator. Nature 432:289–290.

Chung MY, MG Chung. 2005. Pollination biology and breeding systems in the terrestrial orchid *Bletilla striata*. Plant Systematics and Evolution 252:1–9.

Chung MY, JD Nason, MG Chung. 2004. Spatial genetic structure in populations of the terrestrial orchid *Cephalanthera longibracteata* (Orchidaceae). American Journal of Botany 91:52–57.

Claessens J, J Kleynen. 2011. The Flower of the European Orchid: Form and Function. Jean Claessens and Jacques Kleynen Publishers. Koeltz Scientific Books, Germany.

Clayton S, M Aizen. 1996. Effects of pollinia removal and insertion on flower longevity in *Chloraea alpina* (Orchidaceae). Evolutionary Ecology 10:653–660.

Cleland EE, NR Chiariello, R Loarie, HA Mooney, CB Field. 2006. Diverse responses of phenology to global changes in a grassland ecosystem. Proceedings of the National Academy of Sciences of the United States of America 103:13740–13744.

Cleland EE, I Chuine, A Menzel, HA Mooney, MD Schwartz. 2007. Shifting plant phenology in response to global change. Trends in Ecology and Evolution 22:357–365.

Clements MA. 1999. Embryology. Pages 38–58 *in* AM Pridgeon, PJ Cribb, MW Chase, FN Rasmussen, editors. Genera Orchidacearum, vol. 1. Oxford University Press, Oxford, England.

Clements MA, DL Jones, I Sharma, ME Nightingale, MJ Garratt, KJ Fitzgerald, AM MacKenzie, BPJ Molloy. 2002. Phylogenetics of Diurideae (Orchidaceae) based on the internal transcribed spacer (ITS) regions of nuclear ribosomal DNA. Lindleyana 17:135–171.

Coates F, M Duncan. 2009. Demographic variation between populations of *Caladenia orientalis*—a fire-managed threatened species. Australian Journal of Botany 57: 326–339.

Coates F, ID Lunt, RL Tremblay. 2006. Effects of disturbance on population dynamics of the threatened orchid *Prasophyllum correctum* D. L. Jones and implications for grassland management in south-eastern Australia. Biological Conservation 129: 59–69.

Coats P. 1970. Flowers in History. Viking Press, New York, USA.

Coleman E. 1927. Pollination of the orchid *Cryptostylis leptochila* F. v. M. Victorian Naturalist 44:20–22.

———. 1928a. Pollination of an Australian orchid by a male ichneumonid *Lissopimpla semipunctata* Kirby. Transactions of the Royal Entomological Society of London 76: 533–539.

———. 1928b. Pollination of the orchid *Cryptostylis leptochila* F. v. M. Victorian Naturalist 44:333–340.

———. 1929a. Pollination of *Cryptostylis subulata* (Labill.) Reichh. Victorian Naturalist 46:62–66.

———. 1929b. Pollination of an Australian orchid, *Cryptostylis leptochila* F. M. Muell. Journal of Botany, British and Foreign 67:97–99.

———. 1930a. Pollination of *Cryptostylis erecta* R. Be. Victorian Naturalist 46:62–66.

———. 1930b. The pollination of a second Australian orchid by the ichneumon *Lissopimpla semipunctata* Kirby (Hymenoptera, Parasitica). Proceedings of the Royal Entomological Society of London Series A General Entomology 5:15.

———. 1931. Further observations on the fertilization of Australian orchids by the male Ichneumonid *Lissopimpla semipunctata* Kirby. Proceedings of the Royal Entomological Society of London Series A General Entomology 6:22.

———. 1932. Pollination of *Diuris pedunculata* R. Br. Victorian Naturalist 49:179–186.

———. 1933a. Pollination of orchids, genus *Prasophyllum*. Victorian Naturalist 49: 214–221.

———. 1933b. Pollination of *Diuris sulphurea* R. Br. Victorian Naturalist 50:3–8.

———. 1938. Further observations on the pseudocopulation of the male *Lissopimpla semipunctata* Kirby (Hymenoptera, Parasitica) with the Australian orchid, *Cryptostylis leptochila* F. v. M. Proceedings of the Royal Entomological Society of London Series A General Entomology 13:82–83.

Colwell RK, G Brehm, CL Cardelus, AC Gilman, JT Longino. 2008. Global warming, elevational range shifts, and lowland biotic attrition in the wet tropics. Science 322:258–261.

Combes SA, R Dudley. 2009. Turbulence-driven instabilities limit insect flight performance. Proceedings of the National Academy of Sciences of the United States of America 106:9105–9108.

Combs JK, A Pauw. 2009. Preliminary evidence that the long-proboscid fly *Philoliche gulosa* pollinates *Disa karooica* and its proposed Batesian model *Pelargonium stipulaceum*. South African Journal of Botany 75:757–761.

Committee on the Status of Pollinators in North America. 2007. Status of Pollinators in North America. National Research Council. National Academies Press, Washington, DC, USA www.nap.edu.

Comprehensive Scientific Investigation Team of Guangxi Yachang Orchid Nature Reserve. 2007. The comprehensive investigation report of Guangxi Yachang Orchid Nature Reserve. Guangxi Forestry Inventory and Planning Institute, Nanning, Guangxi, China.

Corlett RT. 1990. Flora and reproductive phenology of the rain forest at Bukit Timah, Singapore. Journal of Tropical Ecology 6:55–63.

———. 1993. Reproductive phenology of Hong Kong shrubland. Journal of Tropical Ecology 9:501–510.

———. 2009. The Ecology of Tropical East Asia. Oxford University Press, New York, USA.

Corlett RT, JV Lafrankie Jr. 1998. Potential impacts of climate change on tropical Asian forests through an influence on phenology. Climatic Change 39:439–453.

Correvon H. 1899. Album des orchidées d'Europe centrale et septentrionale. Librairie Georg, Geneva, Switzerland.

Correvon H, M Pouyanne. 1916. Un curieux cas de mimétisme chez les Ophrydées. Journal de la Société Nationale d'Horticulture de France, ser. 4, 17:29–47.

———. 1923. Nouvelles observations sur le mimétisme et la fécondation chez les *Ophrys speculum* et lutea. Journal de la Société Nationale d'Horticulture de France, ser. 4, 24:372–377.

Curtis P, NJ Vereecken, FP Schiestl, MRB Lumaga, A Scrugli, S Cozzolino. 2009. Pollinator convergence and the nature of species' boundaries in sympatric Sardinian *Ophrys* (Orchidaceae). Annals of Botany 104:497–506.

Courtois EA, CET Paine, P-A Blandinieres, D Stein, J-M Bessiere, E Houel, C Baraloto, J Chave. 2009. Diversity of the volatile organic compounds emitted by 55 species of tropical trees: a survey in French Guiana. Journal of Chemical Ecology 35: 1349–1362.

Cozzolino S, S Aceto, P Caputo, A Widmer, A Dafni. 2001. Speciation processes in Eastern Mediterranean *Orchis* s.1. species: molecular evidence and the role of pollination biology. Israel Journal of Plant Sciences 49:91–103.

Cozzolino S, S D'Emerico, A Widmer. 2004. Evidence for reproductive isolate selection in Mediterranean orchids: karyotype differences compensate for the lack of pollinator specificity. Biological Letters 271:259–262.

Cozzolino S, FP Schiestl, A Müller, O De Castro, AM Nardella, A Widmer. 2005. Evidence for pollinator sharing in Mediterranean nectar-mimic orchids: absence of premating barriers? Proceedings of the Royal Society of London B Biological Sciences 272:1271–1278.

Cozzolino S, G Scopece. 2008. Specificity in pollination and consequences for postmating reproductive isolation in deceptive Mediterranean orchids. Philosophical Transactions of the Royal Society of London B Biological Sciences 363:3037–3046.

Cozzolino S, A Widmer. 2005. Orchid diversity: an evolutionary consequence of deception? Trends in Ecology and Evolution 20:487–493.

Cribb P. 1987. The Genus Paphiopedilum. Collingridge, London, England.

———. 1997. The Genus Cypripedium. Timber Press, Portland, OR, USA.

———. 1998. The Genus Paphiopedilum, 2nd edition. Natural History Publications, Kota, Kinabalu, Sabah, and Royal Botanic Gardens, Kew, England.

———. 1999. Phragmipedium. Pages 153–161 *in* AM Pridgeon, PJ Cribb, MW Chase, FN Rasmussen, editors. Genera Orchidacearum, vol. 1: General Introduction, Apostasioideae, Cypripedioideae. Oxford University Press, Bath Press, Avon, England.

Cropper SC, DM Calder. 1990. The floral biology of *Thelymitra epipactoides* (Orchidaceae), and the implications of pollination by deceit on the survival of this rare orchid. Plant Systematics and Evolution 170:11–27.

Cruden RW. 1977. Pollen-ovule ratios a conservative indicator of the breeding systems in flowering plants. Evolution 31:32–46.

Crüger H. 1864. A few notes on the fecundation of orchids and their morphology. Journal of the Linnean Society of London Botany 8:127–135.

Curry K J, LM McDowell, WS Judd, WL Stern. 1991. Osmophores, floral features, and systematics of *Stanhopea* (Orchidaceae). American Journal of Botany 78:610–623.

Dafni A. 1983. Pollination of *Orchis caspia*—a nectarless plant species which deceives the pollinators of nectariferous species from other plant families. Journal of Ecology 71:467–474.

———. 1984. Mimicry and deception in pollination. Annual Review of Ecology and Systematics 15:259–278.

———. 1986. Floral mimicry: mutualism and undirectional exploitation of insect by plants. Pages 81–90 *in* B Juniper, R Southwood, editors. Insect and the Plant Surface. Edward Arnold Publishing, London, England.

———. 1987. Pollination in *Orchis* and related genera: evolution from reward to deception. Pages 79–104 *in* J Arditti, editor. Orchid Biology: Reviews and Perspectives 6. Cornell University Press, Utica, NY, USA.

Dafni A, P Bernhardt. 1989. Pollination of terrestrial orchids in Southern Australia and the Mediterranean region: systematics, ecological, and evolutionary implications. Pages 192–252 *in* MK Hecht, B Wallace, RJ Macintyre, editors. Evolutionary Biology, vol. 24. Plenum Publishing Corporation, New York, USA.

Dafni A, DM Calder. 1987. Pollination by deceit and floral mimesis in *Thelymitra antennifera* (Orchidaceae). Plant Systematics and Evolution 158:11–22.

Dafni A, D Firmage. 2000. Pollen viability and longevity: practical, ecological and evolutionary implications. Pages 113–132 *in* A Dafni, M Hesse, E Pacini, editors. Pollen and Pollination. Springer, Berlin, Germany.

Dafni A, Y Ivri. 1979. Pollination ecology of and hybridization between *Orchis coriophora* L. and *O. collina* Sol. ex Russ. (Orchidaceae) in Israel. New Phytologist 83:181–187.

———. 1981. Floral mimicry between *Orchis israelitica*. Baumann and Dafni (Orchidaceae) and *Bellevalia flexuosa* Boiss (Liliaceae). Oecologia (Berlin) 49:229–232.

Dafni A, PG Kevan. 1996. Floral symmetry and nectar guides: ontogenetic constraints form floral development, colour pattern rules and functional significance. Botanical Journal of the Linnean Society 120:371–377.

Damon A, MA Pérez-Soriano. 2005. Interaction between ants and orchids in the Soconusco region, Chiapas, Mexico. Entomotropica 20:59–65.

Danforth BN, S Sipes, J Fang, SG Brady. 2006. The history of early bee diversification based on five genes plus morphology. Proceedings of the National Academy of Sciences of the United States of America 103:15118–15123.

Darveau C-A, PW Hochachka, KC Welch Jr., DW Roubik, RK Suarez. 2005. Allometric scaling of flight energetics in Panamanian orchid bees: a comparative phylogenetic approach. Journal of Experimental Biology 208:3581–3591.

Darwin C. 1845. Journal of Researches into the Natural History and Geology of the Countries Visited during the Voyage of HMS Beagle round the World, under the Command of Capt Fitz Roy RN, 2nd edition. John Murray, London, England.

———. 1859. On the Origin of Species by Means of Natural Selection, or the Preservation of Favoured Races in the Struggle for Life. John Murray, London, England.

———. 1862. On the Various Contrivances by Which British and Foreign Orchids Are Fertilised by Insects, and on the Good Effects of Intercrossing. John Murray, London, England.

———. 1868. The Variation of Animals and Plants under Domestication. John Murray, London, England.

———. 1869. Notes on the fertilization of orchids. Annals and Magazine of Natural History 4 (series 4): 141–159.

————. 1876. The Effects of Cross and Self-Fertilisation in the Vegetable Kingdom. John Murray, London, England.

————. 1877. The Various Contrivances by Which Orchids Are Fertilised by Insects, 2nd edition. D. Appleton and Company, New York, USA.

————. 1880. The Power of Movement in Plants. John Murray, London, England.

————. 1984. The Various Contrivances by Which Orchids Are Fertilized by Insects, 2nd edition, revised with a new foreword by M Ghiselin. University of Chicago Press, Chicago, IL, USA.

Darwin F., editor. 1887. The Life and Letters of Charles Darwin, including an Autobiographical Chapter, vol. 3. John Murray, London, England.

————, editor. 1896. The Life and Letters of Charles Darwin, including an Autobiographical Chapter. Appleton and Company, New York, USA.

Daumann E. 1971. Zum Problem der Täuschblumen. Preslia 43:304–317.

Daumer K. 1956. Reizmetrische Untersuchung des Farbensehens der Bienen. Zeitschrift für vergleichende Physiologie 38:413–478.

Davies KL. 2009. Food-hair form and diversification in orchids. Pages 159–184 in T Kull, J Arditti, SM Wong, editors. Orchid Biology—Reviews and Perspectives, vol. 10. Comstock Publishers Associates, Ithaca, NY, USA.

Davies KL, M Stpiczyńska. 2011. Comparative labellar micromorphology of Zygopetalinae (Orchidaceae). Annals of Botany (London) 108:945–964.

————. 2012. Comparative labellar anatomy of resin-secreting and putative resin-mimic species of *Maxillaria* sl. (Orchidaceae: Maxillarinae). Botanical Journal of the Linnean Society 170:405–435.

Davies KL, MP Turner. 2004. Morphology of floral papillae in *Maxillaria* Ruiz and Pav. (Orchidaceae). Annals of Botany (London) 93:75–86.

Defrize J, M Théry, J Casas. 2010. Background colour matching by a crab spider in the field: a community sensory ecology perspective. Journal of Experimental Biology 213:1425–1435.

De Jong TJ, PGL Klinkhamer. 1994. Plant size and reproductive success through female and male function. Journal of Ecology 82:399–402.

De Jong TJ, NM Waser, PGL Klinkhamer. 1993. Geitonogamy: the neglected side of selfing. Trends in Ecology and Evolution 8:321–325.

DeLange P, J Rolfe, I St. George, J Sawyer. 2007. Wild Orchids of the Lower North Island. Dept. Conservation. Wellington Conservancy, Wellington, NZ.

Delforge P. 1994. Guide des Orchid'ees d'Europe, d'Afrique du Nord et du Proche-Orient. Delachaux et Niestl'e, Neuchâtel-Paris, Switzerland.

Delle-Vedove R, N Juillet, J-M Bessière, C Grison, N Barthes, T Pailler, L Dormont, B Schatz. 2011. Colour-scent associations in a tropical orchid: three colours but two odours. Phytochemistry 72:735–742.

Delpino F. 1868–1875. Ulteriori Osservazioni Sulla Dicogamia nel Regno Vegetale. Milano, pt. 1 (1868/69), pt. 2, fasc. 1 (1870), fasc. 2 (1875). Estratto dagli Atti della Soc. Ital. delle Sci. Nat., Milano, vols. 11, 12.

Devey DS, RM Bateman, MF Fay, JA Hawkins. 2008. Friends or relatives? Phylogenetics and species delimitation in the controversial European orchid genus *Ophrys*. Annals of Botany 101:385–402.

Devillers P, J Devillers-Terschuren. 1994. Essai d'analyse systématique du genre *Ophrys*. Les Naturalistes belges 75 (Orchidées 7):273–400.

Dickinson M. 2006. Insect flight. Current Biology 16:309–314.

Dickson CR, S Petit. 2006. Effect of individual height and labellum colour on the pollination of *Caladenia* (syn. *Arachnorchis*) *behrii* (Orchidaceae) in the northern Adelaide region, South Australia. Plant Systematics and Evolution 262:65–74.

Dillon ME, R Dudley. 2004. Allometry of maximum vertical force production during hovering flight of Neotropical orchid bees (Apidae: Euglossini). Journal of Experimental Biology 207:417–425.

Dixon KW. 2009. Pollination and restoration. Science 325:571–573.

Dixon KW, SD Hopper. 2009. An introduction to *Caladenia* R. Br.—Australia's jewel among terrestrial orchids. Australian Journal of Botany 57:i–vii.

Dixon KW, RL Tremblay. 2009. Biology and natural history of *Caladenia*. Australian Journal of Botany 57:247–258.

Dixon P, C Weinig, J Schmitt. 2001. Susceptibility to UV damage in *Impatiens capensis* (Balsaminaceae): testing for opportunity costs to shade-avoidance and population differentiation. American Journal of Botany 88:1401–1408.

Dobson HE. 2006. Relationship between floral fragrance composition and type of pollinator. Pages 147–198 *in* NA Dudareva, E Picheresky, editors. Biology of Floral Scent. CRC Press, Boca Raton, FL, USA.

Dockrill AW. 1969. Australian Indigenous Orchids, vol. 1. Society for Growing Australian Plants, Sydney, Australia.

Dodson CH. 1962a. Pollination and variation in the subtribe Catasetinae (Orchidaceae). Annals of the Missouri Botanical Garden 49:35–56.

———. 1962b. The importance of pollination in the evolution of orchids of tropical America. American Orchid Society Bulletin 31:525–534, 641–649, 731–735.

———. 1965. Studies in orchid pollination—the genus *Coryanthes*. American Orchid Society Bulletin 34:680–687.

———. 1966. Studies in orchid pollination—*Cypripedium, Phragmipedium* and allied genera. American Orchid Society Bulletin 35:125–128.

Dodson CH, RL Dressler, HG Hills, RM Adams, NH Williams. 1969. Biologically active compounds in orchid fragrances. Science 164:1243–1249.

Dodson CH, GP Frymire. 1961. Natural pollination of orchids. Missouri Botanical Garden Bulletin 49:133–139.

Donoghue MJ. 1989. Phylogenies and the analysis of evolutionary sequences, with examples from seed plants. Evolution 43:1137–1156.

Dormont L, R Delle-Vedove, JM Bessière, M Hossaert-Mc Key, B Schatz. 2010. Rare white-flowered morphs increase the reproductive success of common purple morphs in a food-deceptive orchid. New Phytologist 185:300–310.

Dötterl S, NJ Vereecken. 2010. The chemical ecology and evolution of bee-flower interactions: a review and perspectives. Canadian Journal of Zoology 88:668–697.

Douzery EJP, AM Pridgeon, P Kores, HP Linder, H Kurzweil, MW Chase. 1999. Molecular phylogenetics of Diseae (Orchidaceae): a contribution from nuclear ribosomal ITS sequences. American Journal of Botany 86:887–899.

Dressler RL. 1968. Pollination by euglossine bees. Evolution 22:202–210.

———. 1976. How to study orchid pollination without any orchids. Pages 534–537 *in* Proceedings of the Eighth World Orchid Conference, Frankfurt, Germany.

———. 1981. The Orchids: Natural History and Classification. Harvard University Press, Cambridge, MA, USA.

———. 1982. Biology of orchid bees (Euglossini). Annual Review of Ecology and Systematics 13:373–394.

———. 1993a. Field Guide to the Orchids of Panama and Costa Rica. Cornell University Press, Ithaca, NY, USA.

———. 1993b. Phylogeny and Classification of the Orchid Family. Dioscorides Press, Portland, OR, USA.

Dressler RL, CH Dodson. 1960. Classification and phylogeny in the Orchidaceae. Annals of the Missouri Botanical Garden 47:25–68.

Dudley R. 2002. Mechanisms and implications of animal flight maneuverability. Integrative and Comparative Biology 42:135–140.

Duffy KJ, G Scopece, S Cozzolino, MF Fay, RJ Smith, JC Stout. 2009. Ecology and genetic diversity of the dense-flowered orchid, *Neotinea maculata*, at the centre and edge of its range. Annals of Botany 104:507–516.

Dukas R. 2008. Evolutionary biology of insect learning. Annual Review of Entomology 53:145–160.

Dukas R, LA Real. 1993a. Learning constraints and floral choice behaviour in bumble bees. Animal Behaviour 46:637–644.

———. 1993b. Effects of recent experience on foraging decisions by bumble bees. Oecologia (Berlin) 94:244–246.

Dyer AG, L Chittka. 2004a. Fine colour discrimination requires differential conditioning in bumblebees. Naturwissenschaften 91:224–227.

———. 2004b. Biological significance of distinguishing between similar colours in spectrally variable illumination: bumblebees (*Bombus terrestris*) as a case study. Journal of Comparative Physiology A 190:105–114.

Dyer AG, C Neumeyer. 2005. Simultaneous and successive colour discrimination in the honeybee (*Apis mellifera*). Journal of Comparative Physiology A 191:547–557.

Dyer AG, AC Paulk, DH Reser. 2011. Colour processing in complex environments: insights from the visual system of bees. Proceedings of the Royal Society of London B Biological Sciences 278:952–959.

Dyer AG, J Spaethe, S Prack. 2008. Comparative psychophysics of bumblebee and honeybee colour discrimination and object detection. Journal of Comparative Physiology A Neuroethology Sensory Neural and Behavioral Physiology 194:617–627.

Dyer AG, HM Whitney, SEJ Arnold, BJ Glover, L Chittka. 2006. Bees associate warmth with floral colour. Nature 442:525.

Eberhard WG. 1997. Grave-robbing by male *Eulaema seabrai* bees (Hymenoptera: Apidae). Journal of the Kansas Entomological Society 70:66.

Eberle G. 1974. Nektarausscheidung im Sporn des Wanzenknabenkrautes *Orchis coriophora* L. Die Orchidee 25:222–225.

Ebert D, C Hayes, RD Peakall. 2009. Chloroplast simple sequence repeat markers for evolutionary studies in the sexually deceptive orchid genus *Chiloglottis*. Molecular Ecology Resources 9:784–789.

Eckert CG, KE Samis, S Dart. 2006. Reproductive assurance and the evolution of uniparental reproduction in flowering plants. Pages 183–203 *in* LD Harder, SCH Barrett, editors. The Ecology and Evolution of Flowers. Oxford University Press, Oxford, England.

Edens-Meier R, M Arduser, E Westhus, P Bernhardt. 2011. Pollination ecology of *Cypripedium reginae* Walter (Orchidaceae): size matters. Telopea 13:327–340.

Edens-Meier R, N Vance, Y Luo, P Li, E Westhus, P Bernhardt. 2010. Pollen-pistil interactions in North American and Chinese *Cypripedium* L. (Orchidaceae). International Journal of Plant Science 171:370–381.

Edens-Meier R, E Westhus, P Bernhardt. 2013. Floral biology of large-flowered *Thelymitra* species (Orchidaceae) and their hybrids in Western Australia. Telopea 15: 165–183.

Eggleton P, RL Vane-Wright, editors. 1994. Phylogenetics and Ecology. Academic Press, London, England.

Eltz T, J Ayasse, K Lunau. 2006. Species-specific antennal responses to tibial fragrances by male orchid bees. Journal of Chemical Ecology 32:71–79.

Eltz T, F Fritzsch, J Ramírez-P, Y Zimmermann, SR Ramírez, J Quezada-Euán, B Bembé. 2011. Characterization of the orchid bee *Euglossa viridissima* (Apidae: Euglossini) and a novel cryptic sibling species, by morphological, chemical, and genetic characters. Zoological Journal of the Linnaean Society 163:1064–1076.

Eltz T, DW Roubik, K Lunau. 2005a. Experience-dependent choices ensure species-specific fragrance accumulation in male orchid bees. Behavioral Ecology and Sociobiology 59:149–156.

Eltz T, DW Roubik, WM Whitten. 2003. Fragrances, male display and mating behavior of *Euglossa hemichlora*—a flight cage experiment. Physiological Entomology 28: 251–260.

Eltz T, A Sager, K Lunau. 2005b. Juggling with volatiles: exposure of perfumes by displaying male orchid bees. Journal of Comparative Physiology A Neuroethology Sensory Neural and Behavioral Physiology 191:575–581.

Eltz T, WM Whitten, DW Roubik, KE Linsenmair. 1999. Fragrance collection, storage, and accumulation by individual male orchid bees. Journal of Chemical Ecology 25: 157–176.

Eltz T, Y Zimmermann, J Haftmann, R Twele, W Francke, JJG Quezada-Euán, K Lunau. 2007. Enfleurage, lipid recycling and the origin of perfume collection in orchid bees. Proceedings of the Royal Society of London B Biological Sciences 274:2843–2848.

Eltz T, Y Zimmermann, C Pfeiffer, JR Pech, R Twele, W Francke, JJG Quezada-Euán, K Lunau. 2008. An olfactory shift is associated with male perfume differentiation and species divergence in orchid bees. Current Biology 18:1844–1848.

Endara L, DA Grimaldi, A Bitty, BA Roy. 2010. Lord of the flies: pollination of *Dracula* orchids. Lankesteriana 10:1–11.

Endress PK. 1996. Diversity and Evolutionary Biology of Tropical Flowers. Cambridge University Press, New York, USA.

Erickson R. 1965a. Insects and Australian orchids. Australian Orchid Review 30: 172–174.

———. 1965b. Orchids of the West, 2nd edition. Paterson Brokensha Pty., Ltd., Perth, Western Australia.

Faast R, JM Facelli. 2009. Grazing orchid: impact of florivory on two species of *Caladenia* (Orchidaceae). Australian Journal of Botany 57:361–372.

Faast R, JM Facelli, AD Austin. 2011. Seed viability in declining populations of *Caladenia rigida* (Orchidaceae): are small populations doomed? Plant Biology 13: 86–95.

Faast R, L Farrington, JM Facelli, A Austin. 2009. Bees and white spiders: unravelling the pollination syndrome of *Caladenia rigida* (Orchidaceae). Australian Journal of Botany 57:315–325.

Fabre J-H. 1879–1907. Souvenirs Entomologiques, vols. 1–10. Librairie Ch. Delagrave, Paris, France.

Faegri K, L van der Pijl. 1979. The Principles of Pollination Ecology, 3rd edition. Pergamon Press, Oxford, England.

Fan C-M, E-C Yang, IM Tso. 2009. Hunting efficiency and predation risk shapes the color-associated foraging traits of a predator. Behavioral Ecology 20:808–816.

Farrington L, P MacGillivray, R Faast, A Austin. 2009. Investigating DNA barcoding options for the identification of *Caladenia* (Orchidaceae) species. Australian Journal of Botany 57:276–286.

Feinsinger P. 1978. Ecological interactions between plants and hummingbirds in a successional tropical community. Ecological Monographs 48:269–287.

Fenster CB, WS Armbruster, P Wilson, MR Dudash, JA Thomson. 2004. Pollination syndromes and floral specialization. Annual Review of Ecology, Evolution, and Systematics 35:375–403.

Ferdy JB, PH Gouyon, J Moret, B Godelle. 1998. Pollinator behavior and deceptive pollination: learning process and floral evolution. American Naturalist 152:696–705.

Ferdy JB, S Loriot, M Sandmeier, M Lefranc, C Raquin. 2001. Inbreeding depression in a rare deceptive orchid. Canadian Journal of Botany 79:1181–1188.

Ferguson CS, K Donham. 1999. Pollinator of clustered lady's-slipper orchid, *Cypripedium fasciculatum* (Orchidaceae) in Oregon. North American Native Orchid Journal 5:180–184.

Ferlan L. 1957. Le "pseudo-accouplement" des Hyménoptères mâles et des *Ophrys*. La Nature 3265:180–185.

Fernandez PC, FF Locatelli, N Person-Rennell, G Delo, BH Smith. 2009. Association conditioning tunes transient dynamics of early olfactory processing. Journal of Neuroscience 29:10191–10202.

Feuerherdt L. 2002. The role of echidnas (*Tachyglossus aculeatus*) in dispersing mycorrhizal fungi associated with the endangered pink-lipped spider orchid (*Caladenia behrii*) north of Adelaide and distribution of these fungi at Warren Conservation Park, South Australia. Honours thesis, University of South Australia Adelaide, South Australia.

Feuerherdt L, S Petit, M Jusaitis. 2005. Distribution of mycorrhizal fungus associated with the endangered pink-lipped spider orchid (*Arachnorchis* [syn. *Caladenia*] *behrii*) at Warren Conservation Park in South Australia. New Zealand Journal of Botany 43:367–371.

Firmage DH, FR Cole. 1988. Reproductive success and inflorescence size in *Calopogon tuberosus* (Orchidaceae). American Journal of Botany 75:1371–1377.

Fisher BL. 1992. Facultative ant association benefits a Neotropical orchid. Journal of Tropical Ecology 8:109–114.

Fisher BL, JK Zimmerman. 1988. Ant/orchid associations in the Barro Colorado National Monument, Panama. Lindleyana 3:12–16.

Fitter AH, RSR Fitter. 2002. Rapid changes in flowering time in British plants. Science 296:1689–1691.

Fitzgerald RD. 1875–1895. Australian Orchids, vols. 1 and 2. Government Printer, Sydney, NSW, Australia.

Flook PK, S Klee, CHF Rowell. 1999. Combined molecular phylogenetic analysis of the Orthoptera (Arthropoda, Insecta) and implications for their higher systematics. Systematic Biology 48:233–253.

Foster DJ, RV Cartar. 2011. What causes wing wear in foraging bumble bees? Journal of Experimental Biology 214:1896–1901.

Francisco A, L Ascensao. 2013. Structure of the osmophore and labellum micromorphology in the sexually deceptive orchids *Ophrys bombyliflora* and *O. tentredinifera*
(Orchidaceae). International Journal of Plant Sciences 174:619–636.

Francke W, W Schröder, G Bergström, J Tengö. 1984. Esters in the volatile secretion of
bees. Nova Acta Regiae Societatis Scientiarum Upsaliensis, Serie V:C 3:127–136.

Franke S, F Ibarra, CM Schulz, R Twele, J Poldy, RA Barrow, R Peakall, FP Schiestl,
W Francke. 2009. The discovery of 2,5-dialkylcyclohexan-1,3-diones as a new
class of natural products. Proceedings of the National Academy of Sciences of the
United States of America 106:8877–8882.

Freudenstein JV, C van den Berg, DH Goldman, PJ Kores, M Molvray, MW Chase. 2004.
An expanded plastid DNA phylogeny of Orchidaceae and analysis of jackknife
branch support strategy. American Journal of Botany 91:149–157.

Frey FM. 2004. Opposing natural selection from herbivores and pathogens may maintain
floral-color variation in *Claytonia virginica* (Portulacaceae). Evolution 58:2426–2437.

Fritz AL. 1990. Deceit pollination of *Orchis spitzelii* (Orchidaceae) on the island of
Gotland in the Baltic: a suboptimal system. Nordic Journal of Botany 9:577–587.

Fritz AL, LA Nilsson. 1996. Reproductive success and gender variation in deceit-
pollinated orchids. Pages 319–338 *in* DG Lloyd, SCH Barrett, editors. Floral
Biology: Studies on Floral Evolution in Animal-Pollinated Plants. Chapman and
Hall, New York, USA.

Galizia CG, J Kunze, A Gumbert, AK Borg-Karlson, S Sachse, C Markl, R Menzel. 2005.
Relationship of visual and olfactory signal parameters in a food-deceptive flower
mimicry system. Behavioral Ecology 16:159–168.

Gandía-Herrero F, F García-Carmona, J Escribano. 2005. Floral fluorescence effect.
Nature 437:334.

Garnet JR. 1940. Observations on the pollination of orchids. Victorian Naturalist 56:
191–197.

Garside S. 1922. The pollination of *Satyrium bicallosum* Thunb. Annals of the Bolus
Herbarium 3:137–154.

Gaskett AC. 2011. Orchid pollination by sexual deception: pollinator perspectives.
Biological Reviews 86:33–75.

———. 2012. Floral shape mimicry and variation in sexually deceptive orchids with a
shared pollinator. Biological Journal of the Linnean Society 106:469–481.

Gaskett AC, ME Herberstein. 2010. Colour mimicry and sexual deception by tongue
orchids (*Cryptostylis*). Naturwissenschaften 97:97–102.

Gaskett AC, CG Winnick, ME Herberstein. 2008. Orchid sexual deceit provokes ejaculation. American Naturalist 171:E206–E212.

Geerts S, A Pauw. 2007. A malachite sunbird pollination guild in the Cape flora, with
implications for the endangered *Brunsvigia litoralis*. South African Journal of
Botany 73:289–289.

———. 2009. Hyper-specialization for long-billed bird pollination in a guild of South
African plants: the malachite sunbird pollination syndrome. South African Journal
of Botany 75:699–706.

Gegear RJ, TM Laverty. 2001. The effect of variation among floral traits on the flower
constancy of pollinators. Pages 1–20 *in* L Chittka, JD Thompson, editors. Cognitive Ecology of Pollination: Animal Behaviour and Floral Evolution. Cambridge
University Press, Cambridge, England.

Gentry AH, CH Dodson. 1987. Diversity and biogeography of Neotropical vascular epiphytes. Annals of the Missouri Botanical Garden 74:203–233.

George A, J Cooke. 1981. *Rhizanthella*: the underground orchid of Western Australia. Pages 77–78 *in* L Lawler, RD Kerr, editors. Proceedings of the Orchid Symposium, 13th International Botanical Congress, Sydney, Australia.

Gerlach G. 2011. The Genus *Coryanthes*: a paradigm in ecology. Lankesteriana 11: 253–264.

Gerlach G, RL Dressler. 2003. Stanhopeinae Mesoamericanae I. Lankesteriana 8:23–30.

Gerlach G, R Schill. 1991. Composition of orchid scents attracting euglossine bees. Botanica Acta 104:379–391.

Gigord LDB, MR Macnair, A Smithson. 2001. A negative frequency dependent selection maintains a dramatic flower color polymorphism in the rewardless orchid *Dactylorhiza sambucina* (L.) Sóo. Proceedings of the National Academy of Sciences of the United States of America 98:6253–6255.

Gigord LDB, MR Macnair, M Stritesky, A Smithson. 2002. The potential for floral mimicry in rewardless orchids: an experimental study. Proceedings of the Royal Society of London B Biological Sciences 269:1389–1395.

Gill DE. 1989. Fruiting failure, pollination inefficiency, and speciation in orchids. Pages 458–481 *in* D Otte, JA Endler, editors. Speciation and Its Consequences. Academy of Natural Sciences Publications, Philadelphia, PA, USA.

Gill FB. 1971. Ecology and evolution of the sympatric Mascarene white-eyes, *Zosterops borbonica* and *Zosterops oliveacea*. Auk 88:35–60.

Giurfa M. 2004. Conditioning procedure and color discrimination in the honeybee *Apis mellifera*. Naturwissenschaften 91:228–231.

Giurfa M, A Dafni, PR Neal. 1999. Floral symmetry and its role in plant-pollinator systems. International Journal of Plant Sciences 160 (series 6): S41-S50.

Giurfa M, L Lehrer. 2001. Honeybee vision and floral displays: from detection to close-up recognition. Pages 61–82 *in* L Chittka, JD Thompson, editors. Cognitive Ecology of Pollination: Animal Behaviour and Floral Evolution. Cambridge University Press, Cambridge, England.

Giurfa M, J Núñez, L Chittka, R Menzel. 1995. Colour preferences of flower-naive honeybees. Journal of Comparative Physiology A 177:247–259.

Giurfa M, M Vorobyev, R Brandt, B Posner, R Menzel. 1997. Discrimination of coloured stimuli by honeybees: alternative use of achromatic and chromatic signals. 180: 235–243.

Giurfa M, M Vorobyev, P Kevan, R Menzel. 1996. Detection of coloured stimuli by honeybees: minimum visual angles and receptor specific contrasts. Journal of Comparative Physiology A 178:699–709.

Given BB. 1954. Evolutionary trends in the Thynninae (Hymenoptera: Tiphiidae) with special reference to feeding habits of Australian species. Transactions of the Royal Entomological Society of London 105:1–10.

———. 1957. Observations on behavior of Australian Thynninae. Proceedings of the Eighth Pacific Science Congress (Manila, 1953) 3A:1265–1274.

Glover BJ, HM Whitney. 2010. Structural colour and iridescence in plants: the poorly studied relations of pigment colour. Annals of Botany 105:505–511.

Godfery MJ. 1925a. The fertilisation of *Ophrys speculum*, *O. lutea* and *O. fusca*. Orchid Review 33:33–40, 67–69, 195.

———. 1925b. The fertilisation of *Ophrys speculum, O. lutea* and *O. fusca*. Journal of Botany, British and Foreign 63:33–40.

———. 1927. The fertilisation of *Ophrys fusca* Link. Journal of Botany, British and Foreign 65:350–351.

———. 1928. Classification of the genus *Ophrys*. Journal of Botany, British and Foreign 66:33–36.

———. 1929. Recent observation on the pollination of *Ophrys*. Journal of Botany, British and Foreign 67:298–302.

———. 1930. Further notes on the fertilisation of *Ophrys fusca*, and *O. lutea*. Journal of Botany, British and Foreign 68:237–238.

Gögler J, J Stökl, A Sramkova, R Twele, W Francke, S Cozzolino, P Cortis, A Scrugli, M Ayasse. 2009. Ménage à trois—two endemic species of deceptive orchids and one pollinator species. Evolution 63:2222–2234.

Gögler J, R Twele, W Francke, M Ayasse. 2011. Two phylogenetically distinct species of sexually deceptive orchids mimic the sex pheromone of their single common pollinator, the cuckoo bumblebee *Bombus vestalis*. Chemoecology 21:243–252.

Goldblatt P, JC Manning. 2000. The long-proboscid fly pollination system in southern Africa. Annals of the Missouri Botanical Garden 87:146–170.

Goldsmith TH. 1994. Ultraviolet receptors and color vision: evolutionary implications and a dissonance of paradigms. Vision Research 34:1479–1487.

Gollan JR, MB Ashcroft, M Batley. 2011. Comparison of yellow and white pan traps in surveys of bee fauna in New South Wales, Australia (Hymenoptera: Apoidea: Anthophila). Australian Journal of Entomology 50:174–178.

Goodwillie C, S Kalisz, CG Eckert. 2005. The evolutionary enigma of mixed mating systems in plants: occurrence, theoretical explanations, and empirical evidence. Annual Review of Ecology, Evolution, and Systematics 36:47–79.

Göth A, CS Evans. 2004. Social responses without early experience: Australian brush-turkey chicks use specific visual cues to aggregate with conspecifics. Journal of Experimental Biology 207:2199–2208.

Gould SJ. 1978. The Panda's peculiar thumb. Natural History 87:105–108.

Goulson D, J Cruise, K Sparrow, A Harris, K Park, M Tinsley, A Gilburn. 2007. Choosing rewarding flowers: perceptual limitations and innate preferences influence decision making in bumblebees and honeybees. Behavioral Ecology and Sociobiology 61: 1523–1529.

Govaerts R, P Bernet, K Kratochvil, G Gerlach, GD Carr, P Alrich, AM Pridgeon, J Pfahl, MA Campacci, DH Baptista, H Tigges, J Shaw, P Cribb, AGK Kreuz, J Wood. 2012. World Checklist of Orchidaceae. Facilitated by the Royal Botanic Gardens, Kew. Published on the Internet; http://apps.kew.org/wcsp/monocots; retrieved January 30, 2014.

Goyret J, RA Raguso. 2008. Why do *Manduca sexta* feed from white flowers? Innate and learnt colour preferences in a hawkmoth. Naturwissenschaften 95:569–576.

Gracie C. 1993. Pollination of *Cyphomandra endopogon* var. *endopogon* (Solanaceae) by *Eufriesea* spp. (Euglossini) in French Guiana. Brittonia 45:39–46.

Gray A. 1862. Fertilization of orchids through the agency of insects. American Journal of Science, series 2, 34:420–429.

Gregg KB. 1983. Variation in floral fragrances and morphology: incipient speciation in *Cycnoches*. Botanical Gazette 144:566–576.

Griffiths KE, JWH Trueman, GR Brown, R Peakall. 2011. Molecular genetic analysis and ecological evidence reveals multiple cryptic species among thynnine wasp pollinators of sexually deceptive orchids. Molecular Phylogenetics and Evolution 59:195–205.

Grim T. 2005. Mimicry *vs.* similarity: which resemblances between brood parasites and their hosts are mimetic and which are not? Biological Journal of the Linnean Society 84:69–78.

Gronquist M, A Bezzerides, A Attygalle, J Meinwald, M Eisner, T Eisner. 2001. Attractive and defensive functions of the ultraviolet pigments of a flower (*Hypericum calycinum*). Proceedings of the National Academy of Sciences of the United States of America 98:13745–13750.

Gumbert A. 2000. Color choices by bumble bees (*Bombus terrestris*): innate preferences and generalization after learning. Behavioral Ecology and Sociobiology 48: 181–187.

Gumbert A, J Kunze. 2001. Colour similarity to rewarding model affects pollination in a food deceptive orchid, *Orchis boryi*. Biological Journal of the Linnaean Society 72: 419–433.

Gurevitch J, SM Scheiner, GA Fox. 2006. The Ecology of Plants, 2nd edition. Sinauer Associates, Inc., Sunderland, MA, USA.

Haas A. 1946. Neue Beobachtungen zum Problem der Flugbahnen bei Hummel-Männchen. Zeitschrift für Naturforschung 1:596–600.

———. 1949. Gesetzmässiges Flugverhalten der Männchen von *Psithyrus silvestris* Lep. und einiger solitären Apiden. Zeitschrift für vergleichende Physiologie 31:671–683.

———. 1952. Die Mandibeldrüse als Duftorgan bei einigen Hymenopteren. Naturwissenschaften 39:484.

Hale RJ. 2000. Nest utilisation and recognition by juvenile gryllacridids (Orthoptera: Gryllacrididae). Australian Journal of Zoology 48:643–652.

Hale RJ, WJ Bailey. 2004. Homing behaviour of juvenile Australian raspy crickets (Orthoptera: Gryllacrididae). Physiological Entomology 29:426–435.

Hale RJ, DCF Rentz. 2001. The Gryllacrididae: an overview of the world fauna with emphasis on Australian examples. Pages 95–110 *in* LH Field, editor. The Biology of Wetas, King Crickets and Their Allies. Cabi Publishing, Wallingford, England.

Hallé N. 1977. Orchidacées. Flore de la Nouvelle-Calédonie et Dependances, vol. 8. Muséum National d'Histoire Naturelle, Paris, France.

Hamilton WD, SP Brown. 2001. Autumn tree colours as a handicap signal. Proceedings of the Royal Society of London B Biological Sciences 268:1489–1493.

Harder LD. 2000. Pollen dispersal and the floral diversity of Monocotyledons. Pages 243–257 *in* KL Wilson, D Morrison, editors. Monocots: Systematics and Evolution. CSIRO Publishing, Melbourne, Australia.

Harder LD, SD Johnson. 2009. Darwin's beautiful contrivances: evolutionary and functional evidence for floral adaptation. New Phytologist 183:530–545.

Hardie RC, K Kirschfeld. 1983. Ultraviolet sensitivity of fly photoreceptors R7 and R8: evidence for a sensitising function. Biophysics of Structure and Mechanism 9: 171–180.

Harvey PH, AJL Brown, JHM Smith, S Nee, editors. 1996. New Uses for New Phylogenies. Oxford University Press, Oxford, England.

Hayward MW. 2009. Conservation management for the past, present and future. Biodiversity and Conservation 18:765–775.

He H, ZN Qin, Y-L Liu, X-P Liao. 2007. Spatial and temporal characteristics of abnormal monthly mean temperature and their changes in Guangxi Province. Advances in Climate Change Research 3:95–99.

Heiling AM, K Cheng, L Chittka, A Goeth, ME Herberstein. 2005. The role of UV in crab spider signals: effects on perception by prey and predators. Journal of Experimental Biology 208:3925–3931.

Heiling AM, ME Herberstein, L Chittka. 2003. Crab-spiders manipulate flower signals. Nature 421:334.

Heinrich B. 1975. Bee flowers: a hypothesis on flower variety and blooming times. Evolution 29:325–334.

Hempel de Ibarra N, M Vorobyev, R Brandt, M Giurfa. 2000. Detection of bright and dim colours by honeybees. Journal of Experimental Biology 203:3289–3298.

Herrera CM. 1991. Dissecting factors responsible for individual variation in plant fecundity. Ecology 72:1436–1448.

———. 1993. Selection on floral morphology and environmental determinants of fecundity in a hawk moth-pollinated violet. Ecological Monographs 63:251–275.

Hietz P. 1999. Diversity and conservation of epiphytes in a changing environment. International Union of Pure and Applied Chemistry, vol 70. http://www.iupac.org/symposia/proceedings/phuket97/hietz.html.

Hills HG, NH Williams, CH Dodson. 1972. Floral fragrances and isolating mechanisms in the genus *Catasetum* (Orchidaceae). Biotropica 4:61–76.

Hines HM, BA Counterman, R Papa, PA de Moura, MZ Cardos, M Linares, J Mallet, RD Reed, CD Jiggins, MR Kronforst, WO McMillan. 2011. Wing patterning gene redefines the mimetic history of *Heliconius* butterflies. Proceedings of the National Academy of Sciences of the United States of America 108:19666–19671.

Hodges SA, ML Arnold. 1995. Spurring plant diversification: are floral nectar spurs a key innovation? Proceedings of the Royal Society of London B Biological Sciences 262:343–348.

Hodges SA, NJ Derieg. 2009. Adaptive radiations: from field to genomic studies. Proceedings of the National Academy of Sciences of the United States of America 106:9947–9954.

Hoehne FC. 1933. Contribuição para o conhecimento do gênero *Catasetum* especialmente hermaphroditismo e trimorphismo das suas flores. Boletin Agricola Brasileira 133–196.

Hooker JD. 1860. The Botany of the Antarctic Voyage of HM Discovery Ships Erebus and Terror in the Years 1839–1843. Volume 1, Flora Antarctica. Part 2, Botany of Fuegia, The Falklands, Kerguelen's Land etc. Reeve Brothers, London, England.

Hopper SD. 2009. Taxonomic turmoil down-under: recent developments in Australian orchid systematics. Annals of Botany 104:447–455.

Hopper SD, AP Brown. 2004. Robert Brown's *Caladenia* revisited, including a revision of its sister genera *Cyanicula*, *Ericksonella* and *Pheladenia* (Caladeniinae: Orchidaceae). Australian Systematic Botany 17:171–240.

———. 2006a. Contributions to Western Australian orchidology: 3. New and reinstated taxa in *Eriochilus*. Nuytsia 16:29–61.

———. 2006b. Australia's wasp-pollinated flying duck orchids revised (*Paracaleana*: Orchidaceae). Australian Systematic Botany 19:211–244.

———. 2007. A revision of Australia's hammer orchids (*Drakaea*: Orchidaceae), with some field data on species-specific sexually deceived wasp pollinators. Australian Systematic Botany 20:252–285.

Hu SJ, H Hu, N Yan, JL Huang, SY Li. 2011. Hybridization and asymmetric introgression between *Cypripedium tibeticum* and *C. yunnanense* in Shangrila County, Yunnan Province, China. Nordic Journal of Botany 29:625–631.

Huang C-B, J-L Chen, C-L Feng, Z-F Lu, J-X Li. 2008. Characteristics of climate vertical distribution in Yachang Orchid Nature Reserve. Journal of Northwest Forestry University 23:39–43.

Huang X-S, H-W Zhou, M-L Huang. 2005. Changes in temperature and rainfall over the past 50 years in Guangxi province. Guangxi Meterology 26:9–11.

Huber FK, R Kaiser, W Sauter, FP Schiestl. 2005. Floral scent emission and pollinator attraction in two species of *Gymnadenia* (Orchidaceae). Oecologia (Berlin) 142: 564–575.

Huda MK, CC Wilcock. 2008. Impact of floral traits on the reproductive success of epiphytic and terrestrial tropical orchids. Oecologia (Berlin) 54:731–741.

———. 2012. Rapid floral senescence following male function and breeding systems of some tropical orchids. Plant Biology 14:278–284.

Hugel S, C Micheneau, J Fournel, BH Warren, A Gauvin-Bialecki, T Pailler, MW Chase, D Strasberg. 2010. *Glomeremus* species from the Mascarene Islands (Orthoptera, Gryllacrididae) with the description of the pollinator of an endemic orchid from the island of Réunion. Zootaxa 2545:58–68.

Hutchings MJ. 2010. The population biology of the early spider orchid *Ophrys sphegodes* Mill. 3. Demography over three decades. Journal of Ecology 98:867–878.

Inda LA, M Pimentel, MW Chase. 2012. Phylogenetics of tribe Orchideae (Orchidaceae: Orchidoideae) based on combined DNA matrices: inferences regarding timing of diversification and evolution of pollination syndromes. Annals of Botany 110: 71–90.

Indsto JO, PH Weston, MA Clements. 2009. A molecular phylogenetic analysis of *Diuris* (Orchidaceae) based on AFLP and ITS reveals three major clades and a basal species. Australian Systematic Botany 22:1–15.

Indsto JO, PH Weston, MA Clements, AM Batley, RJ Whelan. 2006. Pollination of *Diuris maculata* (Orchidaceae) by male *Trichocolletes venustus* bees. Australian Journal of Botany 54:669–679.

Indsto JO, PH Weston, MA Clements, AG Dyer, M Batley, RJ Whelan. 2007. Generalised pollination of *Diuris alba* R. Br. (Orchidaceae) by small bees and wasps. Australian Journal of Botany 55:628–634.

Inouye DW. 2008. Effects of climate change on phenology, frost damage and floral abundance of montane wildflowers. Ecology 89:353–362.

Internicola AI, G Bernasconi, LDB Gogord. 2008. Should food-deceptive species flower before or after rewarding species? An experimental test of pollinator visitation behaviour under contrasting phonologies. Journal of Evolutionary Biology 21: 1358–1365.

Internicola AI, N Juillet, A Smithson. 2006. Experimental investigation of the effect of spatial aggregation on reproductive success in a rewardless orchid. Oecologia (Berlin) 150:435–441.

Internicola AI, PA Page, G Bernasconi. 2009. Carry-over effects of bumblebee associative learning in changing plant communities leads to increased costs of foraging. Arthropods and Plant Interactions 3:17–26.

Internicola AI, PA Page, G Bernasconi, LDB Gigord. 2007. Competition for pollinator visitation between deceptive and rewarding artificial inflorescences: an experimental test of the effects of floral colour similarity and spatial mingling. Functional Ecology 21:864–872.

IPCC (Intergovernmental Panel on Climate Change). 2007. Climate Change 2007: Synthesis Report. Contribution of Working Groups I, II and III to the Fourth Assessment. Report of the Intergovernmental Panel on Climate Change [Core Writing Team, Pachauri, RK and A Reisinger (editors)]. IPCC, Geneva, Switzerland.

Jacobson M. 1965. Insect Sex Attractants. John Wiley and Sons, New York, USA.

Jacquemyn H, R Brys. 2010. Temporal and spatial variation in flower and fruit production in a food-deceptive orchid: a five-year study. Plant Biology 12:145–153.

Jacquemyn H, R Brys, M Hermy, JH Willems. 2007a. Long-term dynamics and population viability in one of the last populations of the endangered *Spiranthes spiralis* (Orchidaceae) in the Netherlands. Biological Conservation 134:14–21.

Jacquemyn H, R Brys, O Honnay. 2009. Large population sizes mitigate negative effects of variable weather conditions on fruit set in two spring woodland orchids. Biology Letters 5:495–498.

Jacquemyn H, R Brys, K Vandepitte, O Honnay, I Roldan-Ruiz, T Wiegand. 2007b. A spatially explicit analysis of seedling recruitment in the terrestrial orchid *Orchis purpurea*. New Phytologist 176:448–459.

Jacquemyn H, C Micheneau, DL Roberts, T Pailler. 2005. Elevation gradients of species diversity, breeding system and floral traits of orchid species on Réunion Island. Journal of Biogeography 32:1751–1761.

Janes JK, DA Steane, RE Vaillancourt. 2010. An investigation into the ecological requirements and niche partitioning of Pterostylidinae (Orchidaceae) species. Australian Journal of Botany 58:335–341.

Janzen DH. 1971. Euglossine bees as long-distance pollinators of tropical plants. Science 71:203–205.

———. 1981a. Bee arrival at two Costa Rican female *Catasetum* orchid inflorescences and a hypothesis on euglossine population structure. Oikos 36:177–183.

———. 1981b. Differential visitation of *Catasetum* orchid male and female flowers. Biotropica 13:77.

Janzen DH, P DeVries, DE Gladstone, ML Higgins, TM Lewinsohn. 1980. Self- and cross-pollination of *Encyclia cordigera* (Orchidaceae) in Santa Rosa National Park, Costa Rica. Biotropica 12:72–74.

Jeanes JA. 2004. A revision of the *Thelymitra pauciflora* R. Br. (Orchidaceae) complex in Australia. Muelleria 19:19–79.

———. 2006. Resolution of the *Thelymitra fuscolutea* R. Br. (Orchidaceae) complex of southern Australia. Muelleria 24:3–24.

———. 2009. Resolution of the *Thelymitra variegata* (Orchidaceae) complex of southern Australia and New Zealand. Muelleria 27:149–170.

———. 2011. Resolution of the *Thelymitra aristata* (Orchidaceae) complex of southeastern Australia. Muelleria 29:110–129.

———. 2013. An overview of the *Thelymitra nuda* (Orchidaceae) complex in Australia including the description of six new species. Muelleria 31: 3–30.

Jeffrey DC, J Arditti, H Koopowitz. 1970. Sugar content in floral and extrafloral exudates of orchids: pollination, myrmecology and chemotaxonomy implication. New Phytol. 69:187–195.

Jentsch A, J Kreyling, C Beierkuhnlein. 2007. A new generation of climate change experiments: events not trends. Frontiers in Ecology and Environment 5:365–374.

Jentsch A, J Kreyling, J Bottcher-Treschkow, C Beierkuhnlein. 2009. Beyond gradual warming—extreme weather events alter flower phenology of European grassland and heath species. Global Change Biology 5:837–849.

Jersáková J, SD Johnson. 2006. Lack of floral nectar reduces self-pollination in a fly-pollinated orchid. Oecologia (Berlin) 147:60–68.

Jersáková J, SD Johnson, A Jürgens. 2009. Deceptive behavior in plants. II. Food deception by plants: from generalized systems to specialized floral mimicry. Pages 223–246 in F Baluška, editor. Plant-Environment Interactions: From Sensory Plant Biology to Active Plant Behavior. Springer-Verlag, Berlin, Germany.

Jersáková J, SD Johnson, P Kindlemann. 2006a. Mechanisms and evolution of deception pollination in orchids. Biological Reviews 81:219–235.

Jersáková J, SD Johnson, P Kindlemann, AC Pupin. 2008. Effect of nectar supplementation on male and female components of pollination success in the deceptive orchid Dactylorhiza sambucina. Acta Oecologia (Berlin) 33:300–306.

Jersáková J, P Kindlemann, SS Renner. 2006b. Is the corolla colour dimorphism in Dactylorhiza sambucina maintained by differential seed viability instead of frequency-dependent selection? Folia Geobotanica 41:61–76.

Jiang D-B, H-J Wang, X-M Lang. 2005. Evaluation of East Asian climatology as simulated by seven coupled models. Advances in Atmospheric Science 22:479–495.

Johns J, B Molloy. 1983. Native Orchids of New Zealand. AH and AW Reed Ltd, Wellington, NZ.

Johnson SD. 1994a. Evidence for Batesian mimicry in a butterfly-pollinated orchid. Biological Journal of the Linnean Society 53:91–104.

———. 1994b. Red flowers and butterfly pollination in the fynbos of South Africa. Pages 137–148 in M Arianoutsou, RH Groves, editors. Groves Plant-Animal Interactions in Mediterranean-Type Ecosystems. Kluwer Academic Publishers, Dordrecht, the Netherlands.

———. 1996. Bird pollination in South African species of Satyrium (Orchidaceae). Plant Systematics and Evolution 203:91–98.

———. 1997. Pollination ecotypes of Satyrium hallackii (Orchidaceae) in South Africa. Botanical Journal of the Linnean Society 123:225–235.

———. 2000. Batesian mimicry in the non-rewarding orchid Disa pulchra, and its consequences for pollinator behaviour. Biological Journal of the Linnean Society 71:119–132.

———. 2003. Specialized pollination systems in southern Africa. South African Journal of Science 99:345–348.

———. 2010. The pollination niche and its role in the diversification and maintenance of the southern African flora. Philosophical Transactions of the Royal Society of London B Biological Sciences 365:499–516.

Johnson SD, R Alexandersson, HP Linder. 2003a. Experimental and phylogenetic evidence for floral mimicry in a guild of fly-pollinated plants. Biological Journal of the Linnean Society 80:289–304.

Johnson SD, S Andersson. 2002. A simple field method for manipulating ultraviolet reflectance of flowers. Canadian Journal of Botany 80:1325–1328.

Johnson SD, WJ Bond. 1992. Habitat dependent pollination success in a cape orchid. Oecologia (Berlin) 91:455–456.

———. 1994. Red flowers and butterfly pollination in the fynbos of South Africa. Pages 137–148 in M. Arianoutsou, RH Groves, editors. Plant-animal interactions in Mediterranean-type ecosystems. Kluwer Academic Publishers, Dordrecht, the Netherlands.

Johnson SD, M Brown. 2004. Transfer of pollinaria on bird's feet: a new pollination system in orchids. Plant Systematics and Evolution 244:181–188.

Johnson SD, PI Craig, J Ågren. 2004. The effects of nectar addition on pollen removal and geitonogamy in the non-rewarding orchid *Anacamptis morio*. Proceedings of the Royal Society of London B Biological Sciences 271:803–809.

Johnson SD, TJ Edwards. 2000. The structure and function of orchid pollinaria. Plant Systematics and Evolution 222:243–269.

Johnson SD, AL Hargreaves, M Brown. 2006. Dark, bitter-tasting nectar functions as a filter of flower visitors in a bird-pollinated plant. Ecology 87:2709–2716.

Johnson SD, A Jurgens. 2010. Convergent evolution of carrion and faecal scent mimicry in fly-pollinated angiosperm flowers and a stinkhorn fungus. South African Journal of Botany 76:796–807.

Johnson SD, WR Liltved. 1997. Hawkmoth pollination of *Bonatea speciosa* (Orchidaceae) in a South African coastal forest. Nordic Journal of Botany 17:5–10.

Johnson SD, HP Linder, KE Steiner. 1998. Phylogeny and radiation of pollination systems in *Disa* (Orchidaceae). American Journal of Botany 85:402–411.

Johnson SD, S Morita. 2006. Lying to Pinocchio: floral deception in an orchid pollinated by long-proboscid flies. Botanical Journal of the Linnean Society 152: 271–278.

Johnson SD, PR Neal, LD Harder. 2005. Pollen fates and the limits on male reproductive success in an orchid population. Biological Journal of the Linnean Society 86: 175–190.

Johnson SD, LA Nilsson. 1999. Pollen carryover, geitonogamy, and the evolution of deceptive pollination systems in orchids. Ecology 80:2607–2619.

Johnson SD, CI Peter, LA Nilsson, J Ågren. 2003b. Pollination success in a deceptive orchid is enhanced by co-occurring rewarding magnet plants. Ecology 84:2919–2927.

Johnson SD, KE Steiner. 1997. Long-tongued fly pollination and evolution of floral spur length in the *Disa draconis* complex (Orchidaceae). Evolution 51:45–53.

———. 2000. Generalization *versus* specialization in plant pollination systems. Trends in Ecology and Evolution 15:140–143.

———. 2003. Specialized pollination systems in southern Africa. South African Journal of Science 99:345–348.

Jones DL. 1970. The pollination of *Corybas diemenicus* (HMR) Rupp and WH Nicholls ex HMR Rupp. Victorian Naturalist 87:372–374.

———. 1972. The pollination of *Prasophyllum alpinum* R. Br. Victorian Naturalist 89: 260–263.

———. 1981. The pollination of selected Australian orchids. Pages 40–43 *in* L Lawler, RD Kerr, editors. Proceedings of the Orchid Symposium Held as a Satellite Function of the 13th International Botanical Congress, Sydney, Australia. The Orchid Society of New South Wales, Harbour Press, Sydney, NSW, Australia.

———. 1988. Native Orchids of Australia. Reed Books Pty Ltd, Port Melbourne, Victoria, Australia.

———. 2001. Tribe Diurideae. Pages 59–213 *in* AM Pridgeon, PJ Cribb, MW Chase, FN Rasmussen, editors. Genera Orchidacearum. Volume 2. Oxford University Press, Oxford, England.

————. 2006. A Complete Guide to Native Orchids of Australia including the Island Territories. Reed New Holland Publishers, Sydney, Australia.

Jones DL, CC Bower. 2001. *Thelymitra*. Pages 205–213 *in* A Pridgeon, P Cribb, M Chase, FN Rasmussen, editors. Genera Orchidacearum, vol. 2. Oxford University Press, Oxford, England.

Jones DL, MA Clements. 2002. A reassessment of *Pterostylis* R. Br. (Orchidaceae). Australian Orchid Research 4:3–63.

Jones DL, MA Clements, IK Sharma, AM Mackenzie. 2001. A new classification system of *Caladenia* R. Br. (Orchidaceae). Orchadian 13:389–419.

Jones DL, H Wapstra, P Tonelli, S Harris. 1999. The Orchids of Tasmania. Miegunyah Press of Melbourne University, Melbourne, Australia.

Jonsson S. 2008. Bertil Kullenberg—in memoriam. Entomologisk Tidskrift 129:121–124.

Juillet N, S Dunand-Martin, LDB Gigord. 2006. Evidence for inbreeding depression in the food-deceptive colour-dimorphic orchid *Dactylorhiza sambucina* (L.) Soo. Plant Biology 9:147–151.

Juillet N, MA Gonzalez, PA Page, LDB Gigord. 2007. Pollination of the European food-deceptive *Traunsteinera globosa* (Orchidaceae): the importance of nectar producing neighbouring plants. Plant Systematics and Evolution 265:123–129.

Juillet N, CC Salzmann, G Scopece. 2011. Does facilitating pollinator learning impede deceptive orchid attractiveness? A multi-approach test of avoidance learning. Plant Biology 13:570–575.

Juillet N, G Scopece. 2010. Does floral trait variability enhance reproductive success in deceptive orchids? Perspectives in Plant Ecology, Evolution and Systematics 12: 317–322.

Junker RR, N Blüthgen. 2010. Floral scents repel facultative flower visitors, but attract obligate ones. Annals of Botany (London) 105:777–782.

Jurgens A, S Dotterl, U Meve. 2006. The chemical nature of fetid floral odours in stapeliads (Apocynaceae-Asclepiadoideae-Ceropegieae). New Phytologist 172:452–468.

Kaiser R 1993. The Scent of Orchids. Editiones Roche, Basel, Switzerland.

————. 2011. Scent of the Vanishing Flora.Verlag Helvetica Chimica Acta, Zürich, Switzerland.

Kaiser R, C Nussbaumer. 1990. Dehydrogeosmin, a novel compound occurring in the flower scent of various species of Cactaceae. Helvetica Chimica Acta 73:133.

Kajobe R, D Roubik. 2006. Honey-making bee colony abundance and predation by apes and humans in a Uganda forest reserve. Biotropica 38:210–218.

Karlson P, M Lüscher. 1959. Pheromones: A new term for a class of biologically active substances. Nature 183:55–56.

Kay, KM, and Schemske, DW. 2003. Pollinator assemblages and visitation rates for 11 species of Neotropical *Costus* (Costaceae). *Biotropica* 35:198–207.

Kelber A, A Balkenius, EJ Warrant. 2003a. Colour vision in diurnal and nocturnal hawkmoths. Integrative and Comparative Biology 43:571–579.

Kelber A, D Osorio. 2010. From spectral information to animal colour vision: experiments and concepts. Proceedings of the Royal Society of London B Biological Sciences 277:1617–1625.

Kelber A, M Vorobyev, D Osorio. 2003b. Animal colour vision—behavioural tests and physiological concepts. Biological Reviews 78:81–118.

Kennedy H. 1978. Systematics and pollination of the "close-flowered" species of *Calathea* (Marantaceae). University of California Publications in Botany 71:1–90.

Kevan PD. 1972. Floral colors in the High Arctic with reference to insect-flower relations and pollination. Canadian Journal of Botany 50:2289–2316.

Kevan PD, WGK Backhaus. 1998. Color vision: ecology and evolution in making the best of the photic environment. Pages 163–183 *in* WGK Backhaus, R Kliegl, JS Werner, editors. Color Vision: Perspectives from Different Disciplines. W. DeGruyter GmbH and Co., Berlin, Germany.

Kevan PD, AM Baker. 1983. Insects as flower visitors. Annual Reviews of Entomology 28:407–453.

Kevan PD, M Giurfa, L Chittka. 1996. What are there so many and so few white flowers? TIPS 8:280–284.

Kevan PD, R Menzel. 2012. The plight of pollination and the interface of neurobiology, ecology and food security. Environmentalist doi:10.1007/s10669-012-9394-5.

Kimsey LS. 1982. Systematics of bees of the genus *Eufriesea* (Hymenoptera, Apidae). University of California Publications in Entomology 95:1–125.

———. 1984. The behavioral and structural aspects of grooming and related activities in euglossine bees (Hymenoptera: Apidae). Journal of Zoology (London) 204: 541–550.

Kindlmann P, J Jersàkovà. 2006. Effect of floral display on reproductive success in terrestrial orchids. Folia Geobotanica 41:47–60.

Klein Tank AMG, FW Zwiers, X Zhang. 2009. Climate Data and Monitoring: Guidelines on Analysis of Extremes in a Changing Climate in Support of Informed Decisions for Adaptation. WCDMP-No. 72, WMO-TD No. 1500. World Meteorological Organizations, Geneva, Switzerland.

Klinkhamer PGL, TJ De Jong. 1993. Attractiveness to pollinators: a plant's dilemma. Oikos 66:180–184.

Knoll F. 1858. Über den Scleudervorgang der männlichen Catasetum-Blüte. Berichte der Bayerischen Botanischen Gesellschaft zur Erforschung der Heimischen Flora? 71:337–348.

Knowles LL. 2009. Estimating species trees: methods of phylogenetic analysis when there is incongruence across genes. Systematic Biology 58:463–467.

Knudsen JT. 2002. Variation in floral scent composition within and between populations of *Geonoma macrostachys* (Arecaceae) in the western Amazon. American Journal of Botany 89:1771–1778.

Knudsen JT, R Eriksson, J Gershenzon, B Ståhl. 2006. Diversity and distribution of floral scent. Botanical Review 72:1–120.

Knudsen JT, SA Mori. 1996. Floral scents and pollination in Neotropical Lecythidaceae. Biotropica 28:42–60.

Knuth P. 1909. Handbook of Flower Pollination, vol. 3. Clarendon Press, Oxford, England.

Kocyan A, PK Endress. 2001. Floral structure and development of *Apostasia* and *Neuwiedia* (Apostasioideae) and their relationships to other Orchidaceae. International Journal of Plant Sciences 162:847–867.

Koivisto AM, E Vallius, V Salonen. 2002. Pollination and reproductive success of two colour variants of a deceptive orchid, *Dactylorhiza maculata* (Orchidaceae). Nordic Journal of Botany 22:53–58.

Koopowitz H. 2008. The Tropical Slipper Orchids Paphiopedilum and Phragmipedium Species and Hybrids. Timber Press, Portland, OR, USA.

Kores PJ, M Molvray, PH Weston, SD Hopper, AP Brown, KM Cameron, M Chase. 2001. A phylogenetic analysis of Diurideae (Orchidaceae) based on plastid DNA sequence data. American Journal of Botany 88:1903–1914. doi:10.2307/3558366.

Kores PJ, PH Weston, M Molvray, W Mark, M Chase. 2000. Phylogenetic relationships within the Diurideae (Orchidaceae); inferences from plastid *mat*K DNA sequences. Pages 449–456 *in* KL Wilson, DA Morrison, editors. Monocots: Systematics and Evolution. CSIRO Publishing, Melbourne, Victoria, Australia.

Kretzschmar H, W Eccarius, H Dietrich. 2007. The Orchid Genera Anacamptis, Orchis, Neotinea. Echinomedia, Bürgel, Germany.

Kroodsma DE. 1975. Flight distances of male euglossine bees in orchid pollination. Biotropica 7:71–72.

Kropf M, SS Renner. 2005. Pollination success in monochromic yellow populations of the rewardless orchid *Dactylorhiza sambucina*. Plant Systematics and Evolution 254:185–197.

———. 2008. Pollinator-mediated selfing in two deceptive orchids and a review of pollinium tracking studies addressing geitonogamy. Oecologia (Berlin) 155:497–508.

Kudo G, H Hiroshi Ishii, Y Hirabayashi, T Ida. 2007. A test of the effect of floral color change on pollination effectiveness using artificial inflorescences visited by bumblebees. Oecologia (Berlin) 154:119–128.

Kull T. 1998. Fruit set and recruitment in populations of *Cypripedium calceolus* L. in Estonia. Botanical Journal of the Linnean Society 126:27–38.

Kull T, J Arditti, SM Wong. 2009. Orchid Biology: Reviews and Perspectives, vol. 10. Comstock Publishers Associates, Ithaca, NY, USA.

Kullenberg B. 1949. A description in Swedish with two photographs of the visit of the male *Campsoscolia ciliata* on the flower of *Ophrys speculum*. Sveriges Natur (Svenska naturskyddsföreningens årsbok): 96–97.

———. 1950a. Investigations on the pollination of *Ophrys* species. Oikos 2:1–19.

———. 1950b. Observations sur *Ophrys* et les insectes. Bulletin de la Société des Sciences Naturelles du Maroc 38:138–141.

———. 1950c. Flugblomstret (*Ophrys insectifera*) och insekterna. Svensk faunistisk Revy 12:21–30.

———. 1951. *Ophrys insectifera* L. et les insectes. Oikos 3:53–70.

———. 1952a. Recherches sur la biologie florale des *Ophrys*. Bulletin de la Société d'Histoire Naturelle d'Afrique du Nord 43:53–62.

———. 1952b. Nouvelles observations sur les rapports entre *Ophrys* et les insectes. Bulletin de la Société des Sciences Naturelles du Maroc 32:175–179.

———. 1953. Some observations on the scents among bees and wasps (Hymenoptera). Entomologisk Tidskrift 74:1–7.

———. 1956a. On the scents and colours of *Ophrys* flowers and their specific pollinators among the Aculeate Hymenoptera. Svensk Botanisk Tidskrift 50:25–46.

———. 1956b. Field experiments with chemical sexual attractants on Aculeate Hymenoptera males I. Zoologiska Bidrag Uppsala 31:253–352.

———. 1961. Studies in *Ophrys* pollination. Zoologiska Bidrag Uppsala 34:1–340.

———. 1973a. New observations on the pollination of *Ophrys* L. (Orchidaceae). Zoon (Uppsala) Suppl. 1:9–14.

————. 1973b. Field experiments with chemical sexual attractants on Aculeate Hymenoptera males II. Zoon (Uppsala) Suppl. 1:31–42.

Kullenberg B, G Bergström. 1973. The pollination of *Ophrys* orchids. Nobel Symposium
25:253–258.

————. 1976a. The pollination of *Ophrys* orchids. Botaniska Notiser 129:11–19.

————. 1976b. Hymenoptera Aculeata males as pollinators of *Ophrys* orchids. Zoologica
Scripta 5:13–23.

Kullenberg B, A-K Borg-Karlson, A-L Kullenberg. 1984. Field studies on the behaviour of
the *Eucera nigrilabris* male in the odour flow from flower labellum extract of *Ophrys
tenthredinifera*. Nova Acta Regiae Societatis Scientiarum Upsaliensis, Serie V:C 3:
79–110.

Kunze J, A Gumbert. 2001. The combined effect of color and odor on flower choice behavior of bumble bees in flower mimicry systems. Behavioral Ecology 12:447–456.

Kurzweil H. 1987. Developmental studies in orchid flowers II: Orchidoid species.
Nordic Journal of Botany 7:443–451.

————. 1988. Developmental studies in orchid flowers III: Neottioid species. Nordic
Journal of Botany 8:271–282.

Kurzweil H, P Linder, L Stern, AM Pridgeon. 1995. Comparative vegetative anatomy
and classification of Diseae (Orchidaceae). Botanical Journal of the Linnean Society
117:171–220.

Kurzweil H, PH Weston, AJ Perkins. 2005. Morphological and ontogenetic studies
on the gynostemium of some Australian members of *Diurideae* and *Cranichideae*
(Orchidaceae). Telopea 11:11–33.

Lammi A, M Kuitunen. 1995. Deceptive pollination of *Dactylorhiza incarnata*: an experimental test of the magnet species hypothesis. Oecologia (Berlin) 101:500–503.

Lande R, DW Schemske. 1985. The evolution of self-fertilization and inbreeding depression in plants. I. Genetic models. Evolution 39:24–40.

Landwehr J. 1977. Wilde orchideën van Europa. s-Graveland, Amsterdam, the Netherlands.

Larsen MW, C Craig Peter, SD Johnson, JM Olesen. 2008. Comparative biology of pollination systems in the African-Malagasy genus *Brownleea* (Brownleeinae: Orchidaceae). Botanical Journal of the Linnean Society 156:65–78.

Lavarack PS. 1976. The taxonomic affinities of the Australian Neottioideae. Taxon 25:
289–296.

Laverty TM. 1992. Plant interactions for pollinator visits: a test of the magnet species
effect. Oecologia (Berlin) 89:502–508.

Leadbeater E, L Chittka. 2007. Social learning in insect-form miniature brain to consensus building. Current Biology 17:703–713.

Lehnebach CA, M Riveros. 2003. Pollination biology of the Chilean endemic orchid
Chloraea lamellata. Biodiversity and Conservation 12:1741–1751.

Lehnebach CA, AW Robertson, D Hedderley. 2005. Pollination studies of four New
Zealand terrestrial orchids and the implication for their conservation. New Zealand
Journal of Botany 43:467–477.

Lemey P, M Salemi, AM Vandamme. 2009. The Phylogenetic Handbook: A Practical Approach to Phylogenetic Analysis and Hypothesis Testing, 2nd edition. Cambridge
University Press, Cambridge, England.

Lentini PE, TG Martin, P Gibbons, J Fischer, SA Cunningham. 2012. Supporting wild pollinators in a temperate agricultural landscape: maintaining mosaics of natural features and production. Biological Conservation 149:84–92.

Levin RA, LA McDade, RA Raguso. 2003. The systematic utility of floral and vegetative fragrance in two genera of Nyctaginaceae. Systematic Biology 52:334–351.

Lev-Yadun S, K Gould. 2007. What do red and yellow autumn leaves signal? Botanical Reviews 73:279–289.

Li J, Z Liu, G Salazar, P Bernhardt, H Perner, Y Tomohisa, X Jin, S Chung, Y Luo. 2011. Molecular phylogeny of *Cypripedium* (Orchidaceae: Cypripedioideae) inferred from multiple nuclear and chloroplast regions. Molecular Phylogenetics and Evolution 61:308–320.

Li P, Y Luo. 2009. Reproductive biology of an endemic orchid *Cypripedium smithii* in China and reproductive isolation between *C. smithii* and *C. tibeticum*. Biodiversity Science 17:406–413.

Li P, Y Luo, P Bernhardt, Y Kou, H Perner. 2008a. Pollination of *Cypripedium plectrochilum* (Orchidaceae) by *Lasioglossum* spp. (Halictidae), the roles of generalist attractants *versus* restrictive floral architecture. Plant Biology 10:200–230.

Li P, Y Luo, P Bernhardt, X Tang, Y Kou. 2006. Deceptive pollination of the lady's slipper *Cypripedium tibeticum* (Orchidaceae). Plant Systematics and Evolution 262: 53–63.

Li P, Y Luo, Y Deng, Y Kou. 2008b. Pollination of the lady's slipper *Cypripedium henryi* Rolfe (Orchidaceae). Botanical Journal of the Linnean Society 156:491–499.

Linder HP, H Kurzweil. 1999. Orchids of Southern Africa. A. A. Balkema, Rotterdam, the Netherlands.

Lindley J. 1826. Orchidearum Sceletos. Typis Ricardi Taylor, London,England.

———. 1846. The Vegetable Kingdom; or the Structure, Classification and Uses of Plants Illustrated upon the Natural System. Bradbury and Evans, London, England.

Linnaeus C. 1737. Genera plantarum. C Wishoff, Leiden, Sweden.

———. 1753. Species plantarum. L Salvius, Stockholm, Sweden.

Lipow SR, P Bernhardt, N Vance. 2002. Comparative rates of pollination and fruit set in widely separated populations of a rare orchid (*Cypripedium fasciculatum*). International Journal of Plant Sciences 163:775–782.

Little RJ. 1983. A review of floral food deception mimicries with comments on floral mutualism. Pages 294–309 *in* CE Jones, RJ Little, editors. Handbook of Experimental Pollination Biology. Van Nostrand Reinhold, New York, USA.

Liu H. 2010. Geodorum eulophioides—challenges in conserving one of the rarest Chinese orchids. Orchids Magazine, March issue. Orchid: The Bulletin of the American Orchid Society 79:161–163.

———. 2011. Conserving orchids in the wild: the expected and unexpected difficulties of conserving a rare, attractive wild orchid. Tropical Gardens Magazine: The Fairchild Tropical Botanic Garden 66:47–48.

Liu H, C-L Feng, B-S Chen, Z-S Wang, X-Q Xie et al. 2012. Overcoming extreme weather events: successful but variable assisted translocations of wild orchids in southwestern China. Biological Conservation 150:68–75.

Liu H, C-L Feng, Y-B Luo, B-S Chen, Z-S Wang, H-Y. 2010. Potential challenges of climate change to orchid conservation in a wild orchid hotspot in Southwestern China. Botanical Review 76:174–192.

Liu KW, Z-L Liu, S-P Lei, L-Q Li, L-J Chen, Y-T Ahang. 2005. Study on pollination biology in *Paphiopedilum armeniacum* (Orchidaceae). Shenzhen Science and Technology 139:171–183.

Liu Z-J, LJ Chen, WH Rao, L-Q Li, Y-T Zhang. 2008. Correlation between numeric dynamics and reproductive behavior in *Cypripedium lentiginosum*. Acta Ecologica Sinica 28:111–121.

Liu Z-J, J-Y Zhang, Z-Z Ru, S-P Lei, L-J Chen. 2004. Conservation biology of *Paphiopedilum purpuratum* (Orchidaceae). Biodiversity Science 12:509–516.

Llandres AL, FM Gawryszewski, AM Heiling, ME Herberstein. 2011. The effect of colour variation in predators on the behaviour of pollinators: Australian crab spiders and native bees. Ecological Entomology 36:72–81.

Lloyd DC. 1952. Biological observations on some Thynnids of Western Patagonia. Bulletin of Entomological Research 42:707–719.

Lloyd DG. 1979. Some reproductive factors affecting the selection of self-fertilization in plants. American Naturalist 113:67–79.

Lopez DF. 1963. Two attractants for *Eulaema tropica* L. Journal of Economic Entomology 56:540.

Losos JB. 2008. Phylogenetic niche conservatism, phylogenetic signal and the relationship between phylogenetic relatedness and ecological similarity among species. Ecology Letters 11:995–1007.

Luangsuwalai K, S Ketsa, A Wisutiamonkul, WG van Doorn. 2008. Lack of visible post-pollination effects in pollen grains of two *Dendrobium* cultivars: relationship with pollinia ACC, pollen germination, and pollen tube growth. Functional Plant Biology 35:152–158.

Lubinsky PM, M Van Dam, A Van Dam. 2006. Pollination of *Vanilla* and evolution of Orchidaceae. Lindleyana 75:926–929.

Luer C. 1975. The Native Orchids of the United States and Canada excluding Florida. New York Botanical Garden, Bronx, New York, USA.

Lunau K. 1996. Unidirectionality of floral colour changes. Plant Systematics and Evolution 200:125–140.

———. 2000. The ecology and evolution of visual pollen signals. Plant Systematics and Evolution 222:89–111.

———. 2007. Stamens and mimic stamens as components of floral colour patterns. Botanische Jahrbücher für Systematik, Pflanzengeschichte und Pflanzengeographie 127:13–41.

Lunau K, H Knüttel. 1995. Vision through colored eyes. Naturwissenschaften 82: 432–434.

Lunau K, EJ Maier. 1995. Innate colour preferences of flower visitors. Journal of Comparative Physiology A Neuroethology Sensory Neural and Behavioral Physiology 177:1–19.

Luterbacher J, MA Liniger, A Menzel, N Estrella, PM Della-Marta. 2007. Exceptional European warmth of autumn 2006 and winter 2007: historical context, the underlying dynamics, and its phenological impacts. Geophysical Research Letter 34: L12704.

Luyt R, SD Johnson. 2001. Hawkmoth pollination of the African epiphytic orchid *Mystacidium venosum*, with special reference to flower and pollen longevity. Plant Systematics and Evolution 228:49–62.

Mabberley D. 1999. Ferdinand Bauer: The Nature of Discovery. Merell Holberton Publishers, London, England.

Maddison WP, DR Maddison. 2011. Mesquite: A Modular System for Evolutionary Analysis. Version 2.75; http://mesquiteproject.org.

Majetic CJ, RA Raguso, SJ Tonsor, T-A Ashman. 2007. Flower color–flower scent associations in polymorphic *Hesperis matronalis* (Brassicaceae). Phytochemistry 68: 865–874.

Manning A. 1967. An Introduction to Animal Behavior. Addison-Wesley Publishing Co., Reading, MA, USA.

Manning JC, P Goldblatt. 1996. The *Prosoeca peringueyi* (Diptera: Nemestrinidae) pollination guild in southern Africa: long-tongued flies and their tubular flowers. Annals of the Missouri Botanical Garden 83:67–86.

———. 1997. The *Moegistorhynchus longirostris* (Diptera: Nemestrinidae) pollination guild: long-tubed flowers and a specialized long-proboscid fly pollination system in southern Africa. Plant Systematics and Evolution 206:51–69.

Mant JG, C Brändli, NJ Vereecken, CM Schulz, W Francke, FP Schiestl. 2005a. Cuticular hydrocarbons as sex pheromone of *Colletes cunicularius* (Hymenoptera: Colletidae) and the key to its mimicry by the sexually deceptive orchid, *Ophrys exaltata*. Journal of Chemical Ecology 31:1765–1787.

Mant JG, GR Brown, PH Weston. 2005b. Opportunistic pollinator shifts among sexually deceptive orchids indicated by a phylogeny of pollinating and non-pollinating thynnine wasps (Tiphiidae). Biological Journal of the Linnean Society 86:381–395.

Mant JG, R Peakall, FP Schiestl. 2005c. Does selection on floral odor promote differentiation among populations and species of the sexually deceptive orchid genus *Ophrys*? Evolution 59:1449–1463.

Mant JG, R Peakall, PH Weston. 2005d. Specific pollinator attraction and the diversification of sexually deceptive *Chiloglottis* (Orchidaceae). Plant Systematics and Evolution 253:185–200.

Mant JG, FP Schiestl, R Peakall, PH Weston. 2002. A phylogenetic study of pollinator conservatism among sexually deceptive orchids. Evolution 56:888–898.

Marloth R. 1895. The fertilization of *Disa uniflora*, Berg, by insects. Transactions of the South African Philosophical Society, vols. 8–9: xciii–xcv.

Martini P, C Schlindwein, A Montenegro. 2008. Pollination, flower longevity, and reproductive biology of *Gongora quinquenervis* Ruiz and Pavón (Orchidaceae) in an Atlantic forest fragment of Pernambuco, Brazil. Plant Biology 5:495–503.

Martins DJ, SD Johnson. 2007. Hawkmoth pollination of aerangoid orchids in Kenya, with special reference to nectar sugar concentration gradients in the floral spurs. American Journal of Botany 94:650–659.

McAlpine DK. 1978. On the status of *Thelymitra truncata* Rogers. Orchadian 5:179–181.

McCook L. 1989. Systematics of *Phragmipedium* (Cypripedioideae: Orchidaceae). PhD diss., Cornell University, New York, USA.

McCrae DP, BPJ Molloy. 1998. The artificial reconstruction of the New Zealand hybrid *Theylmitra × dentata* (Orchidaceae). Pages 121–125 *in* R Lynch, editor. Ecosystems, Entomology and Plants. The Royal Society of New Zealand Miscellaneous Series 48, Wellington, NZ.

McLachlan JS, JJ Hellman, MW Schwart. 2007. A framework for debate of assisted migration in an era of climate change. Conservation Biology 21:297–302.

McLean CA, A Moussalli, D Stuart-Fox. 2010. The predation cost of female resistance. Behavioral Ecology 21:861–867.

McMullen CK. 1987. Breeding systems of selected Galapagos Islands angiosperms. American Journal of Botany 74:1694–1705.

Meisel JE, CL Woodward. 2005. Andean orchid conservation and the role of private lands: a case study from Ecuador. Selbyana 26:49–57.

Memmott J, PG Craze, MW Nickolas, MV Price. 2007. Global warming and the disruption of plant-pollinator interactions. Ecology Letters 10:710–717.

Menzel A. 2005. A 500 year pheno-climatological view on the 2003 heatwave in Europe assessed by grape harvest dates. Meteorologische Zeitschrift 14:75–77.

Menzel A, H Seifert, N Estrella. 2011. Effects of recent warm and cold spells on European plant phenology. International Journal of Biometeorology doi:10.1007/s00 484-011-0466-x.

Menzel A, TH Sparks, N Estrella, E Koch, AAR Ahas, KM Alm-Kublwe, P Bissolli, O Braslavska, A Briede, FM Chimielewski, Z Crepinsek, Y Curnel, A Dahl, C Defila, A Donnelly, Y Filella, K Jatszak, F Mage, A Mestre, O Nordli, H Van Vliet, F Wielgolaski, S Zach, A Zust. 2006. European phenological response to climate change matches the warming pattern. Global Change Biology 12:1969–1976.

Menzel R. 1985. Learning in honey bees in an ecological and behavioral context. Pages 55–74 in B Holldobler, M Lindauer, editors. Experimental Behavioral Ecology. Gustav Fischer, Stuttgart, Germany.

Menzel R, W Backhaus. 1991. Colour vision in insects. Pages 262–293 in P Gouras, editor. The Perception of Colour. Macmillan, London, England.

Menzel R, M Blakers. 1976. Colour receptors in the bee eye: morphology and spectral sensitivity. Journal of Comparative Physiology A Neurethology Sensory Neural and Behavioral Physiology 108:11–13.

Micheneau C. 2005. Systématique moléculaire de la sous-tribu des Angraecinae: perspectives taxonomiques et implications de la relation plantes-pollinisateurs dans l'évolution des formes florales. PhD diss., Université de La Réunion, Réunion, France.

Micheneau C, BS Carlsward, MF Fay, B Bytebier, T Pailler, MW Chase. 2008a. Phylogenetics and biogeography of Mascarene angraecoid orchids (Vandeae, Orchidaceae). Molecular Phylogenetics and Evolution 46:908–922.

Micheneau C, J Fournel, A Gauvin-Bialecki, T Pailler 2008b. Auto-pollination in a long-spurred endemic orchid (*Jumellea stenophylla*) on Reunion Island (Mascarene Archipelago, Indian Ocean). Plant Systematics and Evolution 272:11–22.

Micheneau C, J Fournel, L Humeau, T Pailler. 2008c. Orchid-bird interactions: a case study from *Angraecum* (Vandeae, Angraecinae) and *Zosterops* (white-eyes, Zosteropidae) on Reunion Island. Botany 86:1143–1151.

Micheneau C, J Fournel, T Pailler. 2006. Bird pollination in an angraecoid orchid on Reunion Island (Mascarene Archipelago, Indian Ocean). Annals of Botany 97: 965–974.

Micheneau C, J Fournel, BH Warren, S Hugel, A Gauvin-Bialecki, T Pailler, D Strasberg, MW Chase. 2010. Orthoptera, a new order of pollinator. Annals of Botany 105: 355–364.

Micheneau C, SD Johnson, MF Fay. 2009. Orchid pollination: from Darwin to the present day. Botanical Journal of the Linnean Society 161:1–19.

Michener CD. 2007. The Bees of the World, 2nd edition. Johns Hopkins University Press, Baltimore, MD, USA.

Miller-Rushing AJ, RB Primack. 2008. Global warming and flowering times in Thoreau's Concord: a community perspective. Ecology 89:332–341.

Moeller DA. 2005. Pollinator community structure and sources of spatial variation in plant-pollinator interactions in *Clarkia xantiana* ssp *xantiana*. Oecologia (Berlin) 142:28–37.

Molloy B. 1990. Pollination systems of New Zealand native orchids. Pages 36–56 *in* I St. George, D McCrae, editors. The New Zealand Orchids: Natural History and Cultivation. New Zealand Native Orchid Group, Dunedin, NZ.

Molloy B, MI Dawson. 1998. Speciation in *Thelymitra* (Orchidaceae) by natural hybridism and amphidiploidy: ecosystems, entomology and plants; proceedings of a symposium held at Lincon University to mark the retirement of Bryony Macmillan, John Dugdale, Peter Wardel, and Brian Molloy, September 1, 1995. Royal Society of New Zealand Miscellaneous Series 48:103–113.

Molvray M, P Kores. 1995. Character analysis of the seed coat in the Spiranthoideae and Orchidoideae, with special reference to the Diurideae (Orchidaceae). American Journal of Botany 82:1443–1454.

Moore LB, E Edgar. 1970. Flora of New Zealand, vol. 2: Indigenous Tracheophyta: Monocotyledones except Gramineae. A. R. Shearer, Government Printer, Wellington, NZ.

Morellato LPC. 2003. South America. Pages 75–92 in MD Schwartz, editor. Phenology: An Integrative Environmental Science. Kluwer Academic Publishers, Dordrecht, the Netherlands.

Morse DH. 1981. Modification of bumblebee foraging: the effect of milkweed pollinia. Ecology 62:89–97.

Moya S, JD Ackerman. 1993. Variation in the floral fragrance of *Epidendrum ciliare* (Orchidaceae). Nordic Journal of Botany 13:41–47.

Müller F. 1871. Application of the Darwinian theory to flowers and the insects which visit them. American Naturalist 5: 271–297.

Müller H. 1883. The Fertilisation of Flowers. Macmillan and Co., London, England.

Murren C. 2002. Effects of habitat fragmentation on pollination: pollinators, pollinia viability and reproductive success. Journal of Ecology 90:100–107.

Naug D, HS Arathi. 2007. Receiver bias for exaggerated signals in honeybees and its implications for the evolution of floral displays. Biology Letters 3:635–637.

Neal PR, A Dafni, M Giurfa. 1998. Floral symmetry and its role in plant-pollinator systems: terminology, distribution, and hypotheses. Annual Review of Ecology and Systematics 29:345–373.

Neiland MRM, CC Wilcock. 1995. Maximisation of reproductive success by European Orchidaceae under conditions of infrequent pollination. Protoplasma 187:39–48.

———. 1998. Fruit set, nectar reward, and rarity in the Orchidaceae. American Journal of Botany 85:1657–1671.

Nelson G, N Platnick. 1981. Systematics and Biogeography: Cladistics and Vicariance. Columbia University Press, New York, USA.

Nemésio A. 2005. Fluorescent colors in orchid bees (Hymenoptera: Apidae). Neotropical Entomology 34:933–936.

Neumeyer C. 1980. Simultaneous color contrast in the honeybee. Journal of Comparative Physiology A 139:165–176.

Newman E, B Anderson, SD Johnson. 2012. Flower colour adaptation in a mimetic orchid. Proceedings of the Royal Society of London B Biological Sciences 279: 2309–2313.

Nicholls WH. 1969. Orchids of Australia. Thomas Nelson, Melbourne, Victoria, Australia.

Nicholson CC, JW Bales, JE Palmer-Fortune, RG Nicholson. 2008. Darwin's bee-trap: the kinetics of *Catasetum*, a New World orchid. Plant Signaling and Behavior 3: 19–23.

Nilsson LA. 1979. Anthecological studies on the lady's slippers, *Cypripedium calceolus* (Orchidaceae). Botaniska Notiser 132:329–347.

———. 1980. The pollination ecology of *Dactylorhiza sambucina* (Orchidaceae). Botaniska Notiser 133:367–385.

———. 1983a. Mimesis of bellflower (*Campanula*) by the red helleborine orchid *Cephalanthera rubra*. Nature 305:799–800.

———. 1983b. Anthecology of *Orchis mascula* (Orchidaceae). Nordic Journal of Botany 3:157–79.

———. 1984. Anthecology of *Orchis morio* (Orchidaceae) at its outpost in the North. Nova Acta Regiae Societatis Scientiataum Upsaliensis, Seq.V C 3:166–179.

———. 1988. The evolution of flowers with deep corolla tubes. Nature 334:147–149.

———. 1992. Orchid pollination biology. Trends in Ecology and Evolution 7:255–259.

Nilsson LA, L Jonsson, L Ralison, E Randrianjohany. 1985. Monophily and pollination mechanisms in *Angraecum arachnites* Schltr. (Orchidaceae) in a guild of long-tongued hawk-moths (Sphingidae) in Madagascar. Biological Journal of the Linnean Society 26:1–19.

———. 1987. Angraecoid orchids and hawkmoths in central Madagascar: specialized pollination systems and generalist foragers. Biotropica 19:310–318.

Nilsson LA, E Rabakonandrianina. 1988. Hawk-moth scale analysis and pollination specialization in the epilithic Malagasy endemic *Aerangis ellisii* (Reichenb. fil.) Schltr. (Orchidaceae). Botanical Journal of the Linnean Society 97:49–61.

Nilsson LA, E Rabakonandrianina, B Pettersson. 1992. Exact tracking of pollen transfer and mating in plants. *Nature* 360:666–668.

Norden B, SWT Batra. 1985. Male bees sport black mustaches for picking up parsnip perfume (Hymenoptera: Anthophoridae). Proceedings of the Entomological Society of Washington 87:317–322.

O'Connell LM, MO Johnston. 1998. Male and female pollination success in a deceptive orchid, a selection study. Ecology 79:1246–1260.

Oldroyd BP, S Wongsiri. 2006. Asian Honey Bees—Biology, Conservation, and Human Interactions. Harvard University Press, Cambridge, MA, USA.

Olesen JM, A Valido. 2003. Lizards as pollinators and seed dispersers: an island phenomenon. Trends in Ecology and Evolution 18:177–181.

Ollerton J, SD Johnson, L Cranmer, S Kellie. 2003. The pollination ecology of an assemblage of grassland asclepiads in South Africa. Annals of Botany 92:807–834.

Ollerton J, SD Johnson, AB Hingston. 2006. Geographical variation in diversity and specificity of pollination systems. Pages 283–308 *in* NM Waser, J Ollerton, editors. Plant-Pollinator Interactions: From Specialization to Generalization. University of Chicago Press, Chicago, IL, USA.

Osorio D, M Vorobyev. 2005. Photoreceptor sectral sensitivities in terrestrial animals: adaptations for luminance and colour vision. Proceedings of the Royal Society of London B Biological Sciences 272:1745–1752.

Otten CF. 1979. Ophelia's "long purples" or "dead men's fingers." Shakespeare Quarterly 30:397–402.

Pacek A, M Stpiczyńska, KL Davies, G Szymczak. 2012. Floral elaiophore structure in four representatives of the *Ornithocephalus* clade (Orchidaceae: Oncidiinae). Annals of Botany 110:4809–4820.

Paige KN, TG Whitham. 1985. Individual and population shifts in flower color by scarlet gilia: a mechanism for pollinator tracking. Science 227:315–317.

Pailler T, L Humeau, J Figier, JD Thompson. 1998. Reproductive biology of the functionally dioecious and morphologically heterostylous island endemic *Chassalia coralliodes* (Rubiaceae). Biological Journal of the Linnean Society 64:297–313.

Parmesan C. 2006. Ecological and evolutionary responses to recent climate change. Annual Review in Ecology, Evolution and Systematics 37:637–669.

Parmesan C, N Ryrholm, C Stefanescu, JK Hill, CD Thomas, H Descimon, B Huntley, L Kalla, J Kullberg, T Tammaru, WJ Tennent, JA Thomas, M Warren. 1999. Poleward shifts in geographical ranges of butterfly species associated with regional warming. Nature 399:579–583.

Parmesan C, G Yohe. 2003. A globally coherent fingerprint of climate change impacts across natural systems. Nature 421:37–42.

Parra-Tabla V, MF Vargas. 2007. Flowering synchrony and floral display size affect pollination success in a deceit-pollinated tropical orchid. Acta Oecologia (Berlin): 32:26—35.

Paulus HF. 2006. Deceived males—pollination biology of the Mediterranean orchid genus *Ophrys* (Orchidaceae). Journal Europäischer Orchideen 38:303–353.

———. 2007. Wie insekten-männchen von orchideenblüten getäuscht werden—bestäubungstricks und evolution in der mediterranen ragwurzgattung *Ophrys*. Denisia 20 Neue Serie 66:255–294.

Paulus HF, C Gack. 1990. Pollination of *Ophrys* (Orchidaceae) in Cyprus. Plant Systematics and Evolution 169:177–207.

Pauw A. 2006. Floral syndromes accurately predict pollination by a specialized oil-collecting bee (*Rediviva peringueyi*, Melittidae) in a guild of South African orchids (Coryciinae). American Journal of Botany 93:917–926.

Pauw A, WJ Bond. 2011. Mutualisms matter: pollination rate limits the distribution of oil-secreting orchids. Oikos 120:1531–1538.

Pauw A, JA Hawkins. 2011. Reconstruction of historical pollination rates reveals linked declines of pollinators and plants. Oikos 120:344–349.

Pauw A, J Stofberg, RJ Waterman. 2009. Flies and flowers in Darwin's race. Evolution 63:268–279.

Peakall R. 1987. Genetic systems of Australian terrestrial orchids. Doctoral thesis, Department of Botany, University of Western Australia, Perth, Western Australia.

———. 1989a. A new technique for monitoring pollen flow in orchids. Oecologia (Berlin) 79:361–365.

———. 1989b. The unique pollination of *Leporella fimbriata* (Orchidaceae): pollination by pseudocopulating male ants (*Myrmecia urens*, Formicidae). Plant Systematics and Evolution 167:137–148.

———. 1990. Responses of male *Zaspilothynnus trilobatus* Turner wasps to females and the sexually deceptive orchid it pollinates. Functional Ecology 4:159–167.

Peakall R, AJ Beattie. 1989. Pollination of the orchid *Microtis parviflora* R. Br. by flightless worker ants. Functional Ecology 3:515–522.

————. 1996. Ecological and genetic consequences of pollination by sexual deception in the orchid *Caladenia tentaculata*. Evolution 50:2207–2220.

Peakall R, D Ebert, J Poldy, RA Barrow, W Francke, CC Bower, FP Schiestl. 2010. Pollinator specificity, floral odour chemistry and the phylogeny of Australian sexually deceptive *Chiloglottis* orchids: implications for pollinator-driven speciation. New Phytologist 188:437–450.

Peakall R, SN Handel. 1993. Pollinators discriminate among floral heights of a sexually deceptive orchid: implication for selection. Evolution 46:1681–1687.

Peakall R, FP Schiestl. 2004. A mark-recapture study of male *Colletes cunicularius* bees: implications for pollination by sexual deception. Behavioral Ecology and Sociobiology 56:579–584.

Peisl P, J Forster. 1975. Zur Bestaubungsbiologie des Knaubenkrautes *Orchis coriophora* L. ssp. *fragrans*. Die Orchidee 26:172–176.

Peitsch D, A Fietz, H Hertel, J Souza, DF Ventura, R Menzel. 1992. The spectral input systems of hymenopteran insects and their receptor-based colour vision. Journal of Comparative Physiology A 170:23–40.

Pellegrino G, F Bellusci, A Musacchio. 2008. A double floral mimicry and the magnet species effect in dimorphic co-flowering species, the deceptive orchid *Dactylorhiza sambucina* and rewarding *Viola aethnensis*. Preslia 80:411–22.

Pellegrino G, D Caimi, ME Noce, A Musacchio. 2005. Effects of local density and flower color polymorphism on pollination and reproduction in the rewardless orchid *Dactylorhiza sambucina*. Plant Systematics and Evolution 251:119–129.

Pellissier L, P Vittoz, AI Internicola, LDB Gigord. 2010. Generalized food-deceptive orchid species flower earlier and occur at lower altitude than rewarding ones. Journal of Plant Ecology 3:243–250.

Pellmyr O. 1986. The pollination ecology of two nectarless *Cimicifuga* sp. (Ranunculaceae) in North America. Nordic Journal of Botany 6:713–723.

Pemberton RW. 2010. Biotic resource needs of specialist orchid pollinators. Botanical Review 76:275–292.

————. 2011. Pollination studies in Phragmipediums: flower fly (Syrphidae) pollination and mechanical self-pollination (autogamy) in *Phragmipedium* species (Cypripedioideae). Orchids 80:364–367.

————. 2013. Pollination of slipper orchids (Cypripedioideae): a review. Lankesteriana 13:65–73.

Pemberton RW, GS Wheeler. 2006. Orchid bees don't need orchids: evidence from the naturalization of an orchid bee in Florida. Ecology 87:1995–2001.

Perez F, MTK Arroyo, JJ Armesto. 2009. Evolution of autonomous selfing accompanies increased specialization in the pollination system of *Schizanthus* (Solanaceae). American Journal of Botany 96:1168–1176.

Pérez-Hernández H, A Damon, J Valle-Mora, D Sánchez-Guillen. 2011. Orchid pollination: specialization or chance? Botanical Journal of the Linnaean Society 165: 251–266.

Perner H. 2008. *Sinopedilum*—a new section of the genus *Cypripedium*. Die Orchidee 59: 35–51.

Perraudin F, J Popovici, C Bertrand. 2006. Analysis of headspace-solid microextracts from flowers of *Maxillaria tenuifolia* Lindl. by GC-MS. Electronic Journal of Natural Substances 1:1–5. http://ejns.univ-lyon1.fr.

Peter CI, G Coombs, CF Huchzermeyer, N Venter, AC Winkler, D Hutton, LA Papier,

AP Dold, SD Johnson. 2009. Confirmation of hawkmoth pollination in *Habenaria epipactidea*: leg placement of pollinaria and crepuscular scent emission. South African Journal of Botany 75:744–750.

Peter CI, SD Johnson. 2006. Doing the twist: a test of Darwin's cross-pollination hypothesis for pollination reconfiguration. Biology Letters 2:65–68.

———. 2008. Mimics and magnets: the importance of color and ecological facilitation in floral deception. Ecology 89:1583–1595.

Petit S, CR Dickson. 2005. Grass-tree (*Xanthorrhoea semiplana*, Liliaceae) facilitation of the endangered pink-lipped spider orchid (*Caladenia* syn. *Arachnorchis behrii*, Orchidaceae) varies in South Australia. Australian Journal of Botany 53:455–464.

Petit S, M Jusaitis, D Bickerton. 2009. Effect of pollen load, self-pollination and plant size on seeds and germination in the endangered pink-lipped spider orchid, *Caladenia behrii*. Australian Journal of Botany 57:307–314.

Pettersson B, LA Nilsson. 1993. Floral variation and deceit pollination in *Polystachya rosea* (Orchidaceae) on an inselberg in Madagascar. Opera Botanica 121:237–245.

Pfeifer M, K Wiegand, W Heinrich, G Jetschke. 2006. Long-term demographic fluctuations in an orchid species driven by weather: implications for conservation planning. Journal of Applied Ecology 43:313–324.

Phillips RD, G Backhouse, AP Brown, SD Hopper. 2009a. Biogeography of *Caladenia* (Orchidaceae), with special reference to the South-West Australian Floristic Region. Australian Journal of Botany 57:259–275.

Phillips RD, AP Brown, KW Dixon, SD Hopper. 2011. Orchid biogeography and factors associated with rarity in a biodiversity hotspot, the Southwest Australian Floristic Region. Journal of Biogeography 38:487–501.

Phillips RD, R Faast, CC Bower, GR Brown, R Peakall. 2009b. Implications of pollination by food and sexual deception for pollinator specificity, fruit set, population genetics and conservation of *Caladenia* (Orchidaceae). Australian Journal of Botany 57:287–306.

Phillips RD, T Xu, MF Hutchinson, KW Dixon, R Peakall. 2013. Convergent specialization—the sharing of pollinators by sympatric genera of sexually deceptive orchids. Journal of Ecology 101:826–835.

Pohl M, T Watolla, K Lunau. 2008. Anther-mimicking floral guides exploit a conflict between innate preference and learning in bumblebees (*Bombus terrestris*). Behavioral Ecology and Sociobiology 63:295–302.

Pohl NB, J Van Wyk, DR Campbell. 2011. Butterflies show flower colour preferences but not constancy in foraging at four plant species. Ecological Entomology 36:290–300.

Pohlman CL, AB Nicotra, BR Murray. 2005. Geographic range size, seedling ecophysiology and phenotypic plasticity in Australian *Acacia* species. Journal of Biogeography 32:341–351.

Pokorny J, S Shevell, VC Smith. 1991. Colour appearance and colour constancy. Pages 43–61 *in* P Gouras, editor. The Perception of Colour. Macmillan, London, England.

Potgieter CJ, TJ Edwards. 2005. The *Stenobasipteron wiedemanni* (Diptera, Nemestrinidae) pollination guild in Eastern Southern Africa. Annals of the Missouri Botanical Garden 92:254–267.

Pouyanne M. 1917. La fécondation des *Ophrys* par les insectes. Bulletin de la Société d'Histoire Naturelle d'Afrique du Nord 43:6–7.

Pridgeon AM. 1993. Systematic leaf anatomy of *Caladenia* (Orchidaceae). Kew Bulletin 48:533–543.

————. 1994. Systematic leaf anatomy of Caladeniinae (Orchidaceae). Botanical Journal of the Linnean Society 114:31–48.

Pridgeon AM, RM Bateman, AV Cox, JR Hapemann, MW Chase. 1997. Phylogenetics of subtribe Orchidinae (Orchidoideae, Orchidaceae) based on nuclear ITS sequences. 1. Intergeneric relationships and polyphyly of *Orchis sensu lato*. Lindleyana 12: 89–109.

Pridgeon AM, MW Chase. 1995. Subterranean axes in tribe Diurideae (Orchidaceae): morphology, anatomy, and systematic significance. American Journal of Botany 82:1473–1495.

Pridgeon AM, PJ Cribb, MW Chase, FN Rasmussen, editors. 1999. Genera Orchidacearum, vol. 1. Oxford University Press, Oxford, England.

————, editors. 2001. Genera Orchidacearum, vol. 2. Oxford University Press, Oxford, England.

————, editors. 2003. Genera Orchidacearum, vol. 3. Oxford University Press, Oxford, England.

Priesner E. 1973. Reaktionen von Riechrezeptoren männlicher Solitärbienen (Hymenoptera, Apoidea) auf Inhaltsstoffe von *Ophrys*-Blüten. Zoon Suppl. 1:43–54.

Primack R, E Stacy. 1998. Cost of reproduction in the pink lady's slipper orchid (*Cypripedium acaule*, Orchidaceae): an eleven-year experimental study of three populations. American Journal of Botany 85:1672–1679.

Pringle JWS. 1983. Insect Flight. Carolina Biological Supply Company, Burlington, NC, USA.

Proctor HC. 1998. Effect of pollen age on fruit set, fruit weight, and seed set in three orchid species. Canadian Journal of Botany 76:420–427.

Proctor M, P Yeo, A Lack. 1996. The Natural History of Pollination. Timber Press Inc., Portland, OR, USA.

Quammen D. 2006. The Reluctant Mr. Darwin: An Intimate Portrait of Charles Darwin and the Making of His Theory of Evolution. Great Discoveries. WW Norton and Co., New York, USA.

Raguso RA. 2008. Wake up and smell the roses: the ecology and evolution of floral scent. Annual Review of Ecology, Evolution, and Systematics 39:549–569.

Ramírez SR, RL Dressler, M Ospina. 2002. Abejas euglosinas (Hymenoptera: Apidae) de la región neotropical: listado de especies con notas sobre su biología. Biota Colombiana 3:7–118.

Ramírez SR, T Eltz, MK Fujiwara, G Gerlach, B Goldman-Huertas, ND Tsutsui, NE Pierce. 2011a. Orchid bees and their orchids. Science 333:1676.

————. 2011b. Asynchronous diversification in a specialized plant-pollinator mutualism. Science 333:1742–1746.

Ramírez SR, B Gravendeel, RB Singer, CR Marshall, NE Pierce. 2007. Dating the origin of the Orchidaceae from a fossil orchid with its pollinator. Nature 448:1042–1045.

Ramírez SR, DW Roubik, C Skov, NE Pierce. 2009. Phylogeny, diversification patterns and historical biogeography of euglossine orchid bees (Hymenoptera: Apidae). Biological Journal of the Linnean Society 100:552–572.

Rasmussen FN. 1982. The gynostemium of the neotteoid orchids. Opera Botanica 65: 1–96.

Rasmussen HN. 2008. Terrestrial Orchids: From Seed to Mycotrophid Plant. Cambridge University Press, New York, USA.

Rausher MD. 2008. Evolutionary transitions in floral color. International Journal of Plant Sciences 169:7–21.

Rayment T. 1935. A Cluster of Bees. Endeavour Press, Sydney, NSW, Australia.

Reich PB. 1995. Phenology of tropical forests—patterns, causes, and consequences. Canadian Journal of Botany 73:164–174.

Reisenman C, M Giurfa. 2008. Chromatic and achromatic stimulus discrimination of long wavelength (red) visual stimuli by the honeybee *Apis mellifera*. Arthropod-Plant Interactions 2:137–146.

Ren Z, D Li, P Bernhardt, H Wang. 2011. Flowers of *Cypripedium fargesii* (Orchidaceae) fool flat-footed flies (Platypezidae) by faking fungus-infected foliage. Proceedings of the National Academy of Sciences of the United States of America 108:7478–7480.

Renner SS. 1983. The widespread occurrence of anther destruction of *Trigona* bees in Melastomataceae. Biotropica 15:251–256.

———. 2006. Rewardless flowers in the angiosperms and the role of insect cognition in their evolution. Pages 123–144 in NM Waser, J Ollerton, editors. Plant-Pollinator Interactions: From Specialization to Generalization. University of Chicago Press, Chicago, IL, USA.

Richards AJ. 1986. Plant Breeding Systems. George Allen and Unwin, London, England.

Riches JT. 1876. Fertilization in *Cypripedium* (lady's slipper). Hardwicke's Science-Gossip: an Illustrated Medium of Interchange and Gossip; for Students and Lovers of Nature 192:125–126. Hardwicke and Bogue, London, England.

Ridsdill Smith TJ. 1970a. The behaviour of *Hemithynnus hyalinatus* (Hymenoptera: Tiphiidae), with notes on some other Thynninae. Journal of the Australian Entomological Society 9:196–208.

———. 1970b. The biology of *Hemithynnus hyalinatus* (Hymenoptera: Tiphiidae), a parasite of scarabeid larvae. Australian Journal of Entomology 9:183–195.

Rieseberg LH, JH Willis. 2007. Plant speciation. Science 317:910–914.

Rindal E, AVZ Brower. 2011. Do model-based phylogenetic analyses perform better than parsimony? A test with empirical data. Cladistics 27:331–334.

Rodríguez I, A Gumbert, N de Ibarra, J Kunze, M Giurfa. 2004. Symmetry is in the eye of the "beeholder": innate preference for bilateral symmetry in flower-naïve bumblebees. Naturwissenschaften 91:374–377.

Rodríguez-Robles JA, EJ Meléndez, JD Ackerman. 1992. Effects of display size, flowering phenology, and nectar availability on effective visitation frequency in *Comparettia falcata* (Orchidaceae). American Journal of Botany 79:1009–1017.

Rogers RS. 1913. Mechanisms of pollination in certain Australian orchids. Transactions and Proceedings of the Royal Society of South Australia 36:48–64.

Roguenant A, A Raynal-Roques,Y Sell. 2005. Un Amour d'Orchidée: Le Mariage de la Fleur et de l'Insecte. Belin, Paris, France.

Romero GA, CE Nelson. 1986. Sexual dimorphism in *Catasetum* orchids: forcible pollen emplacement and male flower competition. Science 232:1538–1540.

Rothschild LW, K Jordan. 1903. A revision of the lepidopterous family Sphingidae. Novitates Zoologicae 9:1–972.

Roubik DW. 1982. The ecological impact of nectar-robbing bees and pollinating hummingbirds on a tropical shrub. Ecology 63:354–360.

———. 1989. Ecology and Natural History of Tropical Bees. Cambridge University Press, New York, USA.

———. 1993. Tropical pollinators in the canopy and understory: field data and theory for stratum "preferences." Journal of Insect Behavior 6:659–673.

———. 1998. Grave-robbing by male *Eulaema* (Hymenoptera, Apidae): implications for euglossine biology. Journal of the Kansas Entomological Society 71:188–191.

———. 2000. Deceptive orchids with Meliponini as pollinators. Plant Systematics and Evolution 222:271–279.

———. 2001. Ups and downs in pollinator populations: when is there a decline? Conservation Ecology 5; consecol.org/v015/iss1/art2.

———. 2002. The value of bees to the coffee harvest. Nature 417:708.

———. 2004. Long-term studies of solitary bees: what the orchid bees are telling us. Pages 97–103 *in* BM Freitas, JO Pereira, editors. Solitary Bees—Conservation, Rearing and Management for Pollination. Imprensa Universitaria, Fortaleza, Brazil.

———. 2006. Stingless bee nesting biology. Apidologie 37:124–143.

———. 2012. Ecology and social organisation of bees. *In* eLS. John Wiley and Sons, Ltd, Chichester, England. doi:10.1002/9780470015902.a0023596.

———. 2013. Why they keep changing the names of our stingless bees (Hymenoptera: Apidae; Meliponini): a little history and guide to taxonomy. Pages 1–7 *in* P Vit, D Roubik, editors. Stingless Bees Process Honey and Pollen in Cerumen Pots. Facultad de Farmacia y Bioanálisis, Universidad de los Andes, Mérida, Venezuela. http://www.saber.ula.ve/handle/123456789/35292.

Roubik DW, JD Ackerman. 1987. Long-term ecology of euglossine orchid-bees (Apidae: Euglossini) in Panama. Oecologia (Berlin) 73:321–333.

Roubik DW, JMF Camargo. 2012. The Panama microplate, island studies and relictual species of *Melipona* (*Melikerria*)/(Hymenoptera: Apidae: Meliponini). Systematic Entomology 37:189–199; doi: 10.1111/j.1365-3113.2011.00587.x.

Roubik DW, PE Hanson. 2004. Orchid Bees of Tropical America: Biology and Field Guide. InBio (National Institute of Biodiversity), Heredia, Costa Rica.

Roubik DW, S Sakai, F Gattesco. 2003. Canopy flowers and certainty: loose niches revisited. Pages 360–368 *in* Y Basset, R Kitching, S Miller, V Novotny, editors. Tropical Forest Arthropods, Spatio-Temporal Dynamics and Resource Use in the Canopy. Cambridge University Press, New York, USA.

Roubik DW, D Yanega, M Aluja-S, SL Buchmann, DW Inouye. 1995. On optimal nectar foraging by some tropical bees (Hymenoptera, Apidae). Apidologie 26:197–211.

Rundle HD, P Nosil. 2005. Ecological speciation. Ecology Letters 8:336–352.

Rupp HMR. 1942. The Orchids of New South Wales. Australian Medical Publishing Company Limited, Glebe, NSW, Australia.

Rutishauser T, J Luterbacher, C Defila, D Frank, FH Wanner. 2008. Swiss spring plant phenology 2007: extremes, a multi-century perspective, and changes in temperature sensitivity. Geophysical Research Letters 35:L05703.

Ruxton GD, M Schaefer. 2009. The timing of food-deceptive flowers: a commentary on Internicola et al. Journal of Evolutionary Biology 22:1133–1136.

———. 2011. Alternative explanations for apparent mimicry. Journal of Ecology 99: 899–904.

Sakai AK, SG Weller, WL Wagner, M Nepokroeff, TM Culley. 2006. Adaptive radiation and evolution of breeding systems in *Schiedea* (Caryophyllaceae), an endemic Hawaiian genus. Annals of the Missouri Botanical Garden 93:49–63.

Salazar GA, M Chase, MA Soto Arenas, M Ingrouille. 2003. Phylogenetics of Cranichideae with emphasis on Spiranthinae (Orchidaceae, Orchidoideae): evidence from plastid and nuclear DNA sequences. American Journal of Botany 90:777–795.

Salzmann CC, S Cozzolino, FP Schiestl. 2007a. Floral scent in food-deceptive orchids: species specificity and sources of variability. Plant Biology 9:720–729.

Salzmann CC, AM Nardella, S Cozzolino, FP Schiestl. 2007b. Variability in floral scent in rewarding and deceptive orchids: the signature of pollinator-imposed selection? Annals of Botany (London) 100:757–765.

Sanchez-Azofeifa A, M Kalacska, M Quesada, KE Stoner, JA Lobo, P Arroyo-Mora. 2003. Tropical dry climates. Pages 121–138 in MD Schwartz, editor. Phenology: An Integrative Environmental Science. Kluwer Academic Publishers, Dordrecht, the Netherlands.

Sargent RD. 2004. Floral symmetry affects speciation rates in angiosperms. Proceedings of the Royal Society of London B Biological Sciences 271:603–608.

Sauquet H, PH Weston, CL Anderson, NP Barker, DJ Cantrill, AR Mast, V Savolainen. 2009. Contrasted patterns of hyperdiversification in Mediterranean hotspots. Proceedings of the National Academy of Sciences of the United States of America 106:221–225.

Scanlen E. 2008. Thrips as *Thelymitra* pollinators. New Zealand Native Orchids 108:31.

Scanlen E, I St. George. 2010. Colour Field Guide to the Native Orchids of New Zealand, 2nd edition. New Zealand Native Orchid Group Inc., Wellington, NZ.

Schaefer HM. 2010. Visual communication: evolution, ecology, and functional mechanisms. Pages 3–28 in P Keppeler, editor. Animal Behaviour: Evolution and Mechanisms. Springer, Heidelberg, Germany.

Schaefer HM, G Rolshausen. 2006. Plants on red alert: do insects pay attention? BioEssays 28:65–71.

Schaefer HM, GD Ruxton. 2010. Communication theory and the form of receiver mediated selection. Trends in Ecology and Evolution 25:383–384.

Schaefer HM, V Schaefer, DJ Levey. 2004. How plant-animal interactions signal new insights in communication. Trends in Ecology and Evolution 19:577–584.

Schaefer HM, DM Wilkinson. 2004. Red leaves, insects and coevolution: a red herring? Trends in Ecology and Evolution 19:616–618.

Schiestl FP. 2004. Floral evolution and pollinator mate choice in a sexually deceptive orchid. Journal of Evolutionary Biology 17:67–75.

———. 2005. On the success of a swindle: pollination by deception in orchids. Naturwissenschaften 92:255–264.

———. 2010. The evolution of floral scent and insect chemical communication. Ecology Letters 13:643–656.

Schiestl FP, M Ayasse. 2001. Post-pollination emission of a repellent compound in a sexually deceptive orchid: a new mechanism for maximising reproductive success? Oecologia (Berlin) 126:531–534.

———. 2002. Do changes in floral odor cause speciation in sexually deceptive orchids? Plant Systematics and Evolution 234:111–119.

Schiestl FP, M Ayasse, HF Paulus, D Erdmann, W Francke. 1997. Variation of floral scent emission and postpollination changes in individual flowers of *Ophrys sphegodes* subsp. *sphegodes*. Journal of Chemical Ecology 23:2881–2895.

Schiestl FP, M Ayasse, HF Paulus, C Löfstedt, BS Hansson, F Ibarra, W Francke. 1999. Orchid pollination by sexual swindle. Nature 399:421–422.

———. 2000. Sex pheromone mimicry in the early spider orchid (*Ophrys sphegodes*): patterns of hydrocarbons as the key mechanism for pollination by sexual deception. Journal of Comparative Physiology A 186:567–574.

Schiestl FP, S Cozzolino. 2008. Evolution of sexual mimicry in the Orchidinae: the role of preadaptations in the attraction of male bees as pollinators. BMC Evolutionary Biology 8:27.

Schiestl FP, SD Johnson, RA Raguso. 2010. Floral evolution as a figment of the imagination of pollinators. Trends in Ecology and Evolution 25:382–383.

Schiestl FP, E Marion-Poll. 2002. Detection of physiologically active flower volatiles using gas chromatography coupled with electroantennography. Pages 173–198 in JF Jackson, HF Linskens, editors. Analysis of Taste and Aroma. Springer, Berlin, Germany.

Schiestl FP, R Peakall. 2005. Two orchids attract different pollinators with the same floral odour compound: ecological and evolutionary implications. Functional Ecology 19:674–680.

Schiestl FP, R Peakall, JG Mant, F Ibarra, C Schulz, S Franke, W Francke. 2003. The chemistry of sexual deception in an orchid-wasp pollination system. Science 302: 437–438.

Schiestl FP, DW Roubik. 2002. Odor compound detection in male euglossine bees. Journal of Chemical Ecology 29:253–257.

Schiestl FP, PM Schlüter. 2009. Floral isolation, specialized pollination, and pollinator behavior in orchids. Annual Review of Entomology 54:425–446.

Schlechter R. 1926. Das System der Orchidaceen. Notizblatt der Botanischer Gartens zu Berlin 9:590–591.

Schlumpberger BO, A Jux, M Kunert, W Boland, D Wittmann. 2004. Musty-earthy scent in cactus flowers: characteristics of floral scent production in dehydrogeosmin producing cacti. International Journal of Plant Sciences 165:1007–1015.

Schlüter PM, S Xu, V Gagliardini, E Whittle, J Shanklin, U Grossniklaus, FP Schiestl. 2011. Stearoyl-acyl carrier protein desaturases are associated with floral isolation in sexually deceptive orchids. Proceedings of the National Academy of Sciences of the United States of America 108:5696–5701.

Schmid R. 1969. On the pollination of *Policycnis barbata* (Stanhopeinae) by the euglossine bee *Eulaema speciosa*. Orchid Digest 33:220–223.

Schromburgk RH. 1837. On the identity of three supposed genera of orchidaceous epiphytes. Transactions of the Linnaean Society of London 17:551–552.

Schueller SK. 2004. Self-pollination in island and mainland populations of the introduced hummingbird-pollinated plant, *Nicotiana glauca* (Solanaceae). American Journal of Botany 91:672–681.

Schuh RT, AVZ Brower. 2009. Biological Systematics: Principles and Applications, 2nd edition. Cornell University Press, Ithaca, NY, USA.

Schwartz MD, editor. 2003. Phenology: An Integrative Environmental Science. Kluwer Academic Publishers, Dordrecht, the Netherlands.

Schwartz MD, R Ahas, A Aasa. 2006. Onset of spring starting earlier across the Northern Hemisphere. Global Change Biology 12:343–351.

Scopece G, S Cozzolino, SD Johnson, FP Schiestl. 2010. Pollination efficiency and the evolution of specialized deceptive pollination systems. American Naturalist 175: 98–105.

Scopece G, JM Müller, FP Schiestl, S Cozzolino. 2009. Pollinator attraction in *Anacamptis papilionacea* (Orchidaceae): a food or a sex promise? Plant Species Biology 24: 109–114.

Scopece G, N Musacchio, A Widmer, S Cozzolino. 2007. Patterns of reproductive isolation in Mediterranean deceptive orchids. Evolution 61:2623–2642.

Scopece G, A Widmer, S Cozzolino. 2008. Evolution of postzygotic reproductive isolation in a guild of deceptive orchids. American Naturalist 171:315–326.

Scourse N. 1983. The Victorians and Their Flowers. Timber Press, Portland, OR, USA.

Servettaz O, ML Bino, P Grunanger. 1994. Labellum micromorphology in the *Ophrys bertolinii* agg. and some related taxa (Orchidaceae). Plant Systematics and Evolution 189:123–131.

Shangguan F-Z, J Cheng, Y-X Xiong, Y-B Luo. 2008. Deceptive pollination of an autumn flowering *Eria coronaria* (Orchidaceae). Biodiversity Science 16:477–483.

Shi J, J Cheng, D Luo, Z-S Shangguan, Y-B Luo. 2007. Pollination syndromes predict brood-site deceptive pollination by female hoverflies in *Paphiopedilum dianthum* (Orchidaceae). Acta Phytotaxonomica Sinica 45:551–560.

Shi J, J Cheng, F-Z Shangguan, Y-B Luo, Z-H Deng. 2008. Study of pollination of *Paphiopedilum dianthum* in China. Orchideen Journal Heft 3:100–105.

Shi J, Y-B Luo, P Bernhardt, JC Ran, ZJ Liu, Q Zhou. 2009a. Pollination by deceit in *Paphiopedilum barbigerum* (Orchidaceae): a staminode exploits the innate color preferences of hoverflies (Syrphidae). Plant Biology 11:17–28.

Shi J, Y-B Luo, J Cheng, FZ Shangguan, ZH Deng. 2009b. The pollination of *Paphiopedilum hirsutissimum*. Orchid Review 117:78–81.

Shreeves G, J Field. 2008. Parental care and sexual size dimorphism in wasps and bees. Behavioral Ecology and Sociobiology 62:843–852.

Shuttleworth A, SD Johnson. 2010. The missing stink: sulphur compounds can mediate a shift between fly and wasp pollination systems. Proceedings of the Royal Society of London B Biological Sciences 277:2811–2819.

———. 2012. The *Hemipepsis* wasp-pollination system in South Africa: a comparative analysis of trait convergence in a highly specialized plant guild. Botanical Journal of the Linnean Society 168:278–299.

Siegel C. 2011. Darwin and his love affair with orchids. Orchid Digest 75:60–71.

Silvertown J, K McConway, D Gowing, M Dodd, MF Fay, JA Joseph, K Dolphin. 2006. Absence of phylogenetic signal in the niche structure of meadow plant communities. Proceedings of the Royal Society of London B Biological Sciences 273:39–44.

Singer RB. 2002. The pollination mechanism in *Trigonidium obtusum* Lindl. (Orchidaceae: Maxillariinae): sexual mimicry and trap flowers. Annals of Botany (London) 89:157–163.

———. 2003. Orchid pollination: on recent developments from Brazil. Lankesteriana 7: 111–114.

Singer RB, AA Cocucci. 1999a. Pollination in southern Brazilian orchids which are exclusively or mainly pollinated by halictid bees. Plant Systematics and Evolution 217:101–117.

———. 1999b. Pollination mechanisms in four sympatric southern Brazilian Epidendroideae orchids. Lindleyana 14:47–56.

Singer RB, B Gravendeel, H Cross, SR Ramirez. 2008. The use of orchid pollinia or pollinaria for taxonomic identification. Selbyana 29:6–19.

Singer RB, M Koehler. 2003. Notes on the pollination biology of *Notylia nemorosa* (Orchidaceae): do pollinators necessarily promote cross-pollination? Journal of Plant Research 116:19–25.

Singer RB, AJ Marsaioli, A Flach, MG Reis. 2006. The ecology and chemistry of pollination in Brazilian orchids: recent advances. Pages 569–582 *in* J Teixeira da Silva, editor. Floriculture, Ornamental and Plant Biotechnology. Global Science Books, Isleworth, England.

Singer RB, M Sazima. 2000. The pollination of *Stenorrhynchos lanceolatus* (Aublet) L. C. Rich. (Orchidaceae: Spiranthinae) by hummingbirds in southern Brazil. Plant Systematics and Evolution 223:221–227.

Smith BH, GA Wright, C Daly. 2006. Learning-based recognition and discrimination of floral odors. Pages 263–295 *in* NA Dudareva, E Picheresky, editors. Biology of Floral Scent. CRC Press, Boca Raton, FL, USA.

Smith GE. 1829. Catalogue of Rare or Remarkable Phaenogamous Plants, Collected in South Kent: With Descriptive Notices and Observations. Longman, Rees, Orme, Brown and Green, Paternoster Row, London,England.

Smithson A. 2002. The evolution of rewardlessness in orchids: reward supplementation experiments with *Anacamptis morio*. American Journal of Botany 89:1579–1587.

———. 2006. Pollinator limitation and inbreeding depression in orchid species with and without nectar rewards. New Phytologist 169:419–430.

———. 2009. A plant's view of cheating in plant-pollinator mutualisms. Israel Journal of Plant Sciences 57:151–163.

Smithson A, LDB Gigord. 2001. Are there fitness advantages in being a rewardless orchid? Reward supplementation experiments with *Barlia robertiana*. Proceedings of the Royal Society of London B Biological Sciences 268:1435–1441.

———. 2003. The evolution of empty flowers revisited. American Naturalist 161: 537–552.

Smithson A, N Juillet, MR Macnair, LDB Gigord. 2007. Do rewardless orchids show a positive relationship between phenotypic diversity and reproductive success? Ecology 88:434–442.

Smithson A, MR Macnair. 1996. Frequency-dependent selection by pollinators: mechanisms and consequences with regard to behaviour of bumblebees *Bombus terrestris* (L.) (Hymenoptera: Apidae). Journal of Evolutionary Biology 9:571–588.

———. 1997. Negative frequency-dependent selection by pollinators on artificial flowers without rewards. Evolution 51:715–723.

Snow AA, DF Whigham. 1989. Costs of flower and fruit production in *Tipularia discolor* (Orchidaceae). Ecology 70:1286–1293.

Soto MA, GA Salazar, E Hagsater. 1990. *Phragmipedium xerophyticum*, una nueva especie del sureste de Mexico. Orquidea 12:1–10.

———. 1997. *Phragmipedium xerophyticum*: a new species from southeastern Mexico. North American Native Orchid Journal 3:341–355.

Spaethe J, WH Moser, H Paulus. 2007. Increase of pollinator attraction by means of a visual signal in the sexually deceptive orchid, *Ophrys heldreichii* (Orchidaceae). Plant Systematics and Evolution 264:31–40.

Spaethe J, M Streinzer, HF Paulus. 2010. Why sexually deceptive orchids have colored flowers. Communicative and Integrative Biology 3:139–141.

Spaethe J, J Tautz, L Chittka. 2001. Visual constraints in foraging bumblebees: flower size and color affect search time and flight behavior. Proceedings of the National

Academy of Sciences of the United States of America 98:3898–3903.Sprengel KC. 1793. Das entdekte Geheimniss der Natur im Bau und der Befruchtung der Blumen. Friedrich Vieweg, Berlin, Germany.

———. 1996. Discovery of the secret of nature in the structure and fertilization of flowers. Translated by Peter Haase. Pages 3–43 in DG Lloyd, SCH Barrett, editors. Floral Biology: Studies on Floral Evolution in Animal-Pollinated Plants. Chapman and Hall, New York, USA.

Squirrell J, M Hollingsworth, RM Bateman, MC Tebbitt, ML Hollingsworth. 2002. Taxonomic complexity and breeding system transitions: conservation genetics of *Epipactis leptochila* complex (Orchidaceae). Molecular Ecology 11:1957–1964.

Ställberg-Stenhagen S, E Stenhagen, G Bergström. 1973. Analytical techniques in pheromone studies. Zoon (Uppsala) Suppl. 1:77–82.

Stebbins GL. 1950. Variation and Evolution in Plants. Columbia University Press, New York, USA.

———. 1957. Self-fertilization and population variability in the higher plants. American Naturalist 91:337–354.

———. 1970. Adaptative radiation of reproductive characteristics in angiosperms. I: Pollination mechanisms. Annual Review of Ecology and Systematics 1:307–326.

———. 1974. Flowering Plants: Evolution above the Species Level. Belknap Press of Harvard University Press, Cambridge, MA, USA.

Steiner KE. 1989. The pollination of *Disperis* (Orchidaceae) by oil-collecting bees in southern Africa. Lindleyana 4:164–183.

———. 1998. The evolution of beetle pollination in a South African orchid. American Journal of Botany 85:1180–1193.

———. 2001. Pollination [of subtribe Coryciinae]. Pages 20–21 in AM Pridgeon, PJ Cribb, MW Chase, FN Rasmussen, editors. Genera Orchidacearum, vol. 2. Oxford University Press, Oxford, England.

———. 2010. Twin oil sacs facilitate the evolution of a novel type of pollination unit (meranthium) in a South African orchid. American Journal of Botany 97:311–323.

Steiner KE, B Cruz. 2009. Hybridization between two oil-secreting orchids in South Africa. Plant Systematics and Evolution 277:233–243.

Steiner KE, R Kaiser, S Dötterl. 2011. Strong phylogenetic effects on floral scent variation of oil-secreting orchids in South Africa. American Journal of Botany 98: 1663–1679.

Steiner KE, VB Whitehead. 1988. The association between oil-producing flowers and oil-collecting bees in the Drakenberg of southern Africa. Monographs in Systematic Botany from the Missouri Botanical Garden 25:259–277.

———. 1990. Pollinator adaptation to oil-secreting flowers—*Rediviva* and *Diascia*. Evolution 44:1701–1707.

Steiner KE, VB Whitehead, SD Johnson. 1994. Floral and pollinator divergence in two sexually deceptive South African orchids. American Journal of Botany 81:185–194.

Stephens DW, JR Krebs. 1986. Foraging Theory. Princeton University Press, Princeton, NJ, USA.

Stern WL, MW Morris, WS Judd, AM Pridgeon, RL Dressler. 1993. Comparative vegetative anatomy and systematics of Spiranthoideae (Orchidaceae). Protoplasma 172: 49–55.

Stevens PF. 1984. Metaphors and typology in the development of botanical systematics 1690–1960, or the art of putting new wine in old bottles. Taxon 33:169–211.

Stökl J, HF Paulus, A Dafni, C Schulz, W Francke, M Ayasse. 2005. Pollinator attracting odour signals in sexually deceptive orchids of the *Ophrys fusca* group. Plant Systematics and Evolution 254:105–120.

Stökl J, PM Schlüter, TF Stuessy, HF Paulus, G Assum, M Ayasse. 2008a. Scent variation and hybridization cause the displacement of a sexually deceptive orchid species. American Journal of Botany 95:472–481.

Stökl J, PM Schlüter, TF Stuessy, HF Paulus, R Fraberger, D Erdmann, CSchulz, W Francke, G Assum, M Ayasse. 2009. Speciation in sexually deceptive orchids: pollinator-driven selection maintains discrete odour phenotypes in hybridizing species. Biological Journal of the Linnean Society 98:439–451.

Stökl J, R Twele, DH Erdmann, W Francke, M Ayasse. 2008b. Comparison of the flower scent of the sexually deceptive orchid *Ophrys iricolor* and the female sex pheromone of its pollinator *Andrena morio*. Chemoecology 17:231–233.

Stone R. 2008. Ecologists report huge storm losses in China's forests. Science 319: 1318–1319.

———. 2010. Severe drought puts spotlight on Chinese dams. Science 327:1311.

Stoutamire WP. 1967. Flower biology of the lady's slippers (Orchidaceae: *Cypripedium*). Michigan Botanist 3:107–119.

———. 1983. Wasp-pollinated species of *Caladenia* (Orchidaceae) in South-western Australia. Australian Journal of Botany 31:383–394.

Stpiczyńska M, KL Davies, A Gregg. 2003. Nectary structure and nectar secretion in *Maxillaria coccinea* (Jacq.) L. O. Williams ex Hodge (Orchidaceae). Annals of Botany 93:87–95.

Strasberg D, M Rouget, DM Richardson, S Baret, J Dupont, RM Cowling. 2005. An assessment of habitat diversity and transformation on La Réunion Island (Mascarene Islands, Indian Ocean) as a basis for identifying broad-scale conservation priorities. Biodiversity and Conservation 14:3015–3032.

Streinzer M, T Ellis, HF Paulus, J Spaethe. 2010. Visual discrimination between two sexually deceptive *Ophrys* species by a bee pollinator. Arthropod-Plant Interactions 4:141–148.

Streinzer M, HF Paulus, J Spaethe. 2009. Floral colour signal increases short-range detectability of a sexually deceptive orchid to its bee pollinator. Journal of Experimental Biology 212:1365–1370.

Sugiura N, M Goubar, K Kitamura, K Inoue. 2002. Bumblee pollination of *Cypripedium macranthos* var. *rebunense* (Orchidaceae); a possible case of floral mimicry of *Pedicularis schistostegia* (Orobanchaceae). Plant Systematics and Evolution 235:189–195.

Sultan SE. 2001. Phenotypic plasticity for fitness components in *Polygonum* species of contrasting ecological breadth. Ecology 82: 328–343.

Summerhayes VS. 1966. African orchids XXX. Kew Bulletin 20:165–199.

Sun H, J Cheng, F Zhang, Y-B Luo, S Ge. 2009. Reproductive success of non-rewarding *Cypripedium japonicum* benefits from low spatial dispersion pattern and asynchronous flowering. Annals of Botany 103:1227–1237.

Swarts ND, KW Dixon. 2009. Terrestrial orchid conservation in the age of extinction. Annals of Botany 104:543–556.

Swarts ND, E Sinclair, S Krauss, K Dixon. 2009. Genetic diversity in fragmented populations of the critically endangered spider orchid *Caladenia huegelii*: implications for conservation. Conservation Genetics 10:1199–1208.

Sydes MA, DM Calder. 1993. Comparative reproductive biology of two sun-orchids; the vulnerable *Thelymitra circumsepta* and the widespread *T. ixioides* (Orchidaceae). Australian Journal of Botany 41:577–589.

Tengö J. 1979. Odour-released behaviour in *Andrena* male bees (Apoidea, Hymenoptera). Zoon (Uppsala) 7:15–48.

Tengö J, G Bergström. 1975. All-*trans*-farnesyl hexanoate and geranyl-octanoate in the Dufour gland secretion of *Andrena* (Hymenoptera: Apidae). Journal of Chemical Ecology 1:253–268.

———. 1976. Comparative analyses of lemon-smelling secretions from heads of *Andrena* F. (Hymenoptera, Apoidea) bees. Comparative Biochemistry and Physiology 55:179–188.

———. 1977. Comparative analyses of complex secretions from heads of *Andrena* bees (Hym., Apoidea). Comparative Biochemistry and Physiology 57:197–202.

Théry M, J Casas. 2002. Predator and prey views of spider camouflage. Nature 415:133.

Théry M, M Debut, G Gomez, J Jérôme Casas. 2005. Specific color sensitivities of prey and predator explain camouflage in different visual systems. Behavioral Ecology 16:25–29.

Théry M, D Gomez, J Casas, JS Stephen. 2010. Insect colours and visual appearance in the eyes of their predators. Advances in Insect Physiology 38:267–353.

Thompson JD. 1978. Effect of stand composition on insect visitation in two-species mixtures of *Hieracium*. American Midland Naturalist 100:431–440.

———. 1981. Spatial and temporal components of resource assessment by flower-feeding insects. Journal of Animal Ecology 50:49–59.

Torretta JP, NE Goniz, SS Allscioni, ME Bello. 2011. Biología reproductiva de *Gomesa bifolia* (Orchidaceae, Cymbidieae, Oncidiinae). Darwiniana 49:16–24.

Tovée MJ. 1995. Ultra-violet photoreceptors in the animal kingdom: their distribution and function. Trends in Ecology and Evolution 10:455–460.

Tremblay RL. 1992. Trends in the pollination ecology of the Orchidaceae: evolution and systematics. Canadian Journal of Botany 70:642–650.

Tremblay RL, JD Ackerman. 2007. Floral color patterns in a tropical orchid: are they associated with reproductive success? Plant Species Biology 22:95–105.

Tremblay RL, JD Ackerman, JK Zimmerman, R Calvo. 2005. Variation in sexual reproduction in orchids and its evolutionary consequences: a spasmodic journey to diversification. Biological Journal of the Linnean Society 84:1–54.

Tremblay RL, M-E Perez, M Larcombe, A Brown, J Quarmby, D Bickerton, G French, A Bould. 2009a. Dormancy in *Caladenia*: a Bayesian approach to evaluating latency. Australian Journal of Botany 57:340–350.

———. 2009b. Population dynamics of *Caladenia*: Bayesian estimates of transition and extinction probabilities. Australian Journal of Botany 57:351–360.

Trimen R. 1864. On the fertilization of *Disa grandiflora*. Journal of the Linnean Society of London Botany 7:144–147.

Tripp EA, PS Manos. 2008. Is floral specialization an evolutionary dead-end? Pollination system transitions in *Ruellia* (Acanthaceae). Evolution 62:1712–1737.

Troje N. 1993. Spectral categories in the learning behaviour of blowflies. Zeitschrift für Naturforschung C 48:96–104.

Tso I-M, C-W-L Lin, Y En-Cheng. 2004. Colourful orb-weaving spiders, *Nephila pilipes*, through a bee's eyes. Journal of Experimental Biology 207:2631–2637.

Tucker S. 1997. Floral evolution, development, and convergence: the hierarchical-significance hypothesis. International Journal of Plant Sciences 158 (6) Suppl.: S143-S161.

Tupac OJ, JD Ackerman, P Bayman. 2002. Diversity and host specificity of endophytic *Rhizoctonia*-like fungi from tropical orchids. American Journal of Botany 89:1852–1858.

Tupac OJ, NS Flanagan, EA Herre, JD Ackerman, P Bayman. 2007. Widespread mycorrhizal specificity correlates to mycorrhizal function in the Neotropical, epiphytic orchid *Ionopsis utricularioides* (Orchidaceae). American Journal of Botany 94:1944–1950.

Tyler-Whittle MS. 1970. The Plant Hunters. Chilton Book Co., Philadelphia, PA, USA.

Ulherr J. 1967. A note on the pollination of *Caladenia alba* R. Br. Orchadian 2:94 and 108.

Valdivia CE, MA Cisternas, GS Verdugo. 2011. Reproductive biology aspects of two species of the genus *Gavilea* (Orchidaceae, Chloraeinae) in populations from Central Chile. Gayana Botánica 67:44–51.

Vallius E. 2000. Position-dependent reproductive success of flowers in *Dactylorhiza maculata* (Orchidaceae). Functional Ecology 14:573–579.

Vallius E, A Lammi, M Kuitunen. 2007. Reproductive success of *Dactylorhiza incarnata* ssp. *incarnata* (Orchidaceae): the effects of population size and plant visibility. Nordic Journal of Botany 25:183–189.

Valterová I, J Kunze, A Gumbert, A Luxová, I Liblikas, B Kalinová, A-B Borg-Karlson. 2007. Male bumble bee pheromonal components in the scent of deceit pollinated orchids: unrecognized pollinator cues? Arthropod-Plant Interactions 1:137–145.

van den Berg C, DH Goldman, JV Freudenstein, AM Pridgeon, KM Camerson, MW Chase. 2005. An overview of the phylogenetic relationships within Epidendroideae inferred from multiple DNA regions and recircumscriptions of Epidendreae and Arethuseae (Orchidaceae). American Journal of Botany 92:613–624.

van der Cingel NA. 2001. An Atlas of Orchid Pollination—America, Africa, Asia and Australia. A. A. Balkema Rotterdam, the Netherlands.

van der Niet T, DM Hansen, SD Johnson. 2011. Carrion mimicry in a South African orchid: flowers attract a narrow subset of the fly assemblage on animal carcasses. Annals of Botany 107:981–992.

van der Niet T, SD Johnson. 2009. Patterns of plant speciation in the Cape floristic region. Molecular Phylogenetics and Evolution 51:85–93.

———. 2012. Phylogenetic evidence for pollinator-driven diversification of angiosperms. Trends in Ecology and Evolution 27:353–361.

van der Niet T, SD Johnson, HP Linder. 2006. Macroevolutionary data suggest a role for reinforcement in pollination system shifts. Evolution 60:1596–1601.

van der Niet T, WR Liltved, SD Johnson. 2011. More than meets the eye: a morphological and phylogenetic comparison of long-spurred, white-flowered *Satyrium* species (Orchidaceae) in South Africa. Botanical Journal of the Linnean Society 166: 417–430.

van der Pijl L, CH Dodson. 1966. Orchid Flowers, Their Pollination and Evolution. University of Miami Press, Miami, FL, USA.

Vandewoestijne S, AS Róis, A Caperta, M Baguette, D Tyteca. 2009. Effects of individual and population parameters on reproductive success in three sexually deceptive orchid species. Plant Biology 11:454–463.

Vereecken NJ. 2009. Deceptive behavior in plants. I. Pollination by sexual deception in orchids: a host-parasite perspective. Pages 203–222 *in* Baluška F, editor. Plant-Environment Interactions. Springer-Verlag, Berlin, Germany.

Vereecken NJ, S Cozzolino, FP Schiestl. 2010a. Hybrid floral scent novelty drives pollinator shift in sexually deceptive orchids. BMC Evolutionary Biology 10:103–114.

Vereecken NJ, A Dafni, S Cozzolino. 2010b. Pollination syndromes in Mediterranean orchids—implications for speciation, taxonomy and conservation. Botanical Review 76:220–240.

Vereecken NJ, JG Mant, FP Schiestl. 2007. Population differentiation in female sex pheromone and male preferences in a solitary bee. Behavioral Ecology and Sociobiology 61:811–821.

Vereecken NJ, JN McNeil. 2010. Cheaters and liars: chemical mimicry at its finest. Canadian Journal of Zoology 88:725–752.

Vereecken NJ, FP Schiestl. 2008. The evolution of imperfect floral mimicry. Proceedings of the National Academy of Sciences of the United States of America 105:7484–7488.

———. 2009. On the roles of colour and scent in a specialized floral mimicry system. Annals of Botany 104:1077–1084.

Vereecken NJ, M Streinzer, M Ayasse, J Spaethe, HF Paulus, J Stökl, P Cortis, FP Schiestl. 2011. Integrating past and present studies on *Ophrys* pollination—a comment on Bradshaw et al. Botanical Journal of the Linnean Society 165:329–335.

Vereecken NJ, J Tengö, G Kullenberg, G Bergström. 2009. Bertil Kullenberg (1913–2007) and his *Ophrys* orchids. Journal Europaïscher Orchidëen 41:3–18.

Vermeulen P. 1947. Studies on Dactylorchids. Schotanus and Jens, Utrecht, Sweden.

Vignolini S, MP Davey, RM Bateman, PJ Rudall, E Moyroud, J Tratt, S Malmgren, U Steiner, and BJ Glover. 2012. The mirror crack'd: both pigment and structure contribute to the glossy blue appearance of the mirror orchid, Ophrys speculum. New Phytologist 196:1038–1047.

Vogel S. 1954. Blütenbiologische Typen als Elemente der Sippengliederug, dargestellt anhand der Flora Südafrikas. Fischer, Jena, Germany.

———. 1959. Organographie der Blüten kapländischer Ophrydeen. Akademie der Wissenschaften und der Literatur in Mainz, Abhandlungen der Mathematisch-Naturwissenschaftlichen Klasse 6–7:1–268.

———. 1961. Die bestäubung der kesselfallen-blüten von *Ceropegia*. Beitrage zur Biologie der Pflanzen 36:159–237.

———. 1963a. Das sexuelle anlockungsprinzip der catasetinen- und stanhopeen-blüten und die wahre function ihres sogenannten futtergewebes. Öesterreichische Botanische Zeitschrift 100: 308–337.

———. 1963b. Duftdrüsen im Dienste der Bestäubung: Über Bau und Funktion der Osmophoren. Akademie der Wissenschaften und der Literatur in Mainz, Abhandlungen der Mathematisch-Naturwissenschaftlichen Klasse 10:600–763.

———. 1966. Parfümsammelnde bienen als bestäuber von orchidaceen und *Gloxinia*. Öesterreichische Botanische Zeitschrift 113:302–361.

———. 1972. Pollination von *Orchis papilionacea* L. in den Schwarmbahnen von *Eucera tuberculata* F. Jahresberichte des Naturwissenschaftlichen Vereines in Wuppertal 85: 67–74.

———. 1983. Ecophysiology of zoophilic pollination. Pages 560–611 in OL Lange, PS Nobel, CB Osmond, H Ziegler, editors. Encyclopedia of Plant Physiology, Physiological Plant Ecology, new series, vol. 12c. Springer, Berlin, Germany.

———. 1990. The role of scent glands in pollination: on the structure and function of osmophores. Smithsonian Institution, Libraries and National Science Foundation, Washington, DC, USA.

Vogt CA. 1990. Pollination of *Cypripedium reginae* (Orchidaceae). Lindleyana 5:145–150.

Vorobyev M. 1999. Evolution of flower colors—a model against experiments; reply to comments by Chittka. Naturwissenschaften 86:598–600.

Vorobyev M, R Brandt, D Peitsch, SB Laughlin, R Menzel. 2001. Colour thresholds and receptor noise: behaviour and physiology compared. Vision Research 41:639–653.

Vorobyev M, D Osorio. 1998. Receptor noise as a determinant of colour thresholds. Proceedings of the Royal Society of London B Biological Sciences 265:351–358.

Vöth W. 1975. *Trielis villosa var. rubra*, Bestüber von *Orchis coiophora*. Die Orchidee 26: 170–171.

———. 1982. Die ausgeborgten Bestauber von *Orchis pallens* L. Die Orchidee 33:196–203.

Wakakuwa M, M Kurasawa, M Giurfa, K Arikawa. 2005. Spectral heterogeneity of honeybee ommatidia. Naturwissenschaften 92:464–467.

Wallace AR. 1867. Creation by law. Quarterly Journal of Science 4:470–488.

———. 1871. Contributions to the Theory of Natural Selection, 2nd edition. Macmillan, London, England.

Walther GR, E Post, P Convey, A Menzel, C Parmesan, TJC Beebee, JM Fromentin, O Hoegh-Guldberg, F Bairlein. 2002. Ecological responses to recent climate change. Nature 416:389–395.

Wan M, editor. 1986. Natural Calendar of China I. Science Press, Beijing, China.

———. 1987. Natural Calendar of China II. Science Press, Beijing, China.

Wang L, Q Zhang, Y Chen, D-Y Gong. 2008. Changes of warmer winter and winter temperature over China during 1956–2005. Advances in Climate Change Research 4 Suppl.:18–21.

Warrant EJ. 2008. Seeing in the dark: vision and visual behaviour in nocturnal bees and wasps. Journal of Experimental Biology 211:1737–1746.

Warrant EJ, NA Locket. 2004. Vision in the deep sea. Biological Reviews 79:671–712.

Warren BH, E Bermingham, RP Prys-Jones, C Thebaud. 2006. Immigration, species radiation and extinction in a highly diverse songbird lineage: white-eyes on Indian Ocean islands. Molecular Ecology 15:3769–3786.

Waser NM. 1983. Competition for pollination and floral character differences among sympatric species: a review of evidence. Pages 277–293 in CE Jones, RJ Little, editors. Handbook of Experimental Pollination Biology. Van Nostrand Reinhold Co., New York, USA.

Wasserthal LT. 1997. The pollinators of the Malagasy star orchids *Angraecum sesquipedale, A. sororium* and *A. compactum* and the evolution of extremely long spurs by pollinator shift. Botanica Acta 110:343–359.

———. 1998. Deep flowers for long tongues. Trends in Ecology and Evolution 13: 459–460.

Waterman RJ, MI Bidartondo. 2008. Deception above, deception below: linking pollination and mycorrhizal biology of orchids. Journal of Experimental Botany 59: 1085–1096.

Waterman RJ, MI Bidartondo, J Stofberg, JK Combs, G Gebauer, B Savolainen, TG Barraclough, A Pauw. 2011. The effects of above- and belowground mutualisms on orchid speciation and coexistence. American Journal of Botany 177:E54–E68.

Waterman RJ, A Pauw, TG Barraclough, V Savolainen. 2009. Pollinators underestimated: a molecular phylogeny reveals widespread floral convergence in oil-secreting orchids (sub-tribe Coryciinae) of the Cape of South Africa. Molecular Phylogenetics and Evolution 51:100–110.

Weale JPM. 1869. Notes on the structure and fertilization of the genus *Bonatea*, with a special description of a species found at Bedford, South Africa. Botanical Journal of the Linnean Society 10:470–476.

———. 1873a. Notes on a species of *Disperis* found on the Kagaberg, South Africa. Journal of the Linnean Society of London Botany 7:42–45.

———. 1873b. Notes on some species of *Habenaria* found in South Africa. Journal of the Linnean Society of London Botany 8:47–48.

———. 1873c. Some observations on the fertilization of *Disa macrantha*. Journal of the Linnean Society of London Botany 8:45–47.

Weiss MR. 1995. Floral color change: a widespread functional convergence. American Journal of Botany 82:167–185.

———. 2001. Vision and learning in some neglected pollinators: beetles, flies, moths, and butterflies. *In* L Chittka, J Thomson, editors. Cognitive Ecology of Pollination: Animal Behavior and Floral Evolution. Cambridge University Press, Cambridge, England.

Welsford M, SD Johnson. 2012. Solitary and social bees as pollinators of *Wahlenbergia* (Campanulaceae): single-visit effectiveness, overnight sheltering and responses to flower colour. Arthropod-Plant Interactions 6:1–14.

Wentersdorf KP. 1978. Hamlet: Ophelia's long purples. Shakespeare Quarterly 29: 413–417.

Wertlen AM, C Niggebrügge, M Vorobyev, N Hempel de Ibarra. 2008. Detection of patches of coloured discs by bees. Journal of Experimental Biology 211:2101–2104.

Weston PH, NP Barker. 2006. A new suprageneric classification of the Proteaceae, with an annotated checklist of genera. Telopea 11:314–344.

Weston PH, AJ Perkins, TJ Entwisle. 2005. More than symbioses: orchid ecology, with examples from the Sydney Region. Cunninghamia 9:1–15.

Whigham DF, JP O'Neill, HN Rasmussen, BA Caldwell, MK McCormick. 2006. Seed longevity in terrestrial orchids—potential for persistent *in situ* seed banks. Biological Conservation 129:24–30.

Whitehead VB, KE Steiner. 2001. Oil-collecting bees of the winter rainfall area of South Africa (Melittidae, *Rediviva*). Annals of the South African Museum 108:143–277.

Whitehead VB, KE Steiner, CD Eardley. 2008. Oil collecting bees mostly of the summer rainfall area of Southern Africa (Hymenoptera: Melittidae: *Rediviva*). Journal of the Kansas Entomogical Society 81:122–141.

Whittall JB, SA Hodges. 2007. Pollinator shifts drive increasingly long nectar spurs in columbine flowers. Nature 447:706–709.

Whitten WM, MA Blanco, NH Williams, S Koehler, G Carnevali, RB Singer, L Endara, KM Neubig. 2007. Molecular phylogenetics of *Maxillaria* and related genera (Orchidaceae: Cymbidieae) based on combined molecular data. American Journal of Botany 94:1860–1889.

Whitten WM, NH Williams. 1992. Floral fragrances of *Stanhopea* (Orchidaceae). Lindleyana 7:130–153.

Whitten WM, NH Williams, WS Armbruster, WS Battiste, MA Strekovski, N Linquist. 1986. Carvone oxide: an example of convergent evolution in euglossine pollinated plants. Systematic Botany 11:222–228.

Whitten WM, NH Williams, MW Chase. 2000. Subtribal and generic relationships of

Maxillarieae (Orchidaceae) with emphasis on Stanhopeinae: combined molecular evidence. American Journal of Botany 87:1842–1856.

Whitten WM, AM Young, DL Stern. 1993. Nonfloral sources of chemicals that attract male euglossine bees (Apidae: Euglossini). Journal of Chemical Ecology 19:3017–3027.

Whitten WM, AM Young, NH Williams. 1989. Function of glandular secretions in fragrance collection by male euglossine bees (Apidae: Euglossini). Journal of Chemical Ecology 15:1285–1295.

Widmer A, S Cozzolino, G Pellegrino, M Soliva, A Dafni. 2000. Molecular analysis of orchid pollinaria and pollinaria remains found on insects. Molecular Ecology 9: 1911–1914.

Wiens E. 1978. Mimicry in plants. Evolutionary Biology 11:365–403.

Wignall AE, AM Heiling, K Cheng, ME Herberstein. 2006. Flower symmetry preferences in honeybees and their crab spider predators. Ethology 112:510–518.

Wikelski M, J Moxley, A Eaton-Mordas, MM López-Uribe, R Holland, D Moskowitz, DW Roubik, R Kays. 2010. Large-range movements of Neotropical orchid bees observed via radio telemetry. PLoS ONE, 5, e10738. doi: 10.1371. http://plosone.org/article /info%3Adoi%2F10:1371%2Fjournal.ponc.0010738.

Wille A. 1958. Comparative studies on the dorsal vessels of bees (Hymenoptera: Apoidea). Annals of the Entomological Society of America 51:538–546.

Willems JH, E Dorland. 2000. Flowering frequency and plant performance and their relation to age in the perennial orchid *Spiranthes spiralis* (L.) Chevall. Plant Biology 2:344–349.

Williams NH. 1978. A preliminary bibliography on euglossine bees and their relationships with orchids and other plants. Selbyana 2:345–355.

———. 1982. The biology of orchids and euglossine bees. Pages 119–171 *in* J Arditti, editor. Orchid Biology—Reviews and Perspectives no. 2. Cornell University Press, Ithaca, NY, USA.

Williams NH, WM Whitten. 1982. Identification of floral fragrance components of *Stanhopea embreei* and attraction of its pollinator to synthetic fragrance compounds. American Orchid Society Bulletin 51:1262–1266.

———. 1983. Orchid floral fragrances and male euglossine bees: methods and advances in the last sesquidecade. Biological Bulletin (Marine Biological Laboratory in Woods Hole) 164:355–395.

———. 1999. Molecular phylogeny and floral fragrances of male euglossine bee pollinated orchids: a study of *Stanhopea*. Plant Species Biology 14:129–137.

Willmer P. 2011. Pollination and Floral Ecology. Princeton University Press, Princeton, NJ, USA.

Willson MF. 1983. Plant Reproductive Ecology. Wiley, Chichester, England.

Wittmann D, M Hoffmann, E Scholz. 1988. Southern distributional limits of euglossine bees in Brazil linked to habitats of the Atlantic and subtropical rain forest (Hymenoptera, Apidae, Euglossini). Entomologia Generalis 14:53–60.

Wong BBM, C Salzmann, FP Schiestl. 2004. Pollinator attractiveness increases with distance from flowering orchids. Proceedings of the Royal Society of London B Biological Sciences 271:S212–S214.

Wong BBM, FP Schiestl. 2002. How an orchid harms its pollinator. Proceedings of the Royal Society of London B Biological Sciences 269:1529–1532.

Woodell SRJ. 1979. The role of unspecialized pollinators in the reproductive success of Aldabran plants. Philosophical Transactions of the Royal Society of London B Biological Sciences 286:99–108.

Woods M, AS Warren. 1988. Glasshouses: A History of Greenhouses, Orangeries and Conservatories. Aurum Press, London, England.

Woolcock C, D Woolcock. 1984. Australian Terrestrial Orchids. Thomas Nelson Australia, Melbourne, Victoria, Australia.

Worley AC, L Sawich, H Ghazvini, BA Ford. 2009. Hybridization and introgression between a rare and a common lady's slipper orchid, *Cypripedium candidum* and *C. parviflorum* (Orchidaceae). Botany 87:1054–1065.

Wright M, R Cross, KW Dixon, T Huynh, A Lawrie, L Nesbitt, A Pritchard, N Swarts, R Thomson. 2009. Propagation and reintroduction of *Caladenia*. Australian Journal of Botany 57:373–387.

Xu M, Y-B Luo, Y Xu, P-W Guo, J-W Xu. 2009. Changes in surface air temperature and precipitation over China under the stabilization scenario of greenhouse gas. Advances in Climate Change Research 5:79–84.

Xu S, PM Schluter, FP Schiestl. 2012. Pollinator-driven speciation in sexually deceptive orchids. International Journal of Ecology 2012, Article ID 285081. http://dx.doi .org/10.1155/2012/285081.

Xu S, PM Schluter, G Scopece, H Breitkopf, K Gross, S Cozzolino, FP Schiestl. 2011. Floral isolation is the main reproductive barrier among closely related sexually deceptive orchids. Evolution 65:2606–2620.

Yu X-H, Y-B Luo, D Ming. 2008. Pollination biology of *Cymbidium goeringii* (Orchidaceae) in China. Journal of Systematics and Evolution 46:163–174.

Zhang L, SCH Barrett, J-Y Gao, J Chen, WW Cole, Y Liu, Z-L Bai, Q-J Li. 2005. Predicting mating patterns from pollination syndromes: the case of sapromyiophily in *Tacca chantrieri* (Taccaceae). American Journal of Botany 92:517–524.

Zhang XS, SD O'Neill. 1993. Ovary and gametophyte development are coordinately regulated by auxin and ethylene following pollination. Plant Cell 5:403–418.

Zheng G, P Li, Y Tai, D An, Y Kou, Y-B Luo. 2010. Flowering and fruit set dynamics in *Cypripedium*. Acta Ecologica Sinica 30:3182–3187.

Zhou B-Z, L-H Gu, Y-H Ding, L Shao, Z-M Wu, X-S Yang, C Li, Z Li, X-M Wang, Y-H Cao, BS Zeng, M-K Yu, MY Ang, S-K Wang, H-G Sun, A-G Duan, Y-F An, X Wang, W-J Kong. 2011. The great 2008 Chinese ice storm—its socioeconomic-ecological impact and sustainability lessons learned. Bulletin of the American Meteorological Society 92:48–60.

Zhu K, M Wan. 1983. Phenology. Science Press, Beijing, China.

Ziegler C. 2011. Deceptive Beauties: The World of Wild Orchids. University of Chicago Press, Chicago, IL, USA.

Zimmerman JK, TM Aide. 1989. Patterns of fruit production in a Neotropical orchid: pollinator vs. resource limitation. American Journal of Botany 76:67–73.

Zimmerman JK, DW Roubik, JD Ackerman. 1989. Asynchronous phenologies of a Neotropical orchid and its euglossine bee pollinator. Ecology 70:1192–1195.

Zimmermann Y, SR Ramírez, T Eltz. 2009. Chemical niche differentiation among sympatric species of orchid bees. Ecology 90:2994–3008.

Zimmermann Y, DW Roubik, T Eltz. 2006. Species-specific attraction to pheromonal analogues in orchid bees. Behavioral Ecology and Sociobiology 60:833–843.

Zotz G, MY Bader. 2009. Epiphytic plants in a changing world: global change effects on vascular and non-vascular epiphytes. Progress in Botany 70:147–170.

Peter Bernhardt
Department of Biology
Saint Louis University
Saint Louis, Missouri 63103
USA

Mark A. Clements
Centre for Australian National
    Biodiversity Research
Canberra ACT 2601
Australia

Salvatore Cozzolino
Department of Biology
University of Naples Federico II
Napoli 80126
Italy

Amots Dafni
Institute of Evolution
Department of Evolutionary and
    Environmental Biology
Faculty of Science
Haifa University
Mount Carmel 31905
Israel

Zheng-Hai Deng
Yachang National Orchid Nature
    Reserve
No. 10 Huaping Town
Leye County, Guangxi 533209
China

Retha Edens-Meier
College of Education and Public
    Service
Saint Louis University
Saint Louis, Missouri 63103
USA

Chang-Lin Feng
Experimental Center of Tropical
    Forestry
Chinese Academy of Forestry
Pingxiang City, Guangxi 532600
China

Jacques Fournel
Herbier Universitaire de La Réunion
Faculté des Sciences et Technologies
Saint-Denis 97715
La Réunion
France

Ana Francisco
Departamento de Biologia Vegetal
Faculdade de Ciencias da
    Universidade de Lisboa
Lisboa 1749-016
Portugal

Anne Gaskett
School of Biological Sciences
University of Auckland
Auckland 1142
New Zealand

James O. Indsto
National Herbarium of New South
    Wales
Royal Botanic Gardens and Domain
    Trust
Sydney, New South Wales 2000
Australia

Steven D. Johnson
School of Life Sciences
University of KwaZulu-Natal
Scottsville, Pietermaritzburg 3209
South Africa

Wuying Lin
Department of Earth and
    Environment
Florida International University
Miami, Florida 33199
USA

Hong Liu
Department of Earth and
    Environment Studies, ECS 343
Florida International University
Miami, Florida 331199
USA

Shi-Yong Liu
Yachang National Orchid Nature
    Reserve
No. 10 Huaping Town
Leye County, Guangxi 533209
China

Claire Micheneau
Australian Tropical Herbarium
James Cook University
Cairns, Queensland 4878
Australia

Thierry Pailler
Laboratoire des Peuplements
    Végétaux et Bio-Agresseurs en
    Milieu Tropical
Faculté des Sciences et Technologies
Université de La Réunion
Saint-Denis 97715
La Réunion
France

Robert Pemberton
Florida Museum of Natural History
Gainesville, Florida 32611
USA

Andrew J. Perkins
Western Australian Herbarium
Department of Parks and Wildlife
Perth, Western Australia 6152
Australia

Sophie Petit
Sustainable Environments Research
    Group
School of Natural and Built
    Environments
University of South Australia
Mawson Lakes, South Australia 5095
Australia

David W. Roubik
Smithsonian Tropical Research
    Institute
Ancon, Balboa, Republic of Panama
    34002-9998
USA

Giovanni Scopece
Department of Biology
University of Naples Federico II
Napoli 80126
Italy

Nicolas J. Vereecken
Agroecology and Pollination Group
Landscape Ecology and Plant
    Production Systems (LEPPS/EIB)
Campus de la Plaine CP 264/2—
    Building NO
Université Libre de Bruxelles (ULB)
Blvd. du Triomphe
B-1050 Brussels
Belgium

Xin-Lian Wei
Yachang National Orchid Nature
    Reserve
No. 10 Huaping Town
Leye County, Guangxi 533209
China

Peter H. Weston
National Herbarium of New South
    Wales
Royal Botanic Gardens and Domain
    Trust
Sydney, New South Wales 2000
Australia

Xiao-Qing Xie
Yachang State Forestry Farm
No. 10 Huaping Town
Leye County, Guangxi 533209
China

Luo Yi-Bo
Laboratory of Systematic and
    Evolutionary Botany
Institute of Botany
Chinese Academy of Sciences
Beijing 100093
China

Mormolyca ringens, 298
Mycetophilidae, 123, 136
Myrmechila, 100, 106–7, 109, 111, 122, 130,
    135–36, 138, 140, 159; formicifera, 109;
    trapeziformis, 100
Myrmecia, 131
Myrsine africana, 318
Myrtaceae, 192, 258
Mystacidium, 191, 203, 207, 211–13, 215–17;
    venosum, 191, 211, 213, 215–17

Nectarinia famosa, 78
Nemacianthus, 98, 106–7, 116, 123, 135–36,
    138, 140; caudatus, 98
Nematoceras, 98, 106–7, 115, 123, 136–36,
    138, 140, 303; macranthum, 98
Nemestrinidae, 75, 79, 86, 136
Neobathiea, 204, 212–13, 216; grandidierana,
    213, 216
Neophyllotocus, 125
Neotinea, 24–25; maculata, 25
Neottia ovata, color plate 1
Neozeleboria, 126, 129–30, 143, 160, color
    plate 8
Nephele accentifera, 212, 216
Nephrangis, 203
Nomada, 268
Nomia, 132, 186
Nothofagus, 181
Notylia, 236
Nymphalidae, 86, 136

Obtusipetala (section), 272
Ocyptamus, 269; antiphales, color plate 15
Oeonia, 204, 207
Oeoniella, 204
Oligochaetochilus, 98, 106–7, 110–11; pictus,
    98; rufus, 98
Oncidium, 230, 236, 243
Ophrys, 4, 6–7, 16–17, 29, 33, 45, 47–67,
    143, 158, 160, 175, 254, 295–96, 298–
    99, 304, 331–32; apifera, 6, 16, 49–52,
    175; arachnites, 296; arachnitiformis, 64;
    atlantica, 296; bertolloni, 296; bombyli-
    flora, 61–63; chestermanii, 60; exaltata,
    60–61; ferrum-eqinum, 296; fuciflora,
    49; fusca, 53, 61–63, 296; heldreichii,
    64, 158, 296, 298–99; helenae, 296; in-
    sectifera (syn. Ophyrs muscifera), 4, 7,
    48, 49–51, 53–54, 56–57, 66, 295, 298,
    color plate 2; lutea, 53, 61–62; muscifera,
    4, 6, 10, 48, color plate 2; normanii, 60;

regis-ferdinandii, 296; speculum, 53, 55,
    59–60, 63–64, 66, 296 (see also Ophyrys
    vernixia); sphegodes, 50, 58–60, 62, 296;
    tenthredinifera, 61–62; vernixia, 296
Orchidaceae, 3–4, 12, 75–78, 91–92, 94, 97,
    141, 143, 145–46, 149, 188, 190, 201,
    244, 279
Orchidinae, 29
Orchidoideae, 145, 149, 174, 244
Orchis, 3, 6, 8–12, 16, 23–40, 42–43, 45–46,
    48, 52, color plates 1, 2; anthropophora,
    16; boryi, 28, 36, 40, 42; dinsmorei, 35–
    37; galilaea, 29; italica, 33; latifolia, 27,
    31, 48; mascula, 9, 23, 28, 31, 36, 40, 43;
    morio, 3, 6, 10, 27, 35, 48; pallens, 28;
    pauciflora; 29; provincialis, 40; pyramida-
    lis, 6, 12, 23, 156 (see also Anacamptis
    pyramidalis); spitzelii, 28, 36
Orthoceras, 100, 106–7, 109, 117, 124, 135–
    36, 138, 140; strictum, 100, 109
Orthrosanthus laxa, 187
Oscinosoma, 127
Ossiculum, 203

Pachyplectron, 98, 106–7, 110, 121; arifo-
    lium, 98
Panisea cavalerei, 319
Panogena lingens, 212, 216
Paphiopedilum, 12, 19, 265–67, 269–71, 278–
    86, 296, 298, 300, 303, 317, 320; armin-
    iacum, 279; barbatum, 265; barbigerum,
    269, 279–80, 285; bellatum, 269; callo-
    sum, 269, 280, 285; charlesworthii, 280,
    285; dianthum, 269, 280, 285, 317; hir-
    sutissimum, 269, 317; insigne, 265; mi-
    cranthum, 280; parishii, 269; purpuratum,
    265, 269; rothschildianum, 269, 279–80;
    venustum, 265; villosum, 269, 296
Paracaleana, 153
Partamona, 244
Pedicularis schistostegia, 276
Pelargonium, 86; longicaule, color plate 5
Pergidae, 124, 130–31, 134, 136, 142
Petalochilus, 99, 106–7, 109, 117, 121, 127,
    135–36, 138, 140, 147; carneus, 99; cate-
    natus, 99, 109
Phasmodinae, 223
Pheladenia, 100, 106–7, 109, 118–19, 127,
    135–36, 138, 140, 147; deformis, 100,
    109
Philoliche, 76, 81, 86, 88; aethiopica, 76, 86;
    gulosa, 76, 86; rostrata, 76, 81, 88

Phoridae, 129, 136

Phoringopsis, 100, 106–7, 111, 129, 135–36, 138, 140; *dockrillii*, 100

Phragmipedium, 266, 269, 271, 280–85; *beeseae*, 269; *caudatum*, 269, 282; *lindenii*, 269, 282, 284; *longifolium*; 269, 281–82; *pearcei*, 266, 269, 282, color plate 15; *reticulatum*, 269, 282; *xerophyticum*, 280

Phyllotocus, 126

Phymatothynnus, 126–27

Pimenta dioica, 258

Pinus yunnanensis, 318

Platanthera, 24

Platycarya longipes, 318

Plectrelminthus, 203

Podangis, 203

Pompilidae, 77, 136

Ponthieva, 98, 106–7, 110; *racemosa*, 98

Praecoxanthus, 99, 106–7, 118–19, 125, 135–36, 138, 140, 147; *aphyllus*, 99

Prasophyllum, 99, 106–7, 109, 114–15, 121–22, 128, 135–36, 138, 140, 150; *australe*, 99, 109; *brevilabre*, 99

Primula, 4

Prosoeca, 75, 82–83, 86; *ganglbaueri*, 75, 82–83, 86

Pseudophrys (section), 54, 60, 62

Pseudorchis, 24

Pterocormus, 129

Pterostylis, 93, 154, 165, 175, 298, 303; *longifolia*, 175

Pterygodium, 76–77, 81; *catholicum*, 81

Pterygophorus, 130

Pyrorchis, 100, 106–7, 109, 112, 131, 135–36, 138, 140; *nigricans*, 100, 109

Quercus phillyraeoides, 318

Rangaeris, 203, 212–13, 216; *amaniensis*, 213

Rapanea neriifolia, 318

Raphiolepis indica, 318

Rediviva, 76–77, 79–81, 83, 88–89; *brunnea*, 77, 89; *colorata*, 77, 79, 88; *gigas*, 77, 81; *longimanus*, 77; *macgregori*, 77; *neliana*, 77, 88–89; *pallidula*, 77, 89; *peringueyi*, 76, 79, 81, 83

Rhaesteria, 203

Rhipidoglossum, 203

Rhizanthella, 93, 99, 106–7, 109, 114–15, 129, 135–36, 138, 140, 153; *slateri*, 99, 109

Rhododendron, 5

Rhus chinensis, 318

Rimacola, 100, 106–7, 109, 112–13, 131, 135–36, 138, 140, 153; *elliptica*, 100, 109

Rosa rubus, 318

Rubiaceae, 78, 238

Sarcochilus, 16

Sarcophagidae, 276

Sarcophagus, 276

Satyrium, 73, 76–78, 84, 89, 98, 106–7, 110, 303; *bicallosum*, 73; *longicauda*, color plate 5; *microrrynchum*, 89; *nepalense*, 98; *odorum*, 98

Scabiosa columbaria, 85–86

Scarabaeidae, 125–26, 136

Scathophaga, 269

Sciaridae, 276

Scoliidae, 128, 131, 136, 142, 296

Scrophulariaceae, 75–77, 79, 82–83, 86–87

Selenipedium, 271, 280–81, 283, 286; *aequinoctiale*, 281; *chica*, 281; *palmifolium*, 281

Serapias, 33

Simosyrphus, 126

Simpliglottis, 100, 106–7, 109, 111, 129, 135–36, 138, 140; *valida*, 100, 109

Singularybas, 98, 106–7, 115, 123, 135–36, 138, 140; *oblongus*, 98

Sinopedilum (section), 270, 272, 276, 284

Sobennikoffia, 204

Sobralia, 238, 244–45, 255

Solanaceae, 238, 240

Solenangis, 203, 207

Spathiphyllum, 256

Sphecidae, 129, 136

Sphingidae, 136

Sphyrarhynchus, 203

Spiculaea, 100, 106–7, 112, 130, 135–36, 138, 140; *ciliata*, 100

Spiranthes autumnalis, 7, 11

Spuricianthus, 98, 106–7, 116, 123, 135–36, 138, 140; *atepalus*, 98

Stanhopea, 230–31, 243; *costaricensis*, color plate 11

Stegostyla, 99, 106–7, 109, 117, 127, 135–36, 138, 140, 147; *congesta*, 99; *cucullata*, 109

Stellilabium, 298

Stenobasipteron wiedmannii, 75

Steveniella, 24

Stigmatodactylus, 98, 106–7, 115–16, 123, 135–36, 138, 140, 153; *sikokianus*, 98

Wahlenbergia cuspidata, 82, 85–86
Waireia, 100, 106–7, 112–13, 131, 135–36, 138, 140; *stenopetala*, 100
Watsonia, 86; *densiflora*, 86; *lepida*, 86; *wilmsii*, 86

Xanthopan, 209, 212, 216; *morganii*, 212; *morganii* var. *praedicta*, 209, 216
Xanthorrhoea semiplana, 160
Xanthorrhoeaceae, 85
Xylocopina, 275
Xysmalobium involucratum, 89

Ypsilopus, 203

Zaluzianskya, 86–87; *microsiphon*, 86–87; *natalensis*, color plate 5
Zaprochilinae, 223
Zaspilothynnus, 127, 130, 134; *nigripes*, 134; *trilobatus*, color plates 6, 7
Zeuxine, 98, 106–7, 110, 121; *oblonga*, 98; *strateumatica*, 98
Zingiberaceae, 238
Zosterops, 212, 214, 219–20; *borbonicus*, 212, 214, 219–22; *olivaceus*, 212, 219–21

cleistogamy, 17

climate change, 169, 289, 311–13, 324, 326, 333

clinandrium, 13, 184, 187–88. *See also* mitra

cloud forest, 253

coalescence, 14

cocoon(s), 161

coevolution, 74, 83, 194, 202, 211, 234, 239, 332

cognitive interactions, 299

Coiba Island, 241, 255

cold spell(s), 312–13, 315, 317, 320–22, 324–25

color(s)/coloration/colorful/colored, 8, 27–28, 29, 32, 35, 37–39, 40–41, 48, 55, 57, 64, 67, 75–78, 83, 85, 87, 90, 133, 145, 154–55, 159, 162, 175, 177–78, 180, 183–84, 189, 191–92, 204, 212, 216, 232, 234, 239, 246, 250, 265, 267, 271–73, 276, 279, 280–82, 284, 289, 291–309, 330–32; bi–colored, 279; blue, 64, 75, 77, 144, 177–80, 183, 185–87, 189, 192, 293–94, 298, 301–3, 306–8, 333; change(s) of, 293–94; con-colored, 279; contrast, 292, 294, 298, 305–8; functions of, 292, 295; gloss, 295, 304; green (drab), 4, 76–77, 159, 168, 171, 177–78, 292–93, 298–302, 304, 306–8; high-chroma, 293; low-chroma, 293; mauve, 144, 178–80; morphs, 35, 37, 40, 183; polymorphic, 40, 96, 134; purple, 9, 37, 39, 40, 83, 177, 179–80, 186, 192, 272, 293, 303, 308, 333; red, 53, 72, 77–78, 83, 85–87, 159, 177, 179–80, 183, 189, 196, 219, 234, 272, 279, 293, 301–2; saturation, 293; space, 305–7; white, 4, 40, 64, 77, 139, 141, 154, 159–60, 177–79, 181, 183–84, 186, 192, 204, 208, 211–12, 214, 216–17, 219–23, 248, 272, 281, 293–94, 298–99, 301, 304–6; yellow, 37, 77, 83, 144, 154, 161–62, 176–80, 183, 186, 192, 204, 219, 272–73, 279–80, 293, 297, 300, 303, 308, 333; color plates 3, 6, 13

column(s), 11, 14, 152, 226, 236, 282; anterior, 153; architecture, 330; auricle(s), 154; base of, 161; characters, 150; entrance, 283; evolution, 187; floral, 8; foot, 153; functional, 184; hood, 173–74,176, 185; interconnected, 10, 12; irritable, 12; lateral lobe of, 174; nectar, 215; orchid, 10, 205, 215–16, 219, 221, 330; sample, 241; shape, 206; stamino-dium of, 265; structures, 14; wings, 144, 153–54; color plates 6, 7

Comoros, 204, 206, 208

Congo (Basin), 206

conservation, 3, 164, 168, 171, 229, 261, 287, 312, 326–237, 333

contrivance(s), 6, 18, 23, 30, 45, 47, 51–52, 156–57, 168, 201, 252, 279, 282, 292, 311, 329–32

convergent, 71, 73–76, 78, 82–83, 134, 153, 206, 219, 294, 296–97

Cook, James (1729–1779), 173

corolla(s), 13, 41, 75–76, 78, 82–83, 202, 238

Correvon, Henry (1854–1939), 52, 55

cospeciation, 95

Costa Rica, 253

"Creation by Law," 211

Crüger, Hermann (1818–1864), 231

crypsis, 83, 293, 300

Darwin, Charles (1809–1882), 1, 3–19, 21, 23–27, 30–31, 45–53, 66–67, 69, 72, 88, 91–92, 141, 148–49, 155–57, 159, 161–64, 168, 170, 176, 181–82, 184, 192–94, 196–97, 199, 201–2, 209, 211, 215, 217, 223, 227, 229–36, 238–39, 241–43, 245–48, 250–54, 257–58, 260–61, 265, 267, 270, 278–79, 283, 292, 308, 311, 326, 329–31, 333; *The Origin of Species* (1859), 3, 8, 15, 47; *On the Various Contrivances by Which British and Foreign Orchids Are Fertilised by Insects, and on the Good Effects Of Intercrossing*, First Edition (1862), 6, 23, 47, 201; *The Various Contrivances by Which Orchids Are Fertilised by Insects*, Second Edition (1877), 23, 311 color plates 1, 2

Darwin, Emma (1808–1896; wife), 4

Darwin, Francis (1848–1929; son), 4–5, 18

Darwin, George (1845–1912; son), 6

Darwin, William Erasmus (1839–1914; son), 175

deception, 26–27, 34, 42–44, 255, 301; brood-site, 295; color, 291; evolution of, 24, 42, 44; floral, 42, 45, 255; food, 27–29, 37–38, 45, 103, 123–27, 129, 131–32, 134–35, 141, 144–46, 157, 159, 161–62, 168; nonmimicry, 145; by orchids, 298; organized system of, 48; pollination by, 84, 157, 295; sexual, 45, 54, 84, 94, 103, 125–27, 129–35, 141–43, 146, 156, 159, 161, 166–68, 170, 254, 291, 295, 331; specialized, 28; systems of, 84

diandrous, 12, 14, 263
dioecy, 12
dissection(s), 6, 17
distribution(s), 35, 75–76, 78–79, 271, 277–78, 280–81; allopatric, 60; Angraecoid (distribution of), 203–4, 212; calli, 152–53; of floral scent, 65; geographic, 74, 79–80, 88, 319, 326; global, 321; leaf stomata, 151; matching, 86; natural, 80, 274; of nectaries, 146; Neotropical and Paleotropical, 8;; of oil-collecting bees, 80; of orchids, 80, 192; of photoreceptor types, 302; plant species, 88; pollinator, 79–80, 88; of pollination guilds, 80; range, 66, 87, 319–220, 325; spatial, 41; trichome, 151
dormancy, 170
dorsal sepal, 135, 151–52, 174, 274
Downe Bank, 7; color plates 1, 2
drought, 313, 315, 318–19, 322, 324
dry season(s), 156, 260, 313, 315, 324

echidna, 167
ecology, 13, 16–17, 24, 36, 45, 55–56, 58, 62, 67, 71, 156, 230–31, 257–58, 261, 274, 277–79, 281–82, 286, 304, 325, 333
education, 5, 171
egg(s), 165, 170, 276, 280, 282, 284
ejaculation, 56, 164
electroantennography, 168
embryology, 93
embryo sac, 185
emergence, 35, 37, 133, 161–62, 318
energetic cost(s), 165–66
engagement, 171
England, 5, 48–50, 231; color plates 1, 2
Eocene, 234
epigyny, 14
epiphyte(s)/epiphytic, 230, 247, 278, 317, 325
Erickson, Rica (1908–2009), 182
ethology, 285
eudicot(s)/eudicotyledon(s), 144, 186, 240, 244, 247
Europe/European, 16, 29, 37, 42, 53, 72, 74, 175, 193, 254, 275, 301, 312, 323, 325
evolution(s)/evolutionary, 24, 30, 45, 53, 171, 234, 283; adaptive, 28; adaptations, 244; advantage, 42; advergent, 73; from ancestors, 139; arms race, 167; biology/biologists, 10, 202, 209, 227, 329; botany, 8, 19; of both long nec-

tar tubes and long moth tongues, 202; of breeding systems, 141; of the *Caladenia,* 156; change(s), 83, 139, 142; character, 92; chromosome, 195; constraint, 46; context, 18; convergent, 71, 73, 74, 82–83, 206, 294, 296–97; of corolla tube length, 83; dances, 234; dead end, 48; of deception, 24, 42, 44, 139, 143, 331; of different sets of primary attractants and rewards, 331; of the Diurideae, 147; diversification, 156; door, 239; of early-and late-flowering forms, 81; ecology and, 16, 45, 62, 67, 71, 257, 333; elasticity, 232; of extreme traits, 46; floral, 71, 187, 265, 274, 331–32; of floral color, 301; of floral diversity, 96; of floral traits, 245, 261; of fly proboscis length, 83; history, 148, 201, 239; hypotheses, 94, 146; implications, 173; inferences into, 92; interest in, 209; labellum-column, 187; labile, 65; of large-flowered species, 189, 230; of long-lasting flowers, 32; macro, 67, 88, 90, 142, 147, 202, 283; micro, 90, 142; of mimicry, 297; model, 332; modification, 90; molecular, 101; mysteries, 168; by natural selection, 201; of of nectar production, 139, 141; of new species, 196–97; orchid, 69, 199, 232, 250; of orchid flowering times, 90; of orchid pollination, 333; origins, 94, 146; patterns, 207; perspective, 239; plant, 45; of pollinaria, 233, 243; of pollination strategies, 32; of pollination systems, 91, 94–95, 96, 139, 286; of pollinator shifts, 94; of pollinator syndromes, 293; potential, 218, 227; processes of, 65, 73, 147; race, 239; reciprocal, 202; relationships, 91; reproductive, 261; of reproductive mechanisms, 286; reticulate, 196; scenario, 96, 146; scent, 65; shifts, 87, 95, 218; significance, 26, 45, 188, 324; of small-flowered species, 193; of structures mimicking pollen, 162; success, 157; syndromes, 71; theorists, 141, 148; trajectories, 46; transitions, 227; trend, 270; of the tribe Orchideae, 145; of *trnL-F,* 109; of unisexual flowers, 251; extreme weather events, 313, 319

Fabre, J.-H. (1879–1907), 53
fatty acids, 60

fertilization(s), 16; avoiding, 246; cross-, 51, 148, 196; delayed, 185; of the many seeds, 243; of orchids, 52, 326; self-, 50–51, 142, 196–97, 246

Field Experiments with Chemical Sexual Attractants, 56

Fitzgerald, Robert David (1830–1892), 17–18, 91, 147–48, 175–76, 194, 196, 330

flight (zigzag), 165, 216

floral: attractants, 279; biology, 233, 258; color, 291, 294, 300, 303–4; damage, 244; deceit, 239; evolution, 265; mimicry, 84; odors, 234; phenology, 182; presentation, 183, 271, 279–81; sacs, 77; scent, 65, 213; spurs, 77; syndromes, 82; traits, 38

flower(s), 48, 77, 158, 190, 204, 294, 304; color, 212; deceptive, 26; large, 173, 189; longevity, 32, 219, 223; morphology, 30, 211, 219, 222, 225; nectariferous, 25; orchid, 233; orientation, 151; phenology, 316; pollination, 273; scents, 223; season, 323; showy, 31; small, 173, 193; time, 316; visits, 216

flowering, 19, 260; data, 319; delayed, 322; early, 36; patterns, 182; phenology, 311, 319, 323; stems, 241; time, 316

fly/flies, 75, 276, 284, 303, 308

food deceptive, 26–30, 32–34, 36–37, 40, 45–46, 122, 133, 135, 142–44, 147, 156–57, 159, 165–66, 168, 295–98, 303

food mimic(s)/mimicry/mimicking, 44, 46, 94, 139, 142, 144–45, 187, 190, 234, 277, 279, 280

foragers, 186, 216, 285

Forster, Johann Georg (1754–1794), 173

Forster, Johann Reinhold (1729–1798), 173

Fox, W. D. (1805–1880), 4

fragmentation, 170.

fragrance(s), 30, 56, 61, 184, 229–30, 232–36, 239–42, 246–48, 250, 255–57, 261, 271–73, 276, 279, 287, 331; color plates 13, 14. *See also* scent compounds

fruit/fruiting, 16, 26, 32, 38, 43, 50, 103, 156, 158, 170, 191, 194–95, 213, 215, 217, 219, 220, 222–26, 232, 236, 238–39, 247, 258, 260, 268, 277–78, 280, 282, 284–86, 311

function shift, 18

galea, 152

Gardener's Chronicle, 8

gas chromatography, 55, 58, 168

geitonogamy, 26, 43, 95, 139, 141, 217, 294, 330

genetic load, 43, 141

gland(s): cephalic, 255; column anterior basal paired, 153; Dufour, 57; hammer, 7; labellum nectar secretory, 153; mandibular, 57, 59; nectar, 7, 267; scent, 234, 281, 286; yellow spots

glandular hairs, 267

glasshouse(s), 4, 5, 183, 188, 230, 247, 254. *See also* greenhouse

gnat(s): fungus, 73; mushroom, 276; pollinators, 163

grass tree(s), 160, 169

Gray, Asa (1810–1888), 7, 15, 17–18, 47, 267, 270

gray white-eye birds, 214, 219, 221–22

grazing, 160, 189

greenhouse, 5, 183, 188. *See also* glasshouse(s)

guild(s): co-blooming, 276; composition, 79; concept, 74; ecological, 71, 73, 84; level, 84; members, 71, 79, 81–82, 84, 88; mimic, 144, 186, 191–92, 194; oil-bee guild, 83; of oil-producing orchids, 79, 81; plant, 74–76, 78–80–81, 83–84, 89–90; pollination, 74–76, 78, 80–81, 83, 87–90; of pollinators, 292; pompliliid, 83; shifts, 87; color plate 5

habitat(s), 32, 42, 44, 48, 75–76, 78, 80, 88, 158, 181, 189, 208–9, 222, 227, 252–53, 256, 276

hairs: on the floral labellum, 233; glandular, 267; red, 53; color plate 10

Hamlet, 9

haplodiploid, 156

hawkmoth(s), 72–73, 78, 210–12, 215–18, 225, 294; color plates 4, 5

Hazelhurst, Grant, 7

Henslow, John Stevens (1796–1861), 3, 4

herbarium, 103, 149, 176

herbivory, 159, 161

Himalayas, 5, 271

honey, 23, 26, 27

Hooker, J. D. (1817–1911), 4, 10, 209, 227

horsefly/horseflies, 76

hothouse(s), 4, 230, 247, 254. *See also* glasshouse(s)

human population, 171

humus, 278

orchid(s) (*continued*)

interactions between orchids and oil bees, 80; interest in, 5; labellum/labella, 57, 187, 236, 242; lady's slipper, 7; large-flowered, 173–97, 238, 246; late spider, 49; lineages, 29, 73, 90, 146, 193, 331–32; lip, 216; literature, 254; long lived, 243–44; Mediterranean, 29; members, 74–75, 82; mimicry, 84–86,166, 186, 191, 292; mirror, 64; mischievous, 163; model, 15, 85; modern, 13; modes of floral presentation, 16; monandrous, 13–14, 184, 187; morphology, 23, 45, 47, 66, 222; native, 52–53, 170; natural history of, 309; nectar, 221, 236–37; nectar as bait, 237; nectarless, 32, 186, 243; nectar secretion, 7, 182; negative impact of, 166; Neotropical, 11, 230, 233–34, 244, 254, 261; new, 234; niche, 253; nobility, 72; no fast food in, 233; nonrewarding, 84–85; odors, 234, 258; oil-producing, 79, 81, 83, 88; orange morph, 85; orchid/ant gardens, 237; orchid-bee biological interpretation, 254; orchid bee speciation, 234; orchid bee survival, 234; orchid-bird inter-actions, 221; orchid-free environment, 166; orchid-pollinator interactions, 52, 202, 332; orthopteran pollination of, 223, 225; outcrossing in, 246, 252; ovaries, 13; papers, 18; partitioning pollinators, 233; patch, 166, 224; pet-als and sepals, 186; phenology of, 311; phylogeny of, 92, 205; pigmentation, 186; plant community, 161; pollen, 13, 33–34, 275; pollen dispersal, 167; pol-len grain release, 12; pollen masses, 47, 50; pollen tubes, 185; pollinarium/pol-linaria, 57, 82, 192, 224, 235, 256–57, 330; pollination, 4, 54–55, 66–67, 71, 73, 91, 164, 168, 182, 220, 231, 238, 242–44, 256, 333; pollination biolo-gists, 66; pollination biology, 48, 148; pollination ecology, 17; pollination ex-periments, 230; pollination guilds of wasp, bee, and beetle, 89; pollination mechanics, 243; pollination strategies, 48; pollination studies, 303, 308; polli-nation success, 82; pollination systems, 201–2; pollinator(s), 72, 166, 209, 216, 221, 232–34, 253, 292; pollinia, 163; population of, 7, 169–70, 222, 332; population density, 239; protection for, 160, 169; proto-orchid, 239; publica-tion on, 168; rare, 247; raspy cricket visits, 225; recognition of, 166; recruit-ment, 169; red-lipped, 159; red morphs, 85; relationships between orchids and their pollinators, 91; relationship with sting-less bees, 245; reproduction effi-ciency, 202; reproduction of, 224, 236; reproductive organs, 215; reproductive performance, 32; reproductive success, 169, 311; research, 6, 19, 240, 301; re-searchers, 156; response to fire, 169; reward(s), 27, 32, 37, 146, 233; reward-less, 38; safety or emergency pollination procedure, 161; saprophytic, 232; scent of, 232; seed(s), 80, 243; seed banks, 163; self-pollination, 142, 196, 225–26, 247; services by the same bee, 246; sexually deceptive, 60, 65, 84, 155, 157–59, 164, 162, 164–67, 169–70, 291–92, 296, 299, 301–2, 308, 331–32; shape, 246; sites, 7, 158; slipper, 187, 265, 267, 270, 282–83, 285–86, 329; small flow-ered, 173–97; speciation, 261; species, 4–5, 7, 9, 12, 15–16, 27, 48, 54, 74, 82, 109, 134, 155, 158, 160, 191, 215–16, 225, 234, 253, 283, 313, 318, 322, 324–25; species' survival, 169; specimens, 5; spider orchid(s), 121, 155–56, 158, 161–64, 168–70; spring-blooming, 35; spur, 224; star, 202, 215, 227; stingless bees that have evolved into orchid pol-linators, 244; story of, 19; studies, 5, 17, 71; study of, 156, 333; subdivision of, 8; subsample of, 146; sun orchid(s), 173, 182, 191, 190, 194; survival, 170–71, 333; sympatric, 81, 160; systematics, 202; taxa, 103; taxonomists, 193; tax-onomy, 158; terrestrial, 80, 92–93, 143, 147, 183, 320, 325; thrips, 181; trait tracking by, 83; transitions from out-crossing to selfing, 227; transplanted, 4; treacherous, 158; tropical, 33, 39, 206, 229–30, 233, 253, 332; vision of orchid pollinators, 301; visitors, 255; visual mimicry, 159; volatiles released by, 158, 234, 236; wasp emergence time, 162; weather, 160; white, 219; wild, 315, 321, 323, 326; work, 15

orchidology, 8, 10, 71

orchidophilia/orchidophile, 3–4

osmophore(s), 61–62, 234. *See also* trichome(s)

ovary, 13–14, 152, 194
ovules, 13, 146, 185, 188, 194, 243

Palmer, Irene, 9; color plate 1
Panama, 235, 237, 241–42, 244, 249, 253, 255–56, 259–60, 281; color plates 11, 12, 13, 14
Panglossian adaptationist, 18
papillae, 25, 62–63, 184
Papua New Guinea, 175
parasitic, 6, 60, 168
parasitoid, 167, 276
Patagonia, 167
pattern(s)/patterned/patterning, 30, 34, 38–39, 57, 60, 64–65, 82–84, 87, 90–91, 95, 137, 141, 147–48, 177–78, 180, 182–83, 185, 187, 192, 207, 227, 239, 271–72, 274, 279, 281–85, 296, 299–300, 313, 322–23, 331–32
peduncle(s), 151, 236, 276
pencil experiment, 9–10
perceptual biases, 25, 297
perianth(s), 64, 85, 150, 177–78, 180–84, 187, 189–91, 194, 196, 279, 298
petal(s), 13–14, 64, 152, 155, 179, 184, 186–88, 223, 265, 274, 279, 281, 296
Pfitzer, E. (1846–1906), 270
phenology, 36–37, 41, 160, 169, 182, 232–33, 242, 247, 256, 258, 260, 311–13, 316, 318–19, 322–26
pheromone(s) compounds, 160; of *Bombus* species, 29; diversity, 160; emitted by models, 143; insect, 56, 58, 143; mimic, 156; output, 159; perception of, 158; production, 158, 160; sex, 58–59, 65–66, 157, 292; species-specific, 240; synthetic, 158; visibility, 158; wasp, 157. *See also* scent compounds
Philippines, 175, 278
phylogeny/phylogenies/ phylogenetic(s), 24–26, 34, 60, 65, 83, 88, 91–98, 101, 103–4, 122, 137, 139, 142–50, 160, 174, 204–5, 207, 209, 229, 271, 287, 302, 320, 325
phylogram, 104, 106–7, 109. *See also* trees
pigeons, 8
pigmentation patterns, 185, 192, 272, 279, 282, 331
pistil, 14, 190
pollen, 4, 7–8, 12–14, 17, 31–34, 44–45, 47–48, 50–51, 81–82, 95, 145–46, 149, 154, 160–64, 167, 170, 176, 181–82, 185–92, 194–95, 197, 217, 223–24, 232–36, 239, 241, 243–46, 249–51, 255–57, 261, 265,

267, 270–76, 278–79, 283–85, 294, 300, 330, 332; color plate 10
pollen-to-ovule (P/O) ratio, 163, 143, 194
pollinarium, 9–11, 16, 30–31, 141, 150, 154, 164, 176, 183, 187, 190–91, 193–94, 217, 219, 222, 224, 226, 232, 235, 241, 245–48, 251, 270, 330
pollination(s), 4, 6–7, 9–13, 15–19, 23–27, 29–30, 32–40, 42–48, 50–59, 62, 66–67, 71–75, 79–85, 87–91, 94–96, 102–3, 122–23, 133–34, 137, 139, 141–49, 155–64, 166–67, 170, 175–78, 180–95, 197, 201–2, 206, 208–9, 211–12, 215–27, 229–36, 238–39, 242–44, 246–47, 253–54, 256–59, 261, 265, 267–68, 270–71, 273–86, 292–95, 297–98, 301–3, 308, 312–13, 320, 326, 329–33; color plates 5, 12
pollination guilds, 74–75, 80–81, 83, 87–90; color plate 5
pollinator(s), 7–8, 11–12, 16, 19, 24–32, 34–46, 48–50, 52, 54–55, 57–67, 71–76, 78–82, 84–88, 90–91, 94–95, 102–3, 122–24, 126, 128, 130, 132–37, 139, 141–46, 148–49, 158–70, 181, 185, 187–88, 191–93, 195, 201–2, 208–12, 216–21, 223–27, 229–30, 232–36, 239, 241–49, 251–53, 255–56, 258–61, 267–68, 270–86, 291–302, 304, 307–8, 312–13, 320, 324, 226, 330, 332–33
pollinia/pollinium, 3–4, 6, 9–13, 16–17, 30–31, 48, 50–51, 53, 141, 146, 150, 154, 156, 160–66, 168, 176–78, 180–81, 183, 185–88, 190, 192–94, 196, 204, 216–17, 220, 223, 226, 235, 241–42, 244–50, 270, 276, 283–85, 299, 330
polychromy, 39
polylectic, 186, 189, 191–92, 274–75
polymerase chain reaction (PCR), 97
polyphagic, 275
polymorphism, 38–42, 134, 139, 141
polyphyletic, 92–93, 96, 122
population(s), 4, 7–8, 24, 29, 36, 38–40, 43, 60, 67, 73, 83, 97, 142, 144, 158, 160–61, 164, 167–71, 176, 181–83, 186, 189, 192, 195–97, 202, 209, 218, 222, 225, 227, 229, 239, 243–44, 257, 275, 277–78, 280, 285–87, 311, 313, 316, 319–20, 323–24, 331–33
preadaptation, 18
private channels, 293
proboscis/proboscides, 11–12, 30, 83, 209, 211–12, 215–17

protein(s), 233, 237, 255, 261
pseudobulbs, 237
pseudocopulation, 17, 52, 54, 57, 62, 156
pseudopheromones, 143
pseudopollen, 161, 255
pseudoviscidium, 284

quantum catch, 292, 302, 304–8

raceme(s), 85, 183, 218
rain(y)/rainfall, 77, 80, 161, 162, 169, 190, 238, 252–53, 256, 260, 274, 311–13, 315
rain forest, 242, 252, 261, 323
receptor(s): antennal, 29; blue or green, 298; excitations, 307; long-wavelength, 305; LW (green), 299, 302, 304; noise, 306, 308; photo-, 301–3, 305–6, 308; red, 302; types, 292; UV, 298; UV, blue, and green, 307; UV/SW, 302; viewer's, 305
regurgitation, 157
reproductive cost, 170
reproductive output, 161
resin(s), 233, 255, 261
reticulate evolution, 196. *See also* introgression
rewards, 103, 146, 166, 233, 239–40, 285, 291, 294–95, 331; do not offer, 84; edible,190, 267; false rewards, 331; floral, 84, 291–92; fragrance, 232; inferior, 43; lack, 277, 330; nectar, 95, 145, 149, 233, 291; offering no, 192; oils, 233; pollen, 145, 149, 233, 261; proteins, 233; resins, 233
root(s), 230; non-stolonoid, 150; rootlet, 237–38; stolon-like, 95; stolonoid, 150
rostellum, 13–14, 154, 174, 176, 187, 204, 206, 215, 226, 270, 276
Royal Botanic Gardens and Domain Trust, Sydney, 109, 149
Royal Botanic Gardens at Kew, 4, 149

saponin, 235
scent(s)/ scented, 29, 39, 41–42, 55, 84, 158, 185, 291, 298–99, 304, 308; abdominal, 57; absence of, 41; advertisements, 7; analyses, 331; artificially, 42; blends, 85; caraway seed-like, 242; chemistry, 81, 83–84; chiloglottones, 65, 143; cinnamon, 177, 179, 180; compounds, 30, 58, 65; cues, 83; discernible, 177–78; dummies, 46, 56–57; dung-like, 271; emissions, 90; evening, 78, 225; exotic,

60; faint, 42; female abdominal, 57; food-deceptive strategies, 46; floral, 29, 42, 48, 55–58, 61–62, 65–67, 81, 83–84, 103, 158, 175, 208, 213, 225; flower dummies, 42; head-space analysis, 42; lavenders and lilacs, 179; learned by bees, 41; lemon, 177; models, 41, 331; molecules, 332; musty-fermented, 276; nocturnal, 208, 223; no scent, 76–77; polymorphism, 42; pungent, 77; reward, 250; rosewater, 178; rotting vegetation, 271; scentless, 41, 177; soapy, 76; spicy, 279; strange, 156; strong, 178, 184, 208, 211, 216, 232; sweet, 145, 178, 211, 225; unscented, 78, 156, 219, 281; urine, 279; vanilla, 184, 281
scent compounds: alcohols, 56, 60–61 aldehyde(s), 56, 59–61; alkenes, 59–60, 65–66 benzoids, 242; beta ionone, 257; butyric acid, 56; carboxylic acid, 61; cineole, 256; citronellol, 56; dehydrogeosmin, 65; *(E)*-2,3-dihydrofarnesol, 29; *(E)*-β-farnesene, 29; esters, 59–61; farnesol, 56; fatty acid(s), 60–61, 143, 242; geraniol, 56; hydroxydecanoic acid, 66; ketones, 61; linalool, 56, 61; long-chain hydrocarbons, 59, 61; methyl salicylate, 256; monoterpene, 61, 223–24, 242; nerolidol, 56; oxo-acids, 60, 65; p-dimethoxybenzene, 257; sesquiterpenes, 56, 61, 242; skatole, 256; terpenoids, 242; vanillin, 255–56
scent glands, 234, 281, 286
scentless, 41, 177
scheinsaftblumen, 7, 26
sclerophyll woodlands, 182
seedling recruitment, 169
seeds, 5, 16, 161, 163, 224, 232, 243, 247, 250, 261; banks, 163; bearing embryos, 215; characters, 150; coat, 93, 243; die, 243; dispersal, 167, 260; large, 163; morphology class, 154; mortality, 260; number of, 163, 168; orchid, 80, 243; in orchid capsule, 168; produce/production, 49, 215, 260; reduced output, 170; set, 13, 48, 229, 243, 277, 331; small, 163; survival, 258; tiny, 232, 238, 243; unfertilized, 13; viable, 243
selection, 84, 90, 332; artificial, 8; for crypsis, 83; Darwinian, 239; for floral color change, 294; on flower size, 87;frequency-dependent, 37, 39, 42; for greater floral display, 32; levels, 67;

mate, 13; on morphological characters, 159; natural, 8, 18, 23, 25, 45–46, 72, 192, 194, 197, 201–2, 227, 230, 250, 259, 325, 331–32; pollinator, 46; pollinator-mediated, 46, 85, 159; pressure(s), 139, 143, 260, 331; reciprocal, 202; relaxed, 39, 159, 299; for seedling establishment and seed production, 260; sexual, 331; of terminal taxa, 92; trait, 46; for traits that attract a different pollinator, 88; unidirectional, 202

self-compatibility, 271

self-incompatibility, 185, 271

self-pollination, 6, 11, 16, 26, 43, 51, 91, 95, 142, 162–63, 166–67, 175–78, 180–82, 184, 193–95, 197, 212, 215, 225–26, 246–47, 271, 278–79, 282, 330. *See also* autogamy; geitonogamy

sensory: biases, 297; channels, 292; drive, 297, 309; ecological experiments, 59; modalities, 56, 90; modes, 300; perception, 301; preferences, 82, 297; reaction, 298; systems, 291, 302; trap, 296–97, 300; vulnerabilities, 301

sepaline nectar, 103

sepals, 64, 135, 151–53, 155, 157, 174, 177–80, 184, 186, 223, 236, 274, 279

sex ratio, 166

sexual: attractants/attraction, 29, 56; based pollination, 45; behavior, 29, 56, 58, 292, 295, 298, 300, 308, 332; deception, 29–30, 33, 45, 54, 59–60, 65, 84, 94, 103, 122, 125–27, 129–35, 141–43, 146–48, 155–62, 164–68, 170, 254, 291–92, 295–96, 298–302, 308, 331; excitation degree, 60; exciting effect, 58; imbalance, 30; legacy of mutualist male orchid bees, 261; mechanism, 30; mimics/mimicry, 54, 162, 182, 285, 331; partners, 60; phase, 262; reproduction, 216; selection, 331

Seychelles, 206

Shakespeare, William, 9

silica gel, 97

Soberanía National Park, 256, 259–60

Solomon Islands, 278

South America/American, 5, 7, 16, 72, 93, 143, 167, 211, 251, 253, 281

Sowerby, George Brettingham (1812–1884), 9–12, 14–15

speciation, 195, 201, 246; differential, 142; ecological, 227; by hybridization, 147; introgressive, 195; orchid, 261; orchid

bee, 234; rates, 147; of South African orchids, 88; in *Thelymitra* spp., 147

species, 4, 7–8, 10, 12–13, 23–29, 33–35, 37–39; 48–52, 54, 56, 59–62, 64–67, 71, 73, 79, 81–84, 87–88, 94–97, 102–3, 122–23, 133–34, 142–43, 145, 148–49, 155, 159–63, 165–67, 170, 173–76, 181–84, 186–88, 191–92, 194–96, 201–2, 204, 206–9; 211, 213, 215–17, 219–22, 225–27, 229–31, 234, 237–42, 246, 250, 253–55, 257–58, 260, 265, 267–68, 270–87, 292, 294–96, 298, 300, 302–4, 312–13, 316–26, 332; African, 202, 204, 207; alpine, 188; Amazonian lowland, 254; angiosperm, 191; animal, 144; Asian, 16, 276; Australian, 16, 18, 175; autogamous, 123–24, 126, 128, 130, 132, 142; Batesian mimics of rewarding plant, 86; bearing papillae, 25; bearing spurs, 25; bee, 29, 56–57, 59, 61, 65–66, 79, 144, 192, 233–35, 252–53, 256–58, 273, 275–76; belonging to the monophyletic section, 60; bird, 221–22; bird-pollinated, 219, 222–23, 225; blue-flowered, 185; botanical, 240; boundaries, 209; British, 6, 8–9, 15; canopy, 318; at chemical baits, 256; Chinese, 273; co-flowering/blooming, 35, 37–38, 41, 186, 195, 275–76; color-dimorphic, 28; collared, 184; colored, 40; with color morphs, 37;congeneric, 42, 320; co-occurring, 63; cross-pollinated, 187, 194; cryptic, 158, 168; in cultivation, 71; deceptive, 24–33, 35–45, 48, 82, 122, 133, 143, 147, 156, 164, 168; diversification, 201, 208–9, 283; easily accessible, 148; endangered, 278, 318; endemic, 195, 319–20, 326; epidendroid, 236; epiphytic, 182, 202, 321; euglossine, 236; European, 16, 37; evolution of, 283; exotic, 6; fern, 5; flower, 34, 191; fly, 74, 79; group, 65; Guinea Islands, 204; haplodiploid, 156; hawkmoth-pollinated, 211, 217, 225; heterogeneous, 24; hibernation, highly dispersed, 33; highly specialized, 227; honeybee, 325; hymenopteran, 59, 302, 304; insect, 12, 59, 74, 95, 159; insect-pollinated, 51, 123–24, 126, 128, 130, 132–33; insular, 217–18; invasive, 318; isolation, 246; keystone, 79; large-flower, 176, 185–95, 197; leafless, 97; level, 65; long-spurred, 225; Madagascar, 202, 204, 207; magnet, 43–44, 82;

ultraviolet (UV), 85, 159, 244, 291, 293, 295–98, 300–304, 306–8
unisexual, 251–52

vanilla, 184, 281
vegetative/vegetatively, 95, 150, 161, 176, 182, 189
veins, 4, 13–14, 151, 178–80. *See also* vessels
vessels, 13–14. *See also* veins
vestigial organ, 13, 193
Victorian(s), 5, 8, 18
viscidium, 11, 51, 150, 154, 174, 187–88, 193, 204, 224, 245, 247, 250, 270, 276
vision, 45, 164, 298, 326; achromatic, 305; bee, 83; chromatic, 305; color, 301, 303, 308; field of, 309; fly, 303, 308; human, 303; models, 299; orchid pollinators, 301; tetrachromatic, 302; trichromatic, 301, 306; systems, 292, 297; wasp, 302
visual angle, 299, 304
volatile(s), 29, 42, 55–59, 61–62, 65, 80, 84, 143, 158, 211, 224, 234, 236, 242, 252, 255, 271, 273. *See also individual scent compounds*
vouchers, 97, 176
voyage, 4, 173, 229

Wallace, Alfred Russel (1823–1913), 202, 209–11, 233, 250
wart, 279–80, 284
wasp(s), 16, 177, 230, 269, 279, 332; carrying pollinaria, 330; common, 11; digger, 54; emergence, 162; female, 157–58, 160, 166, 169–70; Ichneumon, 54; male, 7, 49, 54–56, 156–57, 160, 164–65, 167; parasitic, 6; parasitoid, 276; perception, 306; pergid, 144; phenology, 160; pheromones of, 59, 157–58; pollination guilds, 89; pollinators, 170, 245; pompilid, 80, 83, 89; preserved specimens of, 166; scoliid, 55, 59, 66, 142; sex ratio of, 166; solitary, 186; spider-hunting, 77; thynnine, 156, 157, 159, 163, 166–67, 170, 332; tiphiid, 133–34, 143; visibility to predators, 164; vision, 302; wasplike, 166; wasp masses (weight), 165; wasp mimicry by orchids, 162; wasp-pollinated, 84, 155; wild, 56
wavelength(s), 308; absorbed, 293; detect, 301–2; long-, 301, 304–5; mid-, 301; red, 302; reflected, 293, 305–6; short, 301; target, 305; UV, 294, 300, 304; visible, 306; wavelength dependent behaviors, 297
weather, 7, 151, 160–61, 170, 221, 244, 311–13, 315, 319, 323–25, 327
Wells, Herbert George (1866–1946), 19
white winged choughs, 160
whorls, 13
wings: bee, 235; carrier's, 165; column, 144, 153; feathered, 184; forming the clinandrium, 13; insect's, 64, 300; tegulae of, 52; UV-reflectance of, 296

xenogamy, 139, 141–42